THE
HOCKEY
STICK
AND THE
CLIMATE
WARS

THE
HOCKEY
STICK
AND THE
CLIMATE
WARS

DISPATCHES FROM THE FRONT LINES

Michael E. Mann

COLUMBIA UNIVERSITY PRESS

NEW YORK

Columbia University Press
Publishers Since 1893
New York   Chichester, West Sussex
cup.columbia.edu
Library of Congress Cataloging-in-Publication Data
Mann, Michael E., 1965–
The hockey stick and the climate wars : dispatches from the front lines / Michael E. Mann.
p.   cm.
Includes bibliographical references and index.
ISBN 978-0-231-15254-9 (cloth : alk. paper) — ISBN 978-0-231-52638-8 (ebook)
1. Climatic changes—Public opinion.   2. Climatic extremes—Public opinion.   3. Global
warming—Public opinion.   4. Climatology.   5. Environmental policy.   I. Title.
QC903.M36   2012
577.2'2—dc23   2011038813
∞
Columbia University Press books are printed on permanent and durable acid-free paper.
This book is printed on paper with recycled content.
Printed in the United States of America
c 10 9 8 7 6 5 4 3 2 1
References to Internet Web sites (URLs) were accurate at the time of writing. Neither the
author nor Columbia University Press is responsible for URLs that may have expired or
changed since the manuscript was prepared.

This book

is dedicated to

Megan Dorothy Mann

and to

the memory of

Jonathan Clifford Mann

Contents

Abbreviations and Acronyms

AAAS	American Association for the Advancement of Science
AEI	American Enterprise Institute
AGU	American Geophysical Union
AMO	Atlantic Multidecadal Oscillation
AP	Associated Press
API	American Petroleum Institute
AR4	IPCC Fourth Assessment Report (see also IPCC)
CEI	Competitive Enterprise Institute
CEQ	White House Council on Environmental Quality
CFACT	Center for a Constructive Tomorrow
CFCs	Chlorofluorocarbons
CID	Civil Investigative Demand
CO_2	Carbon dioxide
CRU	Climatic Research Unit (of the University of East Anglia in the United Kingdom)
ENSO	El Niño/Southern oscillation
EOF	Empirical orthogonal function
EPA	Environmental Protection Agency of the U.S. government
EPW	Environment and Public Works Committee of the U.S. Senate
FOIA	Freedom of Information Act
GCC	Global Climate Coalition
GFDL	Geophysical Fluid Dynamics Laboratory
GISS	Goddard Institute for Space Studies (see also NASA)
GRL	*Geophysical Research Letters*
IPCC	Intergovernmental Panel on Climate Change

LDEO	Lamont Doherty Earth Observatory (of Columbia University)
MBH	Mann, Bradley, and Hughes
MBH98	Mann, Bradley, and Hughes 1998 article in *Nature*
MBH99	Mann, Bradley, and Hughes 1999 article in *Geophysical Research Letters*
MCA	Medieval climate anomaly (see also MWP)
MWP	Medieval warm period (see also MCA)
NAO	North Atlantic Oscillation
NAS	National Academy of Sciences
NASA	National Aeronautics and Space Administration
NCAR	National Center for Atmospheric Research
NOAA	National Oceanographic and Atmospheric Administration
NSF	National Science Foundation
OSTP	Office of Science and Technology Policy of the White House
PC	Principal component
PCA	Principal component analysis
PNAS	*Proceedings of the National Academy of Sciences*
SEPP	Science and Environmental Policy Project
TAR	IPCC Third Assessment Report (see also IPCC)
UC Berkeley	University of California at Berkeley
U. Mass	University of Massachusetts
U. Va	University of Virginia
WMO	World Meteorological Organization
WR	Wegman Report
WSJ	*Wall Street Journal*

Prologue

What Is the Hockey Stick?

On the morning of November 17, 2009, I awoke to learn that my private e-mail correspondence with fellow scientists had been hacked from a climate research center at the University of East Anglia in the United Kingdom and selectively posted on the Internet for all to see. Words and phrases had been cherry-picked from the thousands of e-mail messages, removed from their original context, and strung together in ways designed to malign me, my colleagues, and climate research itself. Sound bites intended to imply impropriety on our part were quickly disseminated over the Internet. Through a coordinated public relations campaign, groups affiliated with the fossil fuel industry and other climate change critics helped catapult these sound bites onto the pages of leading newspapers and onto television screens around the world. A cartoon video ridiculing me and falsely accusing me of "hiding the decline" in global temperature was released on YouTube and advertised through a sponsored link that appeared with any Google search of my name. The video eventually even made its way onto the *CBS Nightly News*. Pundits dubbed the wider issue of the hacked e-mails "climategate," and numerous investigations were launched. Though our work was subsequently vindicated time and again, the whole episode was a humiliating one—unlike anything I'd ever imagined happening. I had known that climate change critics were willing to do just about anything to try and discredit climate scientists like myself. But I was horrified by what they now had stooped to.

My thoughts turned to an event from a decade earlier. In August 1999, I attended a meeting in Arusha, Tanzania, as a lead author for an upcoming report of the Intergovernmental Panel on Climate Change (IPCC). From my hotel room, I could see one of the world's great wonders, Mount Kilimanjaro, with its magnificent ice cap lying just degrees from the equator. The ice cap, by the end of the twentieth century, had already shrunk to just a third of the area it covered in 1936 when Ernest Hemingway wrote "The Snows of Kilimanjaro," but it was majestic all the same.

After the meeting, I joined a daylong expedition to see one of the world's greatest displays of nature: Serengeti National Park. Here, zebras, giraffes, elephants, water buffalo, hippos, wildebeests, baboons, warthogs, gazelles, and ostriches wander among some of the world's most dangerous predators: lions, leopards, and cheetahs. Among the most striking and curious scenes I saw that day were groups of zebras standing back to back, forming a continuous wall of vertical stripes. "Why do they do this?" an IPCC colleague asked the tour guide. "To confuse the lions," he explained. Predators, in what I call the "Serengeti strategy," look for the most vulnerable animals at the edge of a herd. But they have difficulty picking out an individual zebra to attack when it is seamlessly incorporated into the larger group, lost in this case in a continuous wall of stripes. Only later would I understand the profound lesson this scene from nature had to offer me and my fellow climate scientists in the years to come.

For more than two decades, in their efforts to inform the public about climate change and its potentially disastrous consequences, scientists have run up against powerful vested interests who either deny that such change is occurring or, if it is, that human activity plays much if any role in it. I have been privileged to be part of this scientific effort and, indeed, at times singled out in the ensuing conflict. My story is that of a once-aspiring theoretical physicist, driven by a curiosity about the natural world, who wound up as a central object of attack in what some have characterized as the best funded, most carefully orchestrated assault on science the world has known.

The extent of this assault was recently described by John M. Broder of the *New York Times*: "[T]he fossil fuel industries have for decades waged a concerted campaign to raise doubts about the science of global warming and to undermine policies devised to address it," and they have "created and lavishly financed institutes to produce anti-global-warming studies, paid for rallies and Web sites to question the science, and generated scores of economic analyses that purport to show that policies to reduce emissions of climate-altering gases will have a devastating effect on jobs and the overall economy."[1]

A central figure in this controversy has been the "hockey stick," a simple, easy-to-understand graph my colleagues and I constructed to depict changes in Earth's temperature back to A.D. 1000. The graph was featured in the high-profile "Summary for Policy Makers" of the 2001 report of the Intergovernmental Panel on Climate Change, and it quickly became an icon in the debate over human-caused (anthropogenic) climate change.

The hockey stick's prominence in the climate change debate would secure its status as a principal bête noir for those who denied the importance or even the existence of climate change. Climate change deniers went on to wage a public—and very personal—assault against my coauthors and me in the hope that somehow they might discredit all of climate science, the fruit of the labors of thousands of scientists from around the world, by discrediting us and our work. The Serengeti strategy writ large.

Why did our work stir such passions among the deniers of human-caused climate change? Perhaps because it addressed in such a graphic way the critical question of whether there was truly anything unusual about modern global warming. Instrumental records from around the globe indicate that Earth has warmed by almost 1 degree Celsius (about 1.5°F) over the past century. That may seem a small amount, but it is already noticeable in glacier retreat, rising sea level, more frequent heat waves, and more intense hurricanes, among many other phenomena. If the trend continues, the warming will have large and in some cases horrific consequences for the world's agricultural production, seacoast settlements, ocean health, and biodiversity.

Few records of thermometer readings reach further back in time than a century, however. Is it possible, then, that we have simply managed to catch a glimpse, through the myopic window of modern observations, of what is in reality a larger natural cycle of periodic cooling and warming and that in the decades to come, the average temperature will, by itself, stabilize or even begin to drop?

The field of paleoclimatology seeks to address such questions by placing modern evidence of climate change in a longer-term context. Paleoclimatologists make use, among other things, of the indirect evidence provided by so-called proxy climate data—natural archives of information that record, either physically, chemically, or biologically, some attribute of the climate back in time. Scientists have found, for example, that the shifting ratio of oxygen isotopes in the frozen layers of ice cores, some of which date back thousands of years, can be used to estimate temperature changes back in time at the site of the core.[2]

Some proxy records, such as the layers contained in deep ocean sediments, record only the coarsest changes, such as the coming and going of the major ice ages over eons. Other proxy records, such as tree rings, ice cores, corals, and lake sediments, can potentially tell us about climate conditions such as temperature, rainfall, or wind patterns for a single year or season. By using an array of such records, we can establish a year-by-year chronology of the climate changes of past centuries.

In the late 1990s, my coauthors and I published an attempt to use such paleoclimate proxy data to obtain a quantitative assessment of how Earth's surface temperature had varied in past centuries. We published our original findings, which spanned the past six hundred years, in 1998,[3] and the following year we were able to extend the analysis back over the entire past millennium.[4] The picture that emerged was a wiggly curve documenting past temperature changes over the entire Northern Hemisphere (the hemisphere with the most data) and indicating a sharp rise in temperature over the past century. The graph we drew looked fuzzy because for each point on the curve it included the margin of error, reflecting the simple fact that ice cores, corals, tree rings, and other proxies are useful but rather imperfect thermometers.

Despite the uncertainties, my coauthors and I were able to draw certain important conclusions. We deduced that there had been a decline in temperature from the period running from the eleventh century through the fourteenth—a period sometimes referred to as the medieval warm period—into the colder Little Ice Age of the fifteenth to nineteenth centuries. Think of this as the shaft of a hockey stick laid on its back. This long-term gradual decline was followed by an abrupt upturn in temperatures over the past century. Think of this as the blade. In the original draft, our reconstruction ended in 1980, as relatively few long-term proxy records had been updated since the early 1980s. Yet much of the observed warming had actually taken place since then. An anonymous reviewer suggested that we bring the curve up to date by including the observational temperature data, that is, modern thermometer measurements available through to the present. That led to a sharpened blade. The warmth at the end of the record rose well above that of any period of the past millennium, even taking into account the increasing margin of error as one goes back in time.

Thus was born the hockey stick—though the term itself was actually coined later by a colleague in Princeton.[5] It didn't take long for the hockey stick to become a central icon in the climate change debate. It

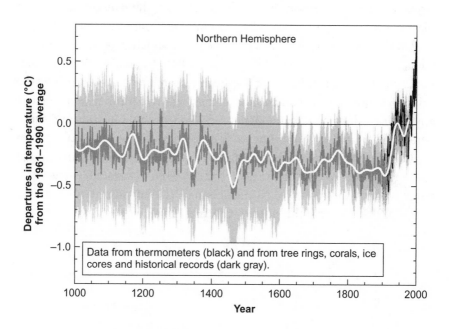

Figure P.1: The Hockey Stick
The thin dark grey curve denotes the estimated Northern Hemisphere temperature for each year (relative to a baseline [0.0] marking the late twentieth-century average) as estimated from proxy data available through 1980. The gray-shaded region indicates the uncertainty in the annual temperature estimates (there is an estimated 95 percent certainty that the temperatures for any given year lie within the shaded region). The thick, smoother curve highlights the long-term variations. The modern instrumental temperature-gauge record through the end of the twentieth century (thin black curve) is shown for comparison. [Adapted from IPCC *Third Assessment Report*, 2001; a similar version of this graphic appeared in our original 1999 publication.]

told an easily understood story with a simple picture: that a sharp and highly unusual rise in atmospheric warming was occurring on Earth. Furthermore, that rise seemed to coincide with dramatic changes in human activity heralded by industrialization and increased use of fossil fuels. The controversy that the hockey stick would ultimately generate, however, had little to do with the depicted temperature rise in and of itself. Rather, it was a result of the perceived threat this simple graph represented to those who are opposed to governmental regulations or other social restraints aimed at protecting our environment and the long-term prospects for the health of our planet.

In this book, I attempt to tell the real story behind the hockey stick. I reflect on the emphasis, and indeed at times the overemphasis, that players on both sides of the climate change debate have often placed on this work. I explore the controversy associated with the hockey stick, some of it real, much of it specious. I use my own story, more than anything else, as a vehicle for exploring broader issues regarding the role of skepticism in science, the uneasy relationship between science and politics, and the dangers that arise when special interests and those who do their bidding attempt to skew the discourse over policy-relevant areas of science. In short, I attempt to use the hockey stick to cut through the fog of disinformation that has been generated by the campaign to deny the reality of climate change. It is my intent, in so doing, to reveal the very real threat to our future that lies behind it.

THE
HOCKEY
STICK
AND THE
CLIMATE
WARS

Chapter 1

Born in a War

The balance of evidence suggests that there is a
discernible human influence on climate.

—The contentious sentence in "Summary for Policy Makers,"
IPCC Third Assessment Report (1995)

It is November 27, 1995, several years before my colleagues and I published our "hockey stick" study. Bill Clinton has been president for nearly three years. The Dow Jones Industrial Average just passed the 5,000 mark for the first time. The TV series *E.R.*, created by novelist Michael Crichton, is the top-rated show on television.

In Madrid, Spain, the Intergovernmental Panel on Climate Change (IPCC) is holding the final plenary meeting for the Second Assessment Report, the purpose of which is to summarize the consensus among scientists regarding the extent of humanity's impact on Earth's climate. At a nearly identical latitude on the other side of the Atlantic, I am working on my Ph.D. dissertation in New Haven, Connecticut. I am oblivious to what is taking place in Madrid; I'm just trying to finish up my research in time to defend my dissertation the coming spring and begin a career as a professional climate researcher.

My research at the time focused on the importance of natural variability—that is, the role of nature, not man—in explaining changes in Earth's climate. The one scientific article I had submitted for publication that touched on the topic of human-caused climate change would, ironically, a few months later be hailed by those who contest the proposition that humans play a significant role in observed climate changes. (That article simply demonstrated a relatively minor inconsistency between theoretical climate model predictions and actual climate observations.)[1] I was especially interested in the role that natural oscillations in the climate system might have played in observed changes in climate during the modern observational era of the nineteenth and twentieth centuries. My first-ever article analyzing climate proxy records had just been published a week earlier.[2] In that article, my coauthors and I showed that these natural oscillations persist over many centuries and might be more important than many scientists had acknowledged in explaining certain modern climate trends. This work, in another twist of irony, would also be celebrated by contrarians in the climate change debate, who were ostensibly unaware that both natural and human influences on climate can and almost certainly do coexist. I myself did not doubt that humans were changing the climate; the extent of evidence was already significant. I had simply chosen to focus in my research on the issue of natural climate variability.

Meanwhile, back in Madrid at the IPCC plenary, a fierce argument had broken out between the scientists crafting the report and government delegates representing Saudi Arabia, Kuwait, and some other major oil exporting nations that profit greatly from societal dependence on fossil fuel energy and, according to the *New York Times*, had "made common cause with American industry lobbyists to try to weaken the conclusions" of the report.[3] The tussle was over whether one could state with confidence that human-caused climate change was already observable. The scientists argued that "the balance of evidence suggests an *appreciable* human influence on climate" because only when human impacts were included could the rise in temperature over the past century be accounted for. The Saudi delegate complained that the word *appreciable* was too strong. He demanded weaker wording.

For two whole days, the scientists haggled with the Saudi delegate over this single word in the "Summary for Policy Makers." They debated, by one estimate, nearly thirty different alternatives before IPCC chair Bert Bolin finally found a word that both sides could accept: "the balance of evidence suggests a *discernible* human influence on climate." The term *discernible* established a middle ground by suggesting that climate change was indeed detectable, as the scientists argued, while acknowledging that humanity's precise role in that change and its magnitude were still subject to dispute—a concession that no doubt pleased the Saudi delegate. This sentence would go on to become famous or, in some circles, infamous. The fact that two entire days at the final plenary were devoted to debating a single word in the report's summary gives you some idea of how contentious the debate over the reality of human-caused climate change had become by 1995.

Why did the scientists care so much about the wording? What would be the harm, after all, if the wording were weakened a bit? I suppose it comes down to how deeply scientists care about getting things right. Details matter, and we argue passionately with each other about them. We don't suffer perceived inaccuracies lightly, and more than anything else, we don't like being misrepresented. The fact that the science in this case might have deep real-world consequences only amplified these natural inclinations.

Among the scientists who fought hard against any watering down of the report's key conclusion was Ben Santer, a climate specialist who works for the Department of Energy's Lawrence Livermore National Laboratory in California. The recipient of a coveted McArthur "genius" award in recognition of his groundbreaking contributions to our understanding of climate change, Santer was a primary author on a series of important papers establishing the human role in observed climate change. As such, Santer was in a better position than anyone—and certainly than a bureaucrat with a political agenda—to assess the level of scientific confidence in concluding that human activity was changing the climate.

As it happens, I had met Santer for the first time a little more than a year earlier, in July 1994, at a two-week workshop on climate science at the National Center for Atmospheric Research. I was attending the workshop as a graduate student invitee, and Santer was one of the invited speakers. I asked him a question about certain details of his analysis following his presentation. His response came across as a bit defensive, as if he perceived my question as an attack. Only later would I understand why.

Santer's work on climate change detection, unbeknownst to me, had been under increasing attack from contrarians in the climate change debate. In 1994, for example, his findings regarding the match between observed and model-predicted surface temperature changes was criticized[4] by Patrick Michaels, a University of Virginia climate scientist who edited the *World Climate Report*,[5] a newsletter with fossil fuel industry funding[6] that featured criticisms of mainstream climate change research.

The attacks against Santer were ratcheted up dramatically following the November 1995 IPCC plenary. In February 1996, for example,

S. Fred Singer, the founder of the Science and Environmental Projection Project and a recipient over the years of substantial fossil fuel funding,[7] published a letter attacking Santer in the journal *Science*.[8] Singer disputed the IPCC finding that model predictions matched the observed warming and claimed—wrongly—that the observations instead showed cooling. Singer went further. He claimed that inclusion of Santer's work in the report violated IPCC rules because the work hadn't yet been published. In fact, the IPCC rules did not require a work cited to be published at the time of the report; if it did, the lag time involved in getting a publication to print would essentially render the report obsolete on arrival. The IPCC requirement was simply that the work be available to reviewers upon request, which Santer's work was. Moreover, a substantial component of the research in question had been published.

Meanwhile, the Global Climate Coalition (GCC), a group also funded by the fossil fuel industry,[9] circulated a report to Washington, D.C. insiders accusing Santer of abusing the peer review system and of "political tampering" and "scientific cleansing"—a charge that was especially distasteful given that Santer had lost relatives in Nazi Germany.[10] The purported basis of these allegations? At the request of the IPCC leadership after the Madrid plenary, Santer, as lead author on an IPCC chapter, had removed a redundant summary so that his chapter's structure would conform to that of the other chapters, all of which had summaries only at the beginning.

A few months later, Frederick Seitz, the founding chairman of another industry-funded group,[11] the George C. Marshall Institute, published an op-ed in the *Wall Street Journal* repeating the same charges. While the paper's editors did eventually publish Santer's rejoinder, they in effect neutered his response by editing his words beyond recognition and removing the names of the forty colleagues who had cosigned the letter, thus leaving *Journal* readers with the misleading impression that Santer stood alone in his defense against the specious charges.[12]

With the help of sympathetic media outlets such as the *Wall Street Journal*, climate change deniers were able to spread false charges about Santer faster than he—or his colleagues—could possibly hope to refute them. The practice of isolating someone like Santer to make an example of an individual scientist—what I call the "Serengeti strategy"—is a tried-and-true tactic of the climate change denial campaign. The climate change deniers isolate individual scientists just as predators on the

Serengeti Plain of Africa hunt their prey: picking off vulnerable individuals from the rest of the herd.

The Santer episode encapsulates the toxic and incendiary environment that existed, largely unbeknownst to me, at the time that I was finishing my Ph.D. and preparing to enter the world of climate research. Little did I know that similar attacks might be made against me just a few years hence, when my work, like Santer's, would be featured as a major pillar of evidence by the IPCC.

Tricks and Treats

It is late December 1974. I've just turned nine, and, as usual, my family and I are celebrating my birthday with relatives in Philadelphia. For more than a year now, I've been pestering my Uncle Paul—an artist and successful entrepreneur to whom I'd always looked for wisdom on all matters of life—to explain what it means to go faster than the speed of light. I was intrigued by such "gee whiz"—but ultimately scientific—concepts. For my birthday that year, Uncle Paul had given me a copy of a popular novel considered inspirational at the time, with an inscription indicating that it would answer my questions. I enjoyed the book, though to this day I can't figure out what it had to do with warp speed, time travel, or any related topics. But I know that already by that age I was fascinated with the world of science.

Math and science were the subjects that had always come most easily to me; perhaps having a father who was a college math professor had something to do with it. In high school, when other kids were partying on Friday nights, I was hanging out with my computer buddies writing programs to solve challenging problems. In fall 1983, after having seen the movie *War Games*, I became determined to write a self-learning tic-tac-toe computer program, just as in the movie, a program that could learn from its mistakes, a rudimentary type of artificial intelligence. The movie carried a thinly veiled lesson about the futility of global thermonuclear war: There can be no winner in a tic-tac-toe game expertly played; if neither player makes a mistake, the game will always result in a tie. Perhaps if the computer—in the movie, it had seized control of America's missile program and was preparing to launch a massive nuclear attack—could be brought to understand this paradox, it could recognize the futility of nuclear war. For me at the time, however, it was just an interesting and challenging computer problem to tackle.

Machine learning of this sort was in principle relatively straight-forward. The real challenge was in how to go about constructing an algorithm—a set of operations or calculations, here in the form of a computer program—to solve the problem as efficiently and elegantly as possible. I had the computer play itself, just like in the movie. That was the easy part. In the beginning, I simply had it make random moves every turn. When it lost to itself, however, I would store both the final and previous configurations of the tic-tac-toe board in a "blacklist"—moves that would no longer be available to the computer.[13] The black-list was used to ensure that the computer, while it continued to make random moves, would not make the same losing move again; in this way it would gradually "learn" how to play tic-tac-toe.

In practice, it might take a very long time for the computer to be-come skilled enough to avoid losing because there are so many possible sequences of moves, and the program gets slower and slower as it has to scan an increasingly long list of disallowed moves before each turn. But I discovered a "trick"—the term scientists and mathematicians often use to denote a clever shortcut to solving a vexing problem—to get the computer program to learn much faster. The trick was to exploit the concept of symmetry. A tic-tac-toe game is the same no matter how you rotate the board, whether you flip it vertically or horizontally, or whether you switch the role of Xs and Os. When you take that symmetry into account, there are actually many fewer truly unique board configura-tions and many fewer losing moves that need to be stored in a blacklist. Now I could get the computer to become unbeatable in tic-tac-toe far more readily. The adrenaline rush, for the scientist, comes from finding tricks that make a problem easier to crack. That—and eating pizza with my friends—was my idea of a fun Friday night.

A Random Walk

A year later, in August 1984, as Ronald Reagan was completing his first term in office and Michael Jackson's *Thriller* was the top-selling record album, I headed off to college at the University of California, Berkeley (UC Berkeley). In part, I must confess, I was looking to get away from the harsh Amherst, Massachusetts, winters I'd endured for the first eighteen years of life, but I was also attracted by the school's reputation as one of the world's leading scientific research institutions. I chose to major in physics, with a second major in applied math. The summer following my freshman year, I began doing research in theoretical physics that emphasized the computational approaches to problem solving I so enjoyed.

The project I was working on bore some resemblance to the tic-tac-toe problem that had captivated me in high school; it too involved the concept of randomness. The project employed what is known as a Monte Carlo method, named for its resemblance to a casino game—a method that I would make use of years later in my climate research. Much as gamblers in Monaco's famous casino town engage in random rolls of the dice in hope of monetary reward, scientists generate random numbers on a computer in hope of simulating processes in nature that have a random component.

One example is the molecular interactions that govern the behavior of a solid or liquid. While the fluctuations of the individual molecules are random in nature, external conditions—the ambient temperature, in particular—influence the collective behavior of the molecules. The warmer the temperature, for example, the more energetic the random fluctuations. Thus low temperatures favor relatively ordered states (e.g., ice crystals), while high temperatures favor relatively disordered states (e.g., water vapor). Shifts between these states are typically abrupt. There is a critical temperature at which the system, when warmed, will suddenly undergo a phase transition from the ordered state to the disordered state, or vice versa in the case of cooling. One can explore phase transitions by representing the interactions between molecules in a computer model simulation, generating random molecular perturbations in the model to mimic the real-world random fluctuations of molecules.

I was using this type of Monte Carlo approach to investigate the theoretical behavior of liquid crystals—the materials used in liquid crystal displays (LCDs) employed in laptops, TVs, and digital watches. My research was aimed at determining how the critical temperature of the transition between the ordered and disordered phases of liquid crystals might vary under different conditions.[14] My adviser was a theoretical physical chemist named Tony Haymet, who much later—coincidentally enough—went on to direct one of the world's premier climate research institutions, the Scripps Institution for Oceanography at the University of California, San Diego.

When Tony left UC Berkeley a couple of years after my arrival, I continued my undergraduate research with Didier de Fontaine, a professor of materials science studying the properties of an exciting new material—a high-temperature superconductor. A superconductor is a material that conducts an electric current with no resistance, a property with profound real-world applications such as in the operation of super-fast bullet trains. Conventional (metallic) superconducting materials need to be cooled nearly to a temperature of absolute zero, making them expensive to maintain. In the mid-1980s, scientists discovered that certain ceramic materials had a remarkable property; they superconducted at much higher temperatures, above even the temperature of liquid nitrogen (a very inexpensive coolant).

When I joined de Fontaine and his group in late 1987, they had been working for months to model the behavior of just such a material: yttrium barium copper oxide (YBCO). They were using Monte Carlo computer approaches similar to those I'd been using to study liquid crystals, so the project was a natural fit. Over the next two years, I worked with de Fontaine's group on modeling the transitions between ordered and disordered phases of YBCO. We were stymied by one vexing problem. At very low temperatures, the simulation would get stuck in a temporary state and wouldn't settle into the true equilibrium state. My key insight—my *ansatz*—was in recognizing that the fact that the material was organizing itself into parallel, variably spaced, linear chains of copper and oxygen atoms was telling us something fundamental about the symmetry of YBCO at very low temperatures. The material was behaving not as a two-dimensional planar system, but as a much simpler one-dimensional linear system. Unlike the two-dimensional problem, the one-dimensional problem had an analytical solution; we could solve it with paper and pencil, allowing us to fully map YBCO's phase behavior.[15] Disappointingly, this behavior turned out to have little if anything to do with the superconductivity of the material. Nonetheless, I had once again found a satisfying trick to solve another challenging scientific puzzle, and that's all that mattered to me at the time.

After graduating from UC Berkeley, I went off to Yale University in fall 1989 to pursue a Ph.D. in physics. I was planning to study theoretical condensed matter physics—the theory behind the behavior of solids and liquids. After two years, I'd completed my coursework and passed my exams, and it was time to select a Ph.D. topic and adviser. Funding was tight, however, and, like other graduate students in physics, I was being steered toward increasingly applied areas of research—in my case, the study of semiconductor devices. This wasn't quite the "big-picture" science I'd had in mind when I decided to go into physics in the first place. I was somewhat disillusioned and felt as if I'd lost my bearings. Facing a vocational crisis, I opened up the university catalog one day to see what other scientific research was being done at Yale where I might be able to apply my math and physics interests to working on a big-picture problem of significance.

As I scrolled through the catalog, I discovered that a professor in the Department of Geology and Geophysics, Barry Saltzman, was using the tools of physics to simulate Earth's climate. That sounded like a big-picture problem to me, and an important one at that. I scheduled an appointment to meet with Saltzman that week. As I talked with him, I became even more excited about the possibility of doing research in this area. He gave me a few articles to read and told me to get back to him if this subject matter seemed to be my cup of tea. I scoured the papers over the weekend. While the terminology and mathematical conventions were a bit different from those I'd encountered in my physics training, I got the gist of the articles—much as someone who speaks Spanish can roughly understand a person speaking Portuguese. On Monday, I told Barry that I was indeed interested in pursuing research in this area. He had some additional funding and could support me on a trial basis for the summer. If all worked out, I could stay on with him to do my Ph.D.

That summer, I assisted a postdoctoral researcher of Barry's with a project aimed at simulating the climate of the Cretaceous period with a state-of-the-art computer model. The Cretaceous period ended about 65 million years ago—with a bang, in fact. That's when the dinosaurs went extinct, due to—it is now generally accepted—the impact of a large asteroid that struck Earth, sending massive amounts of dust into the atmosphere, blocking out much of the incoming sunlight, and sending Earth—in essence—into a quasi-perpetual years-long winter. What we were interested in, however, was the period of peak Cretaceous warmth roughly 35 million years earlier. We were trying to figure out how the high-latitude continents could have been as warm as they were. Fossil evidence indicates that dinosaurs back then were wandering Antarctica! Geological evidence suggests that greenhouse gas concentrations were higher than modern levels by perhaps a factor of four or more—enough to account for the overall extent of apparent global warmth at the time. What the models could not explain, however, was the paleodata indicating that the tropics warmed up only a little, while the poles warmed up a lot—a mystery that is still unsolved today, though some intriguing hypotheses have been proposed.[16]

I spent most of my waking hours that summer of 1991 reading climate science textbooks and papers and soaking up all the knowledge I could about the field of climate research. With the beginning of the fall term, I switched into the Department of Geology and Geophysics, and I went on to take an additional two years of coursework to acquire knowledge in a whole new range of subjects, including fluid dynamics, meteorology, oceanography, climate dynamics, and statistical analysis methods. After passing yet another set of exams, I finally was ready to pursue Ph.D. research with Barry. I had no idea where my seemingly random scientific path would end up taking me, though much of the focus in the field of climate research at the time had to do with two central topics: natural climate variability and human-caused climate change. It was therefore likely that I would end up doing research related to one of these topics.

Chapter 2

Climate Science Comes of Age

The ice we skate is getting pretty thin
The waters getting warm so you might as well swim*

—Smashmouth, "All Star" (1999)

When I entered the field of climate research in the early 1990s, the science was just coming of age. Major research centers around the world were using some of the fastest supercomputers available to run ever-more sophisticated models of Earth's climate. Important new observations were coming in. Thermometer measurements showed that the globe—both land and ocean—had by that time warmed approximately 1 degree Fahrenheit over the past century. The accumulated loss of ice during the previous four decades from melting glaciers around the world could fill Lake Huron. An increasing number of climate measurements were painting a picture of a climate that was changing in much the way models had been predicting it would if we continued to emit greenhouse gases. By the mid-1990s, enough evidence had accumulated to convince the IPCC, as we've seen, that there was a "discernible human influence on climate." While the jury had been out when I began my studies in the early 1990s, it had come in with a judgment by the time I was completing them in the mid-1990s. What had led to that verdict can be described in five easy steps.

Climate Change in Five Easy Steps

By the mid-1990s, scientists were able to connect the dots when it came to establishing a human impact on our climate.

(1) We knew for one thing that human activity—primarily the burning of fossil fuels—had increased carbon dioxide (CO_2) concentrations in the atmosphere. The legendary atmospheric scientist Charles Keeling first began to make direct measurements of atmospheric CO_2 in 1958 at a pristine location far from any pollution sources—nearly three miles above sea level at the top of Mauna Loa in Hawaii. Thanks to Keeling's work, we had an instrumental record of CO_2—the so-called

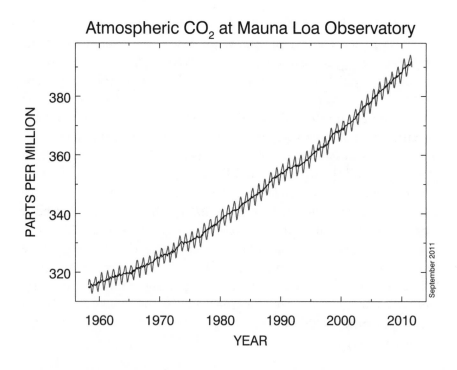

Figure 2.1: The Keeling Curve
Monthly measurements of atmospheric CO_2 concentrations from Mauna Loa, Hawaii, dating back to 1958. The original measurements were made by Charles Keeling. The record is maintained and updated by scientists with the Scripps Institution of Oceanography. [From Scripps Institute of Oceanography/NOAA Earth System Research Laboratory.]

Keeling curve—going back nearly half a century. But we could go back even farther by analyzing CO_2 content in samples of the ancient atmosphere recovered from the layers of ice in ice cores stretching back hundreds of thousands of years.

These records indicate a steady two-century–long rise in CO_2 concentrations coincident with the Industrial Revolution, culminating in modern levels that appear unprecedented in hundreds of thousands of years at least, long before modern humans arrived on the scene. The CO_2 that was building up in the atmosphere as the Industrial Revolution progressed, moreover, had humanity's fingerprints all over it; analysis of the carbon isotopes indicated that the source lay in the burning of fossil fuels. At current rates of fossil fuel burning, CO_2 concentrations, it was estimated, would reach twice preindustrial levels within about four decades. We'd potentially need to reach back tens of millions of years, halfway to the age of dinosaurs, to find previous levels that high.

(2) Scientists also knew that this increase in atmospheric CO_2 (and other trace gases produced by human activity, such as methane) must have a warming effect on Earth's surface. In fact, that had been known for nearly two centuries. The esteemed French scientist and mathematician Joseph Fourier (1768–1830) is generally credited with discovery of the warming effect of these gases. Other great nineteenth-century scientists such as England's John Tyndall (1820–1893) and Sweden's Svante Arrhenius (1859–1927) helped work out the basic physics and chemistry.

Earth is heated from above by the Sun. The only thing keeping the planet from getting hotter and hotter is its ability to cool off by emitting its own invisible form of radiation (infrared radiation) to space. Certain gases in our atmosphere such as carbon dioxide, however, impede this heat loss mechanism by absorbing a fraction of that radiation and re-radiating some of it back down toward the surface, rather than allowing it to escape to space. This requires Earth's surface to send even more infrared radiation out to space. And that it can only do by warming up.

This effect, which in some respects resembles how a greenhouse works, is known as the "greenhouse effect," and the gases responsible for it are thus termed "greenhouse gases." The greenhouse effect is hardly controversial. Indeed, without a natural greenhouse effect, Earth would be a frozen planet lacking life as we know it. By increasing the concentrations of these gases in the atmosphere, it was only logical that we should be further warming Earth's surface.

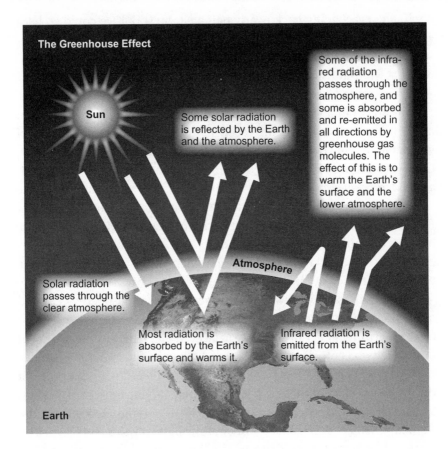

Figure 2.2: The Atmospheric Greenhouse Effect
Earth is warmed by the incoming radiation from the Sun, and its temperature can stabilize only by producing its own invisible outgoing (infrared) radiation. The greenhouse effect involves the absorption of some of that outgoing radiation by greenhouse gases in our atmosphere. The greenhouse effect warms the surface by sending some of this radiation back toward Earth rather than allowing it to escape to space.

(3) Indeed, as mentioned earlier, thermometer measurements told us that by the mid-1990s Earth had already warmed a little more than a degree Fahrenheit (roughly 0.6°C) since the dawn of industrialization. The globe was in fact warming. This observation alone may not seem that decisive; after all, the warming might have been at least partly natural in origin. However, the observation did not exist in isolation. There was now evidence as to the probable cause.

(4) By the mid-1990s, it was possible to investigate the causal mechanisms behind changes in Earth's climate using relatively sophisticated mathematical models of Earth's climate. These models solved the same complex equations of atmospheric physics that numerical weather prediction models did. But they also took into account components of the climate system other than the atmosphere, including the oceans, the continental ice sheets, and even life on Earth (collectively known as the "biosphere"), and they attempted to account for the physical, chemical, and biological interactions among these components. Of course, no theoretical model is ever perfect; even the best model is only an idealization of the actual world. There are always real-world processes that cannot be captured—for example, in the case of a numerical climate model, individual clouds or small-scale air currents like dust devils—that are simply too small for the model to resolve. The key question is, can the model be shown to be useful? Can it make successful predictions?

Climate models had passed that test with flying colors by the mid-1990s. James Hansen, in the late 1980s, successfully predicted the continuing warming that would be observed by the mid-1990s.[1] Even something the model couldn't have predicted in advance—the 1991 eruption of Mount Pinatubo in the Philippines—provided yet another key test. As soon as the eruption occurred, Hansen put what was known about the reflective qualities of volcanic sulfur particulates (known as "sulfate aerosols") into the simulations. The aerosols cooled surface temperatures for several years in the model by shielding the surface from a fraction of incoming sunlight, leading Hansen to make what turned out to be a successful prediction of the temporary cooling that was seen over the ensuing few years.[2]

(5) Finally, perhaps most significant of all, only when human factors were included could the models reproduce the observed warming—both its overall magnitude and, equally important, its geographical pattern over Earth's surface and its vertical pattern in the atmosphere. The primary such human factor was increasing greenhouse gas concentrations due to fossil fuel burning and other human activities. A secondary human factor, sulfate aerosols emitted from industrial smokestacks, also played a role, however. Like volcanic sulfate aerosols, these industrial aerosols have a cooling effect. Unlike volcanic aerosols, which reach the lower stratosphere, allowing them to spread out into a layer covering the globe, industrial aerosols remain confined to the lower atmosphere,

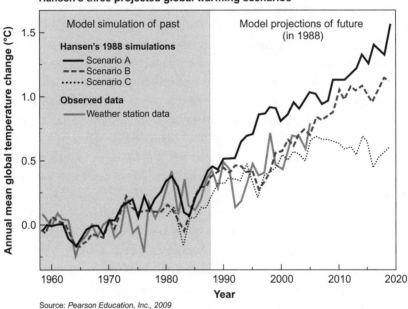

Source: *Pearson Education, Inc., 2009*

Figure 2.3: Testing Climate Models
A comparison of three different simulations of global warming through 2020 made by James Hansen in 1988. The curve made up of weather station observations (available through 2005 in this analysis) closely matches the curve of the middle scenario (B)—the one that is based on the trajectory of an emissions scenario that most closely matches actual greenhouse gas emissions over the preceding twenty years. The upper and lower curves correspond to scenarios A and C, which assume higher and lower emissions respectively.

leading to localized patterns of cooling that offset global warming in some regions. The pattern of warming predicted by the models from the combination of these two human effects on the climate provide a unique "fingerprint" of what the human influence on climate should look like if the models were correct—and the fingerprint matched. The surface and lower atmosphere showed an irregular pattern of warming, while the atmosphere aloft was cooling, just as the models indicated it should.[3] The fingerprint predicted for natural factors alone—for example, from fluctuations in solar output—on the other hand, failed to match the observations. It was the work of Ben Santer and other climate scientists during the mid-1990s in establishing this fingerprint of human influence that provided the "smoking gun," the fifth and final link in the chain of reasoning that allowed the IPCC to declare in 1995 that there was indeed at least a "discernible" human influence on climate.

So the case could be made by the mid-1990s—in just five easy steps—that human activity was changing our climate. The case became stronger over the next decade and a half as increasingly more sophisticated models were developed and a wider array of data was collected that confirmed unprecedented changes taking place in our climate. It is fair to say, though, that even by the mid-1990s there was no longer reason for real scientific debate over the proposition that humans had warmed the planet and changed the climate. That conclusion was now supported by the efforts of thousands of scientists around the world whose work contributed to the various pillars of evidence detailed above. What scientists were still debating with each other at scientific meetings and in the professional journals was the precise balance of human versus natural causes in the changes observed thus far, and just what further changes might loom in our future.

Answers to these more specific questions were far less clear. Climate scientists could surmise that if human civilization continued to follow its current upward trajectory of fossil fuel burning, we would likely see a near doubling of preindustrial atmospheric CO_2 levels by the mid-twenty-first century. Furthermore, we could estimate that such an increase would lead to an additional warming of anywhere between 1.5 and 4.5° C (roughly 3–8°F).

The large spread in estimated temperature increase arose primarily from uncertainty about the effects of so-called feedbacks, responses of the system that can either further amplify or diminish the warming. Certain feedbacks are almost certainly positive, amplifying the given effect; for example, a warmer atmosphere holds more water vapor, and water vapor is itself a greenhouse gas, further warming the surface. The melting of ice as Earth warms exposes more of the ground and ocean surface. These surfaces absorb sunlight more effectively than does ice, which further amplifies the warming. But other factors, such as how clouds change under warmer conditions and to what effect, were highly uncertain, and remain so still.

Most models indicate a tendency for more low clouds in a warmer climate; low clouds mimic the effect of surface ice, for example, reflecting solar radiation back to space. Thus, in these models, clouds play the role of a negative feedback, diminishing the warming. Yet in other credible climate models, clouds behave differently, effectively enhancing the overall greenhouse effect and acting as yet another positive feedback. Representing cloud effects is perhaps the most daunting challenge for climate models, because they occur at scales too small to capture explicitly in the models, and their effects must therefore be represented only through approximations.

What's All the Fuss About?

By the mid-1990s, larger questions regarding the potential societal and environmental impacts of climate change were beginning to receive more attention as well. Did climate change pose a threat to the future welfare of our civilization, and even possibly to our species? And if so, what, if anything, should we do about it, and when? Science alone could not, of course, answer many of these questions. They are as much matters of policy (and risk management, economics, and ethics) as they are matters of science. The science could, however, *inform* matters of policy.

There was increasing recognition by the mid-1990s that another 2°C (3.5°F) warming beyond current levels (for a total of 3°C or 5°F warming relative to preindustrial times) could represent a serious threat to our welfare.[4] Precisely what limitations in global greenhouse gas emissions would be required to avoid that amount of warming remained uncertain, and still does, because of the spread of predictions among models. If we choose to take the midrange model estimates as a best guess, avoiding another 2°C of warming would require stabilizing atmospheric CO_2 concentrations at no higher than about 450 parts per million (ppm).

Preindustrial levels were about 280 ppm, reflecting a long-term balance between natural processes that produce (sources) and those that take up (sinks) CO_2 from the atmosphere. Humans, through extensive fossil fuel burning and other practices, have upset that natural balance, causing CO_2 concentrations to rise steadily. Indeed, those concentrations will continue to rise until human emissions are brought essentially to zero.[5] The carbon we emit into the atmosphere today has an extended legacy; it will potentially reside there for centuries.

Levels, as of 2011, are nearly 390 ppm and are increasing by 2 to 3 ppm per year as a result of annual carbon emissions. The average

American, through various actions and activities, emits roughly 20 tons—the weight of two very large adult male African elephants—of carbon per year. Globally, human beings emit the equivalent of more than 400 million of those elephants—roughly 8.5 billion tons—of carbon per year. A 450 ppm stabilization target would require greenhouse gas emissions be brought to a peak of no more than about 9 billion tons (450 million elephants) per year within the next decade, be lowered to mid-twentieth-century levels of roughly 1 billion tons (50 million elephants) per year by midcentury, and brought to near zero by the end of the century, to avoid breaching 450 ppm. That is a daunting task, as global population continues to increase, developing nations such as China and India continue to ramp up their own emissions, and industrial nations like the United States continue with business as usual. Given the enormity of the challenge, it was convenient for some to simply deny that climate change was happening at all, especially those who were profiting handily from civilization's addiction to fossil fuels.

Some leading climate scientists such as NASA's James Hansen have argued that CO_2 not only needs to be stabilized below 450 ppm, but in fact must be brought back to a level even lower than present. Based on geological evidence regarding ice amounts and sea levels that prevailed in past warm climates, Hansen argues that we need to bring CO_2 down to levels lower than those that persisted when I first entered into climate research in the early 1990s—to 350 ppm, to be specific.[6] In the December 2009 climate change negotiations in Copenhagen, Denmark, a consortium of low-lying island nations already threatened by rising sea levels lobbied for such a target.[7] This target has even been incorporated into the name of the grassroots climate change campaign founded a few years ago by environmental writer Bill McKibben: 350.org. Lowering atmospheric CO_2 concentrations from current levels would require not only dramatically reducing emissions, but in fact making them negative— that is, actively taking ambient CO_2 from the air through expensive and, as yet, largely untested technologies such as open air carbon capture (which attempts to suck the CO_2 out of the air, mimicking what plants do naturally, but at a greatly accelerated rate and without releasing carbon into the atmosphere as plant matter does when it dies and decomposes).

Suppose we were instead to continue with business as usual, shunning efforts to curtail carbon emissions. The impacts on our civilization and environment could be profound. By doing so, we might well be committing ourselves to the melting of the major ice sheets, resulting in a sea level rise as much as six feet by the end of this century[8] and, eventually, twenty feet or more, thus ensuring extensive loss of coastal settlements around the world, including the East and Gulf Coasts of the United States, and the potential disappearance of many low-lying island nations. Many coastal regions, including the U.S. East and Gulf Coasts, might feel the double whammy of inundation from sea level rise and increased erosion and destruction from potentially more powerful hurricanes fueled by warmer oceans. Increasingly widespread and severe droughts would likely take hold over the major continents, including North America, as precipitation became increasingly intermittent and moisture evaporated more readily from warmer soils. Many regions would likely also see increased flooding from more intense rainfall events.

Among the potential impacts would be greater social conflict resulting from movements of large numbers of environmental refugees and increased competition for available resources within and among nations, more widespread famine due to declining agricultural productivity in developing tropical nations already struggling with hunger and malnutrition, and threats to human health and even mortality from potential increases in the spread of infectious disease and stress-related deaths from more frequent and extreme heat waves. Key ecosystems may be lost, including coral reefs and the summer Arctic sea ice environment critical to the survival of polar bears.

Of course, there could be potential benefits in some cases. Agricultural productivity, for example, might increase in some midlatitude regions owing to longer growing seasons, as long as freshwater supply remains available—an important caveat. However, when all the various potential impacts of the climate changes are taken into account, the weight of impacts have been shown to be decidedly negative, and increasingly so as warming progresses.[9] It is possible that the models that indicate a temperature rise of only 3°F or so for CO_2 doubling are right, that the changes will be modest enough that we, and many other living things, might be able to adapt. That appears unlikely, however, given the evidence, and in the worst case scenario, where considerably greater temperature rises occur, Earth will, as NASA's James Hansen bluntly put it, resemble "another planet."[10] Environmental author Bill McKibben has even given a name to this planet: Eaarth.[11] Ultimately the question boils down to this: Are we willing to roll the dice, with Earth lying in the balance? And is it within our rights to imperil future generations should we be wrong?

Given the wealth of scientific evidence amassed by the mid-1990s, one might rightly wonder how there could be a viable opposing position on controlling our carbon emissions. It was already difficult for any scientist to credibly argue that Earth wasn't warming, or that there was no impact on our climate by human activity (though a few still did, nonetheless, and still do). However, even among those who accepted the facts of global warming, there was still an awareness that much uncertainty exists, as we have seen. How much warming would there be? How much of the warming that had occurred could we confidently attribute to human activity? And precisely what impacts would the forecasted changes have on our daily lives? These were still wide-open questions. And while we continue to refine our understanding of climate change, many of these questions remain open to this day.

Taking steps to reduce emissions to levels that would avoid breaching 450 ppm would have been far easier in 1995 than it is now, given that we have emitted more than 100 billion tons of carbon into the atmosphere in the meantime. But even then, it would have been challenging and potentially costly. Those opposed to action could point to that uncertainty for justification. Why should we engage in potentially expensive measures to reduce greenhouse gas emissions, they could say, when the benefits are unclear? The impacts of continued fossil fuel burning in the decades ahead might be mild, some asserted; they might even be favorable. Back when I was working on my Ph.D., I would sometimes encounter such an argument from friends and acquaintances who knew that my research, at least vaguely, had something to do with the topic of global warming.

In fact, I wasn't completely unreceptive to this argument at the time. It was at least an honest, if somewhat flawed, line of reasoning. The flaw, as I would gently point out, is that the logic could just as easily work the other way. What if the problem was actually worse than our current prevailing best estimates? What if the true response of the climate instead lay at the high end of the uncertainty range? The effects in that case could be catastrophic and the costs to civilization and our environment incalculable. In fact, the argument was not my own. I had seen it advanced by Stanford University climate scientist Stephen Schneider in an article that had left an impression on me. Schneider had used the analogy of buying an insurance policy.[12] We don't purchase fire insurance for our homes because we believe our homes are going to burn down. We purchase it because if our house did burn down, it could ruin our lives. We purchase fire insurance to hedge against a perhaps quite low-probability, but undeniably catastrophic, potential outcome. It is useful to think of climate change mitigation the same way. I find Schneider's analogy as compelling today as I did then.

But even in the mid-1990s, as the scientific case had become persuasive, some critics weren't content to engage in the worthy debate to be had over climate change policy, cost-benefit analysis, and risk management. They were instead intent on preempting that debate by continuing to argue that climate change itself, if not a massive and deliberate hoax, was based on bad science. Perhaps they were afraid that general acceptance of the facts behind global warming and the risks it poses would lead the public to demand action to protect the future. Whatever their motive, they sought to deny the science altogether.

The Six Stages of Climate Change Denial

A leaked 2002 memo from leading Republican consultant Frank Luntz warned that the party had nearly "lost the environmental communications battle" and urged its politicians to double down in their efforts to deny the scientific consensus behind global warming.[13] Luntz sounded an alarm: "The scientific debate is closing [against those who deny the reality of climate change] but not yet closed. There is still a window of opportunity to challenge the science. . . . Voters believe that there is no consensus about global warming within the scientific community. Should the public come to believe that the scientific issues are settled, their views about global warming will change accordingly." Luntz suggested a full frontal attack: "you need to continue to make the lack of scientific certainty a primary issue in the debate."

Luntz has been heavily criticized for the now-infamous memo, but in his defense, he was simply the messenger. He was merely communicating the wisdom derived from careful polling and focus groups. From a purely pragmatic standpoint, he was also likely correct; the best tactic for those advocating inaction on climate change seemed to be to continue to attack the science supporting a human influence on climate, as they had for well over a decade.

The climate change denial campaign has always seemed to enjoy the same advantage as the defense in a criminal trial. Those opposed to limiting carbon emissions recognized long ago they need only cast "reasonable doubt" to convince members of the public that it is too expensive to take action. They need not present a logically consistent case. It suffices for them to attempt to simply pick holes in the scientific evidence, however inconsequential. The greater burden lies with those making the scientific case. They must present a case so persuasive that even the most skilled artists of sophistry cannot undermine it. Critics frequently argue that until science is able to offer proof of the reality of human-caused climate change, it is too early to act. Yet this is a red herring. Science can only ever offer weights of evidence, degrees of confidence, and estimated risk. "Proof" is reserved for mathematical theorems and alcoholic beverages.

While there has been little consistency over the years among the various arguments climate change contrarians have made, there is nonetheless a hierarchy to the denialist canon—what I refer to as the "six stages of denial." It goes something like this:

1. CO_2 is not actually increasing.
2. Even if it is, the increase has no impact on the climate since there is no convincing evidence of warming.
3. Even if there is warming, it is due to natural causes.
4. Even if the warming cannot be explained by natural causes, the human impact is small, and the impact of continued greenhouse gas emissions will be minor.
5. Even if the current and projected future human effects on Earth's climate are not negligible, the changes are generally going to be good for us.
6. Whether or not the changes are going to be good for us, humans are very adept at adapting to changes; besides, it's too late to do anything about it, and/or a technological fix is bound to come along when we really need it.

Contrarians have tended to retreat up the ladder of denial as the scientific evidence has become more compelling. With the ever upwardly

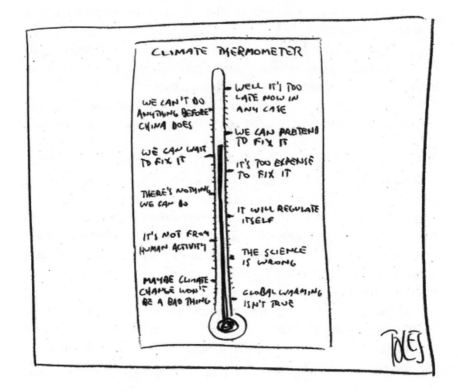

Figure 2.4: The Stages of Denial
A cartoon that uses a thermometer to gauge one conception of the various stages of climate change denial. [TOLES ©2009 The Washington Post. Reprinted with permission of UNIVERSAL UCLICK. All rights reserved.]

trending Keeling curve of CO_2 levels plain for anyone to see, few were calling into question the rise in atmospheric CO_2 by the time I had entered climate science in the early 1990s. But you could still find claims that there was no evidence of warming. John Christy and Roy Spencer of the University of Alabama at Huntsville, for example, argued in a series of papers in the early 1990s that the satellite temperature measurements they had analyzed demonstrated an absence of warming in the atmosphere.[14] Problems with their estimation procedures were established by the late 1990s, however.[15] When their original findings were found to have been almost entirely in error, they did acknowledge that Earth is warming. Spencer still contends, nonetheless, that humans are not to blame for the increase,[16] while Christy accepts that there is a detectable human contribution to the warming, but argues that future warming will be less than standard climate models project.[17]

Contrarians in the climate change debate, oddly enough, have been known to jump several rungs of the ladder of denial at once. For example, S. Fred Singer appeared to leap from warming is not occurring (stage 2) all the way to it is warming but there is nothing we can do about it anyway (stage 6) with the book *Unstoppable Global Warming: Every 1500 Years* that he coauthored in 2006.[18]

Yet one can still find some clinging to that lowest rung of the ladder, the rung I term "CO_2 denial." In a 2007 article in the social science journal *Energy and Environment*, a high school teacher in Germany, Ernst-Georg Beck, argued against there having been any increase in atmospheric CO_2 over the past two centuries.[19] It was an extraordinary claim, given the overwhelming evidence from the work of Keeling and others. What was the evidence that Beck pointed to? A hodgepodge of CO_2 records that were either compromised by systematic errors or taken from heavily polluted urban locations, where CO_2 levels are not representative of the overall ambient global levels. Keeling's son Ralph, a respected atmospheric scientist in his own right, charitably assessed Beck's paper as having "serious conceptual oversights that would have been spotted by any reasonably qualified reviewer."[20] Other commentators weren't as kind.[21]

To return to where things stood when I was completing my Ph.D. in the mid-1990s: No serious, well-credentialed, actively publishing climate scientists could be found clinging to the lower rungs of the ladder of denial. But you could find quite a few legitimately standing on the middle and upper rungs: Yes, there is warming, and some of it almost certainly is anthropogenic in nature. But just how much of it is due to human activity? How much of it might simply be due to natural variability? The implications of these very legitimate questions were potentially far-reaching. If the "noise"—that is, the natural variability—was large enough to have explained a substantial share of twentieth-century warming, it might mean that a relatively small amount of warming could be attributed to human influence. That would in turn imply that the sensitivity of the climate to increasing greenhouse gas concentrations might be at the lower end of the range of uncertainty. Some of this uncertainty persists today.

Chapter 3

Signals in the Noise

One man's signal is another man's noise.

—Origin unknown

When I first began to work with my Ph.D. adviser Barry Saltzman in the early 1990s, he, like many other climate researchers at the time, remained unconvinced that there was yet a detectable human influence on the climate. You might say that Barry was skeptical. Scientists should in fact strive to be skeptics—in the truest sense of the word. That is to say, they should always apply healthy scrutiny to any new claim or finding. True skepticism, however, demands that one subject all sides of a scientific contention or dispute to equal scrutiny and weigh the totality of evidence without prejudice. That should not be conflated with contrarianism or denialism, which is a kind of one-sided skepticism that entails simply rejecting evidence that challenges one's preconceptions. Unfortunately, the term *skeptic* has at times been co-opted by those who are not skeptics at all, but are instead contrarians or deniers, predisposed to the indiscriminate rejection of evidence supporting a human influence on the climate.

In the early 1990s, after carefully weighing all the evidence, scientists could honestly disagree with each other over whether there was a detectable human influence on the climate. They could legitimately be skeptical about whether the human climate change signal had yet emerged. The evidence was not as extensive as it would soon become, and the theoretical models that scientists then employed to study Earth's climate system were still rather primitive. For these reasons, scientists like my adviser were holding out for more evidence, while other scientists, such as NASA's James Hansen and Stanford University's Stephen

Schneider, were convinced by the evidence already in hand that human-caused climate change was indeed now upon us. I myself was closer to Barry's position than to Hansen's or Schneider's. In particular, I felt that natural climate variability might be more important than some scientists thought. Indeed, it was that very assumption that motivated my Ph.D. research topic.

Emerging from the Noise?

Natural climate variability, in the view of many climate scientists at the time, was still a plausible competing mechanism for explaining observed climate trends. There were, in fact, two fundamentally different types of natural climate variability that could potentially explain observed trends. One is external to the climate system, relating to changes in the factors that govern Earth's climate. On timescales relevant to modern global warming, the key natural external factors are the small but measurable changes over time in the output of the Sun (a fraction of a percent, but large enough to have a detectable effect on surface temperatures) and the effect of explosive volcanic eruptions, which place cooling sulfate aerosols into the stratosphere, where they can reside for some time, blocking some of the incoming sunlight and cooling the planet for several years. The problem in trying to explain the observed warming with these external natural factors is that they should have led to cooling, not warming, in recent decades: Solar output shows no increase over the latter half of the twentieth century, and two large cooling volcanic eruptions occurred near the end of the twentieth century: El Chichon in Mexico in 1982 and Mount Pinatubo in the Philippines in 1991.

There is another source of natural variability, however, that could in principle have been responsible for recent trends: internal variability—oscillatory variations of the climate system that take place without any particular external cause such as changing solar output. These climate variations are analogous to weather, the chaotic, seemingly random[1] everyday variation in the behavior of the atmosphere. On longer timescales, the atmosphere doesn't operate in isolation, but instead interacts with other, more sluggish components of Earth's climate system, such as the oceans and the ice sheets. These interactions lead to oscillations with a considerably longer timescale than typical weather fluctuations.

The most familiar example is what we know as the El Niño phenomenon, an oscillation in the climate system that arises from the way atmospheric winds and ocean temperatures in the tropical Pacific influence each other. While El Niño's origins lie in the tropical Pacific, the phenomenon triggers, in turn, changes in wind patterns, temperature, rainfall, and drought around the world. El Niño (and its flip side, La Niña) comes and goes every few years, but there are other types of natural, internally generated climate oscillations that occur over timescales of decades and even centuries.

Work in the 1980s by climate scientists such as Klaus Hasselmann of the Max Planck Institute in Germany and Tom Wigley of the Climatic Research Unit (CRU) of the University of East Anglia in the United Kingdom demonstrated that the climate system was capable of generating sizable oscillations in global temperature on timescales of a century or longer. Such long-term natural variations might potentially have been large enough to explain the overall warming trend of the past century, which was just under 1°F at that time.

By the early 1990s, various climate scientists were trying to determine whether there was specific evidence for natural long-term oscillations in the climate system that could be competing with—or even masquerading as—apparent anthropogenic (human-caused) climate change. In 1994, for example, while I was working on my Ph.D., Michael Schlesinger of the University of Illinois and coauthor Navin Ramankutty published findings suggesting that such oscillations do indeed exist.[2]

They employed a simple theoretical climate model fed with data on anthropogenic impacts (the warming effect of greenhouse gas increases partially offset by the cooling impact of industrial sulfate aerosols) to estimate how the globe should have warmed in the absence of natural variability. Then they subtracted this estimate from the actual temperature record to see what residual variability, presumably natural, was left unexplained. They found some evidence in that leftover variability for the existence of natural temperature oscillations occurring in fifty- to eighty-year cycles. These oscillations could explain why some regions had warmed more than other regions, and why some areas of the North Atlantic had actually cooled in the latter half of the twentieth century. The oscillations, however, could not explain the overall warming of the globe since the nineteenth century. That warming was accounted for in their analysis primarily by an increase in greenhouse gas concentrations from fossil fuel burning—anthropogenic effects. Meanwhile, in my own work I had been employing both climate modeling and data analysis methods in an attempt to understand and identify these multi-decadal climate oscillations.

My collaborator Jeffrey Park, a seismologist, and I used an entirely different statistical approach than that employed by Schlesinger and Ramankutty, one borrowed from the field of seismology.[3] Earthquakes and other seismological disturbances yield waves—oscillations—that travel through solid earth. These waves have a complex spatial pattern as they reach Earth's surface. The surface disturbances are detected by a seismograph, which responds to the subtle vibrations of the ground and registers these vibrations in the form of a trace called a seismogram. Much like medical technicians use electromagnetic oscillations (X-rays, to be precise) in the form of CAT scans to examine the interior of the human body, seismologists study seismograms to infer the structure of Earth's interior. Jeff and I realized that we could apply the same techniques to search for oscillatory signals in the climate record. Just as with seismograms, we had the benefit of a large array of climate measurements around the globe. Much as seismological signals consist of oscillations obscured by random local effects (noise), the climate signals we were looking for were climate oscillations buried in their own type of noise: instrumental biases, episodic climate events like volcanic eruptions, and other competing impacts, including anthropogenic climate change. Our method attempted to identify global-scale climate oscillations with a particular periodicity, buried in the noise, and it tested whether the putative oscillations were statistically significant (i.e., that the apparent signal was sufficiently unlikely to arise from the chance random fluctuations of the noise). Just as in my childhood tic-tac-toe discovery, we had found what scientists and mathematicians call a "trick," a shortcut in systematically attempting to solve a problem, in this case adapting a method originally designed for seismological signals to search instead for climate oscillations.

To me, this kind of work is what made science so exciting. We could now apply our method to search for oscillatory signals in a dataset of global surface temperature records spanning the past century. Our findings, published around the same time as Schlesinger and Ramankutty's study, provided further evidence for oscillations in the surface temperature record.[4] Some of them had periodicities in the three- to seven-year range and were related to El Niño, while others had period-icities in the interdecadal (ten- to twenty-year) range and appeared to be related to the interaction of the atmosphere with the sluggish sub-tropical ocean gyres—ocean current systems that include the well-known Gulf Stream in the Atlantic and the Kurushio current of the Pacific.

The longest oscillation detected in our study, however, was centered in the North Atlantic region and had a periodicity of sixty to eighty years, consistent with the multidecadal oscillation Schlesinger and Ramankutty had found independently. What's more, other scientists were finding evidence that such oscillations were being produced in theoretical climate models by the interactions between the ocean and atmosphere in the North Atlantic, with the long timescales set by the very sluggish subpolar North Atlantic ocean current systems.[5]

None of these studies suggested that such multidecadal oscillations could explain away modern global warming, though. In fact, those os-cillations were only of secondary importance in characterizing long-term variations in the global surface temperature record. The primary pattern was instead one of long-term warming of the globe, and that long-term warming had already been attributed to human influences.

The multidecadal oscillation I'd helped discover would nonetheless become a cause célèbre among climate change contrarians. It would even get a name: the "Atlantic multidecadal oscillation" (AMO)—a moniker I coined off the cuff in a phone interview with science writer Dick Kerr.[6] The AMO appeared to be real, and at least partly responsible for certain phenomena, such as the acceleration of recent warming in parts of the Arctic, that some had attributed to anthropogenic climate change. Other phenomena that have been blamed on the AMO, such as the increase in Atlantic hurricane activity in recent decades, arguably have nothing to do with it all.[7] That hasn't stopped climate change contrarians, how-ever, from dragging out the AMO as a favorite catch-all explanation for just about any observed climate trend. At times I have felt like I helped create a monster.

A Century Is Not Enough

The evidence for AMO-like oscillations in the climate system seemed reasonably strong. What remained vexing to us, however, was that the short instrumental temperature record simply wasn't adequate for any scientifically confident characterization of such a signal. Covering little more than a century, instrumental global temperature observations could capture at most one complete cycle of the oscillation we were studying—not enough to detect a persistent pattern. The apparent oscillation could simply be an artifact of having such a short and noisy record from which to tease out a natural climate oscillation from long-term climate trends.

It was the search for more compelling evidence of these oscillations that first led me to examine longer-term proxy records of climate (records that, as you may recall, stand in past centuries as a proxy for the direct measurements of climate that modern instruments make possible). Well, that, and a bit of serendipity. My parents were attending an event in my hometown of Amherst, Massachusetts. They happened to strike up a conversation with a University of Massachusetts professor over a bottle of wine (*in vino veritas?*). As it turned out, he was a climate researcher. My parents said something to the effect of "What an interesting coincidence! Our son is doing his Ph.D. in climate research." The end result was that the professor, Raymond Bradley, and I were set up on a sort-of scientific blind date. We would meet to talk science the next time I was in town.

It was evident upon our first meeting that there were synergies between our respective research interests. Ray's research specialty involved the use of climate proxy data such as that derived from tree rings, ice cores, corals, lake and ocean sediments, and other natural archives of past climate. These were to become central to our collaboration and to many other studies of changing climate conditions.

Most widespread of all these climate proxies are tree ring data. Depending on the region and environment, tree growth may largely be controlled by growing season conditions such as the warmth of the growing season (typically summer) or the precipitation that falls during the rainy season. One can therefore analyze annual growth rings (the width of the rings and, in some cases, the density of the wood formed) for insights into past climate changes.

One obvious limitation with tree ring data is that they are restricted to the continental regions where trees grow. That is a major problem for the mostly water-covered Southern Hemisphere, but it's also problematic for the Northern Hemisphere. Neither are tree ring data available in polar regions, where tundra and permanent ice cover prevail. And tropical tree species typically do not have annual growth rings (look at a palm tree stump sometime if you don't believe this), limiting the usefulness of tree ring data as tropical climate proxies. Climatically useful tree ring proxies are thus primarily restricted to the midlatitude continental regions, leaving much of the globe un-sampled by this method. Fortunately, other types of proxy data can help fill the gaps.

Corals, for example, are found in oceanic environments, generally in tropical regions, making them complementary to tree rings in the regions of the globe they sample. Corals assimilate oxygen atoms from the ocean in the form of the carbonate ion, CO_3^{-2}, as their layers of calcium carbonate ($CaCO_3$) skeletons are formed. The ratios of the two main stable isotopes of oxygen (the common, lighter ^{16}O versus the less common, heavier ^{18}O) in the corals' annual growth bands, as it turns out, depend on the temperature and salinity (in turn influenced primarily by rainfall) of the near surface ocean waters they grow in. Other chemical measurements from corals such as the ratio of the elements strontium and calcium in the annual skeletal layers provide complementary climate information (the ratio tends to decrease as temperature increases).

In the polar regions, where neither tree ring nor coral records are available, yet another important climate proxy, ice cores, proves useful. The ratios of the ^{16}O and ^{18}O oxygen isotopes in the frozen water (H_2O) that constitutes the annual ice layers recovered from ice cores can tell us about past atmospheric temperatures at the time the snow or ice was laid down. Other information from ice cores, such as dust layers and trace chemical constituents, can provide further hints about both past climate variations and what may have driven them, such as volcanic eruptions and solar output changes. Sediment cores recovered from closed basin lakes in high-latitude environments can also tell us about past seasonal temperature variations, as these variations influence the amount of sediment flushed into the lake during a particular snowmelt season.

These are but a few examples of climate proxy records that can potentially be analyzed to assess past climate phenomena. As the different types of proxy records provide information from complementary regions, and often for complementary seasons, a diverse (multi-proxy) network of these data proves particularly useful in characterizing past climate changes.[8] Since my Ph.D. research was aimed at isolating and understanding long-term climate oscillations, such a network of proxy records could yield a desperately needed longer-term dataset to analyze. It was a logical scientific marriage. Ray and I proceeded to collaborate, even as I was still finishing up my Ph.D. research.

Although proxy records are admittedly indirect and imperfect measures of climate, they could allow us to do something that the brief instrumental record simply could not: to go back centuries in time. Now we could see whether those multidecadal oscillations discovered in the data were real. Were they evident several centuries ago, or were they just figments of the modern record? To answer this question, we applied the oscillation detecting machinery that Jeffrey Park and I had developed to a considerably longer multi-proxy dataset that Ray had used in previous paleoclimate research. We concluded that these oscillations were indeed real.[9] They continued to occur throughout a period stretching over six centuries, and they followed the regional pattern we expected, the signal seeming to be especially large in the North Atlantic.

That wasn't all we found in the data, however. We were able to detect an even longer-term apparent oscillation in the proxy data with a periodicity of roughly 250 years. This oscillation, too, had greatest amplitude in the region surrounding the North Atlantic, and it described a transition in climate conditions that occurred sometime prior to A.D. 1500, coinciding approximately with what is now generally accepted as the boundary, at the global scale, between the so-called Medieval Warm Period and the Little Ice Age. Could our longest-term oscillation, we wondered, be telling us something about that transition?

At the time, we couldn't really address that question. First, our proxy dataset only went back to A.D. 1400, perhaps close to the transition point between the two intervals, but not firmly back into the interval generally considered to constitute the medieval warm period. Second, we were not reconstructing temperature patterns *per se*, but simply evaluating whether the various proxy records were varying in a coherent oscillatory fashion. While most of the records were believed to reflect local temperature changes, we had not explicitly calibrated them against an actual quantitative temperature scale. To identify the relationship between the behavior in these proxy data and the long-term changes they implied in surface temperature patterns—in other words, to reconstruct past temperature patterns from the proxy data—we would need a different approach.

The Medieval Warm Period

Seemingly quite relevant to the issue of whether modern warming might be natural in cause, at least in substantial part, was the purported existence of a period in our not-so-distant preindustrial past characterized by warmth rivaling that of the present day. Evidence of such warmth would not, in and of itself, necessarily contradict a human role in the current warming; after all, that proposition, as we have seen, is based on multiple lines of evidence. However, if warmth less than a thousand years ago rivaled modern warmth, it might seem to support a far larger role for natural climate variability, and the possibility that a large fraction of the current warming could itself be natural.

Investigations into the evidence for a medieval warm period begin with the work of Hubert Lamb (1913–1997), a prominent British scientist who founded the University of East Anglia's Climatic Research Unit. Lamb's seminal work on the climate changes of past centuries has, unfortunately, been among the most misunderstood and most misrepresented in all of paleoclimatology.

In the 1960s, Lamb attempted to estimate past temperature changes in central England from various forms of qualitative and, in some cases, rather anecdotal information (such as reports of vineyards in southern Britain and Vikings in Greenland during medieval times, reports of frost fairs on the Thames, and icebergs off the coast of Norway during later centuries).[10] From the 1960s to the 1980s, Lamb produced various versions of a curve depicting temperature variations over the past thousand years. The curve was at best a rough approximation and was certainly not offered by Lamb as a quantitative reconstruction of past temperature changes (with the exception of the portion of the record since the seventeenth century, which was based on actual thermometer measurements from central England). Nor was the curve ever supposed to represent temperature variations outside one small subregion of Europe. All of that notwithstanding, the Lamb curve was the only one available purporting to depict temperature changes over past centuries at the time of the IPCC First Assessment Report in 1990, wherein it had been featured.

The Lamb curve appeared to suggest a warm period between roughly A.D. 1000 and 1400 during which temperatures in central England were rather warm, followed by an interval of relatively cold conditions from roughly 1500 to 1900, the Little Ice Age. The medieval warm period was depicted as even warmer than the present by a few tenths of a degree in Lamb's estimate. This latter feature would be celebrated for decades to come by climate change contrarians who would continue to point to the Lamb curve as evidence that modern warming is not unusual, even as Lamb's work was supplanted by an increasing number of more robust and quantitative studies suggesting otherwise.

But what was the "present" in the context of the Lamb curve anyway? The record Lamb compiled captured temperature averages only over fifty-year blocks, and the end of the record roughly corresponded to the mid-twentieth century. It did not reflect most of the recent warming, which has taken place during the past half-century. In a recent reassessment of Lamb's original work, Phil Jones—the current director of CRU—and a large group of other paleoclimate researchers (including me) tried to answer the question of whether Lamb's reconstruction actually indicated greater than present-day warmth.[11] They overlaid Lamb's original estimate (as depicted in the 1990 IPCC report) with an up-to-date record of central England temperatures. This comparison made it clear that Lamb's original estimate (if it is indeed reliable; Jones and others have argued otherwise) implies medieval warmth in central England that may have indeed rivaled mid-twentieth-century warmth. The warmth did not, however, reach current levels.

As more widespread climate proxy records were developed through the early 1990s and researchers were able to begin to piece together a picture of the larger-scale patterns of climate during medieval times, a more nuanced view of this period emerged. While various regions of the globe—including parts of Europe, China, western North America, and Australia—appear to have been relatively warm by modern standards during some part of the medieval era, the warmth did not appear synchronous among the various regions. Other regions, including, for example, the southeastern United States and the Mediterranean, showed no evidence of warmth that rivals modern levels.[12] Moreover, many of the more profound changes in regional climate that paleoclimate

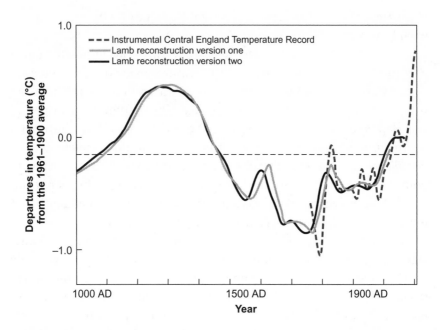

Figure 3.1: The Medieval Warm Period
The graph shows two slightly different versions of Lamb's early qualitative reconstruction of temperatures in central England over the past thousand years,
compared with more modern instrumental temperatures beginning in the seventeenth century. One vertical tick mark spacing corresponds roughly to a 1°C
temperature variation. [Adapted from Jones et al., "High-Resolution Paleoclimatology of the Last Millennium: A Review of Current Status and Future Prospects,"
Holocene (2009): 34.]

researchers were discovering were tied to shifts in atmospheric circulation and rainfall patterns, rather than to changes in temperature. For
these reasons, paleoclimate researchers have increasingly favored the
use of the term "medieval climate anomaly" (MCA) over the potentially
misleading "medieval warm period" moniker.[13]

Toward the Abyss

As I was completing my Ph.D. in 1996, I remained interested in using
proxy climate records to study the longer-term patterns of natural cli-

mate variability. I remained intrigued not only about the role of what I had termed the Atlantic Multidecadal Oscillation, but also how the El Niño phenomenon had behaved in centuries past. I was also interested in teasing out the role of natural "forcing" factors (that is, natural external factors that force the climate to change), such as volcanoes and changes in solar output, in explaining past climate changes. Though I accepted the mainstream view that there was now a detectable human influence on the climate and I was vaguely aware of some of the contentiousness that surrounded that conclusion, anthropogenic climate change was barely on my radar screen. That was about to change, however. If I had avoided entanglement in the thorny issue of humanity's role in global warming thus far, I would soon collide with it head on.

Chapter 4

The Making of the Hockey Stick

The scientists tell us the 1990s were the hottest decade
of the entire millennium.

—President Bill Clinton, State of the Union Address (January 27, 2000)

Although scientific revolutions in how we see the world do occur, the bulk of our scientific understanding comes from the cumulative impact of numerous incremental studies that together paint an increasingly coherent picture of how nature works. The hockey stick was no different. To employ a mixed sports metaphor, the hockey stick did not suddenly appear out of left field. Rather, it arose as a logical consequence of decades of work by paleoclimate researchers that led to increasingly rich networks of climate proxy data and the introduction of new ways to use such data to reconstruct past climates. My colleagues and I were the beneficiaries of this substantial body of past work.

The Reconstruction Zone

A half-century ago, when the British climatologist Hubert Lamb set out to trace temperature trends over the past millennium, he obviously wasn't attempting to address the issue of anthropogenic climate change. It wasn't considered an issue at the time. Reconstructing past climate change was interesting in its own right, however. It was an engaging piece of puzzle-solving, the sort of thing that excites scientists. How could one deduce, from the sparse and imperfect clues available, how climate varied in the past? Could one perhaps even explain some key historical events, such as the rise or collapse of various civilizations in terms of climate changes? Big picture, gee-whiz stuff.

These sorts of questions, rather than the threat of human-caused climate change, drove dozens of paleoclimate researchers around the world to attempt to piece together, over several decades, the riddle of how Earth's climate had varied over time. By the mid-1990s, it was becoming possible to use proxy records such as tree rings and ice cores to build year-by-year chronologies of climate change at many locations around the globe, reaching back centuries and in some cases millennia. These proxy records could be used to address, for example, the question of how cold it really was during the "year without a summer" of 1816 that followed the explosive Tambora volcanic eruption of April 1815. How did that eruption influence rainfall and atmospheric circulation patterns around the world? Was there a relationship between the 1791–1792 El Niño and the deadly failure of the Indian monsoon during that period, which led to drought, famine, and the death of millions?

Thanks to decades of curiosity-driven work by paleoclimatologists, extensive global networks of long-term proxy data were available for analysis by the early 1990s. Of course, by that time, climate change itself and what it might portend for the future were far more prominent issues than they had been in Lamb's day. By featuring the Lamb curve in the First Assessment Report in 1990, the IPCC had suddenly made proxy reconstructions of past temperature policy-relevant. The Lamb curve, as we've seen, implicitly raised the issue of whether modern warming really was unusual; it seemed at the time to resolve that proposition in the negative. Like any controversial and high-profile scientific finding, the curve became a straw man for other scientists to either confirm, refine, or reject. And indeed, other scientists would soon approach the issue using more up-to-date data and more rigorous approaches.

The first truly quantitative reconstruction of past temperature changes at the hemispheric or global scale was attempted by Ray Bradley and Phil Jones in 1993.[1] Their approach was perhaps primitive by today's standards, but their contribution to our understanding was significant. The two researchers assembled a set of about two dozen proxy records representing temperature variations in distinct regions of the Northern Hemisphere (largely during the summer season), supplemented with the few long historical temperature records available. They formed a composite of the proxy records representing the average over all regions and then scaled the composite to match the scale of the modern instrumental temperature record[2] to produce an estimate of Northern

Hemisphere average temperature back in time. The Bradley and Jones reconstruction stretched back to A.D. 1500 and was adopted as the new standard in the 1995 IPCC Second Assessment Report, replacing—to the chagrin of climate change contrarians—the considerably less quantitative or reliable Lamb curve of the 1990 report.

The Bradley and Jones reconstruction did not encompass the medieval period, but it did, for the first time, characterize the extent of Northern Hemisphere average cooling during the period known as the Little Ice Age. Despite the existence of greater cooling in some regions (e.g., Europe) at certain times (e.g., the seventeenth century), the temperature changes recorded in the various proxy records were not synchronous and, in some cases, were even of opposite sign. As a result, the average cooling over the entire Northern Hemisphere at the height of the Little Ice Age was modest—less than 1°C cooler than the late twentieth century, nearly a factor of two smaller than what Lamb had originally estimated for central England.

Numerous efforts were made in the ensuing years to advance the science further. In 1995, Bradley collaborated with solar physicists Judith Lean and Juerg Beer to investigate the relative roles that both natural and human factors might have played in the long-term temperature changes documented by the Bradley and Jones reconstruction.[3] The group built on earlier studies by scientists such as John Eddy,[4] who correlated estimated changes in solar output (derived from historical sunspot measurements and radiocarbon data)[5] with Lamb's estimates of past temperature changes in an attempt to deduce the impact on temperatures of an apparent lowering of solar output during the peak of the European Little Ice Age in the latter half of the seventeenth century. During this period—termed the "Maunder minimum," after British astronomer Edward W. Maunder (1851–1928) who had first studied this anomalous period—no sunspots were observed at all, implying a substantial reduction in solar output at the time. Eddy concluded that there was a statistical connection between the Maunder minimum and coincident changes in climate, including not only cooling temperatures in Europe, but also shifting drought patterns in North America. With the data available at the time, he was unable to provide a meaningful quantitative estimate of the effect.

Bradley and his collaborators revisited the problem with aid of the more extensive quantitative estimates that were now available. They confirmed that the dip in solar output during the Maunder minimum appeared responsible for a cooling of a bit less than 0.5°C during the Little Ice Age. A modest increase in solar output might also have played a role in the early twentieth-century warming, they concluded, but solar impacts could not explain most of the warming of recent decades. Paleoclimatologist Jonathan Overpeck and collaborators reached similar conclusions in a subsequent 1997 study of Arctic temperature trends.[6]

These more recent studies still suffered some limitations. The temperature reconstructions only resolved decadal timescale variations, not individual years, so nothing could be said about, say, the "year without a summer" of 1816 or the apparent mother of all El Niño events in 1791–1792. Moreover, these studies reconstructed only a single time series representative of hemispheric mean temperatures and thus could not establish precisely which regions were warm and which cold in a given year; in other words, they didn't produce a spatial pattern of relative temperature. Finally, these studies didn't provide any margin of error, or "error bars," to indicate the uncertainty in the estimated temperature changes given the imperfect and uncertain nature of the proxy records. While the recent decades were nominally the warmest in these reconstructions, without knowing how much uncertainty there was in the estimates, it was impossible to rule out, with any degree of confidence, that past temperatures might have been as warm as today's.

It was at this point that I stepped onto the paleoclimate reconstruction scene. After defending my Ph.D. dissertation in the spring of 1996, I was funded on a Department of Energy postdoctoral fellowship[7] to continue my paleoclimate work with Ray Bradley at the University of Massachusetts (U. Mass) in Amherst, the town I grew up in. I would also be working closely with Ray's colleague Malcolm Hughes from the University of Arizona, a specialist in the use of tree ring data.

After having left for college more than a decade earlier, I was quite literally returning home. I moved into an upstairs apartment in the house I'd grown up in, just down the street from the center of town. I now constantly ran into old friends and acquaintances, including some of my old high school and elementary school teachers. My father had recently retired as a math professor from U. Mass, and we only narrowly missed being colleagues. Indeed, some of my U. Mass colleagues were parents of kids I'd grown up with. It was all a bit odd. But Amherst was a nice place to live as a "thirty-something," and I couldn't have been happier with my newfound academic home in the U. Mass Department of Geosciences. It was a lively, friendly department. And it also happened to back up on the U. Mass faculty tavern, a favorite gathering place on Friday afternoons and evenings.

The Game Begins

My postdoctoral research was aimed at developing and applying a new statistical approach to the problem of proxy climate reconstruction. Seeking to improve upon previous efforts, I wanted to reconstruct surface temperatures not just for individual decades, but also for individual years. Moreover, I was interested in reconstructing the underlying spatial patterns of temperature variation, not simply average trends over large regions like the Northern Hemisphere or the Arctic. Reconstructing these spatial patterns would not only tell us where it was warm or cold in any particular year, but also would give us insight into the workings of the climate system. It was these patterns that could tell us about the long-term role of the El Niño phenomenon, for example, or the wiggles in the track of the jet stream from year to year called the North Atlantic Oscillation (NAO). How were these patterns influenced by external factors such as volcanic eruptions and changes in solar output? It was questions such as these, rather than the effects of humanity on climate, that I was seeking to address.

There was already a rich history of using sophisticated statistical methods in certain subfields of paleoclimatology, such as dendroclimatology, the study of tree ring data to infer past climate change. Over the previous few decades, researchers such as Hal Fritts, the godfather of dendroclimatology, Ed Cook of Columbia University's Lamont Doherty Earth Observatory (LDEO), and Keith Briffa of the University of East Anglia's Climatic Research Unit had developed various approaches to reconstructing patterns of temperature, rainfall, surface pressure, and drought from tree ring data. In dendroclimatology, one is often lucky enough to be dealing with a set of proxy data with similar attributes, such as a network of precipitation-sensitive tree ring records from, say, the western United States. And there were well-established methods for relating the patterns in the tree ring data to the patterns of climate.

A climate reconstruction could thus be formed by establishing a statistical relationship between the two datasets over their common years of overlap (typically the twentieth century), called the "calibration" or "training" period, which could then be projected back in time for periods long ago when instrumental climate records were not available.[8]

The circumstances of the problem my collaborators and I were attacking were somewhat different. Unlike the typical situation encountered in past work, we were making use of diverse proxy data; our dataset included regional networks of tree ring data, but it also included data from ice cores, corals, and lake sediments, and a smattering of historical climate records. It wasn't appropriate to lump all of the proxy records together. We had far more tree ring data than other types of proxy records, yet the tree ring data represented only a restricted region of the globe, the midlatitude continents. The smaller amount of data drawn from corals, ice cores, and lake sediments represented the other key regions: the oceans, the tropics, and the poles. Allowing the sheer amount of tree ring data to overwhelm the less abundant information from other proxy records would weight our results primarily toward the midlatitude continents, whereas we were seeking to reconstruct patterns over the entire surface of the globe, land and ocean, from equator to pole.

We used a statistical procedure called "principal component analysis" (PCA) to get around the problem. PCA can be used to represent a very large dataset (be it modern thermometer-based temperature measurements distributed over the globe or a set of tree ring records reflective of temperatures distributed across a region such as North America) in terms of a small number of patterns in space and time that describe the most variation in the data. Each pattern can be described by a combination of its temporal history (the principal component or PC time series) and its spatial variations (the empirical orthogonal function or EOF).

This widely used statistical tool would become a major bone of contention among hockey stick critics, so it is worth a special effort to understand how and why it is used. To that end, let us consider a simple synthetic example (see figure 4.1 on page 46) where only two regions make up the world: the hemisphere west of the Greenwich meridian (the west) and the hemisphere east of it (the east). Temperatures, furthermore, are always uniform within each of the two regions. In our example, surface temperatures overall warm in the course of a century (top row), but they fluctuate differently for the west (panel a) and east (panel b). These fluctuations can be measured relative to the average temperature over the century as a whole—a number that defines the zero of the temperature scale (y-axis) and, thus, determines the vertical centering of the data. In the west, temperatures start out cold, about 1 degree below the long-term average. There is substantial warming in the century's first decades, bringing temperatures nearly 0.5 degrees above the long-term average by midcentury. That above-average warming is offset by slight cooling during subsequent decades, bringing temperatures back down roughly to their long-term average by century's end. In the east, we see the converse of this pattern. Slight cooling occurs during the first half-century, when temperatures spend much of the time below their long-term average. But that cooling is more than overcome by a substantial warming during the second half, with temperatures ending up nearly 1 degree above the long-term average by the end of the century.

The variations in surface temperature over the globe as a whole can be fully described mathematically in terms of just two spatiotemporal patterns of variation in the data. One of these two patterns (bottom row) is characterized by a linear increase in temperature over the hundred-year period of just under 1°C in amplitude. In this characterization, the warming is uniform across both west (panel e) and east (panel f) and is described by a PC series that is a simple upward ramp (for purposes of this example, you could think of this pattern as global warming).[9] The other pattern (middle row) is a roof-shaped temperature variation of 1°C amplitude that plays out oppositely in the west (panel c) and east (panel d): The west warms during the first half-century and cools during the later half-century. The east does just the opposite, yielding what looks like an inverted roof. This second pattern in the temperature record can be described, for the west hemisphere, by a PC series that increases toward the middle of the century and decreases thereafter, with precisely the reverse evolution for the east.[10] This pattern could be, say, the imprint on the global temperature record of a single cycle of a multidecadal climate oscillation like the Atlantic Multidecadal Oscillation (AMO) described in chapter 3, which acts to redistribute heat from one part of the globe to another (e.g., from west to east), but doesn't change the average temperature of the globe. Adding together the two patterns yields the total pattern of temperature variation (i.e., top row; adding c and e, gives a, while adding d and f gives b).

We can rank the two contributing patterns by what fraction of the total variation in the overall data (temperature, in this case) they explain. These rankings will depend, however, on what baseline we choose for the data, that is, how we center the data with respect to the y-axis. It is conventional to center the data about their long-term average, as we have done in the above example. By this convention, the relative temperature departures average to zero over the full hundred-year dataset. With the data centered that way, the second of the two patterns (c and d) of warming/cooling (in the west) and cooling/warming (in the east) explains 60 percent of the variation and thus constitutes PC#1. The global warming pattern (e and f) in our example turns out to be the second most prominent pattern, explaining 40 percent of the variation in the data. It is PC#2.

In more realistic examples, there are generally a greater number of significant patterns of variation in the data. Moreover, rarely can they be extracted as cleanly as in this example, since there is typically some degree of contamination by noise, be it random disorganized temperature variations or biases and errors in the data. This example nonetheless demonstrates how PCA can efficiently describe the few leading patterns of variation in a larger dataset, a step that is essential if one is interested, as with paleoclimate reconstructions, in establishing robust relationships between patterns in potentially quite large and noisy datasets. Applying PCA in that case can help sort out the climate signals (the key, most robust patterns of variation in the datasets) from the noise in which they are immersed.

Using PCA, we were able to represent the main information in each of the various regional networks of tree ring data (e.g., North America, Eurasia) in terms of a small number of representative PC series. With the tree ring data that dominate the midlatitude continents[11] represented by a few leading patterns, the handful of coral series from the tropics and ice core series from the Arctic would have commensurate representation in the dataset. As far as each proxy's opportunity to help tease out the past patterns of climate variation was concerned, it was now a fair fight.

Our statistical method established a relationship between the proxy data (which extend several centuries back in time) and the modern surface temperature record during the period of the twentieth century where both datasets overlap. This procedure required finding a way to

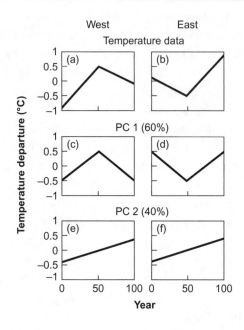

Figure 4.1: PCA Example: Spatial and Temporal Variations in Temperature Data
This example shows how PCA can be used to characterize efficiently a global
temperature dataset (top) in terms of a small number of patterns in space and
time. In this case, just two patterns (middle and bottom) characterize the data.
The horizontal axis is time in years, and the vertical axis depicts the relative de-
parture of temperature in degrees from the average over the baseline, 0 (in this
case, the average over the full hundred-year period).

represent efficiently the information from the modern instrumental
record. Spanning a century, the instrumental surface temperature
dataset contained more than a thousand months of data at more than a
thousand locations around the globe.[12] But only a handful of patterns
could potentially be captured by our noisy proxy dataset. The temperature data thus also needed to be represented in terms of a modest number of its most prominent underlying patterns. We recognized that the
same tool we had used to simplify the representation of dense tree ring
data networks—PCA—could again prove useful. Using PCA, we could
represent the key information in the instrumental temperature dataset
with just a dozen or fewer distinct patterns. The leading temperature
pattern related to the overall warming of the globe, while subsequent

patterns related to phenomena such as El Niño, the North Atlantic Oscillation, and the Atlantic Multidecadal Oscillation.[13]

With the aid of our statistical method, we simply needed to figure out the combination of these dozen or so temperature patterns that most closely matched the behavior of the proxy series we had during the twentieth-century period of overlap. Once those relationships had been established, they could be tested through a process known as "validation" or "verification." In common scientific parlance, these terms describe the process of independent confirmation of a previous finding. In the present context, unlike say, performing a new experiment that backs up previously reported results, one is instead seeking to demonstrate that a statistical model for a phenomenon can successfully predict independent data that were not used in establishing the model in the first place. In other words, it is a way of demonstrating that the statistical relationship is real, and not just a fluke of statistics. Rather than using the full available instrumental record for the calibration of the proxy data, one leaves some subinterval of the instrumental record aside for testing purposes. The proxy data are then calibrated over the shortened calibration interval, and the resulting statistical model is used to predict variations in climate over the remaining subinterval that was not used in the calibration process. Since no information from that part of the instrumental record was used to calibrate the proxy data, the extent to which the climate variations predicted by the statistical model match the climate changes that were actually observed over that time interval provides a true test of the reliability of the climate reconstruction.

This process of validation or verification is essential in establishing the credibility of a proxy-based climate reconstruction. There is, however, an important compromise that must be struck: The shorter the calibration period, the less robust the resulting statistical model, but the shorter the validation period, the less reliable the independent validation test.[14] The information from the calibration and validation process can also be used to estimate statistical uncertainties in the reconstructions, yielding the important margins of error (variously referred to as "error bars" or "confidence intervals") that characterize the envelope of uncertainty surrounding the climate reconstructions.

The instrumental data become increasingly sparse as one goes back in time before the twentieth century; indeed, only a handful of records are available as far back as the early nineteenth century. On the other hand, proxy data are available only up through the time they were collected. Many of the key proxy records were obtained during the 1970s or early 1980s, and proxy data become increasingly sparse after that period. We consequently chose the data-rich interval of 1902 to 1980 for our calibration period, leaving aside the earlier nearly half-century of sparser instrumental data for the validation tests.[15] These tests established that the statistical reconstructions were skillful—that is, meaningful in a statistically rigorous sense[16]—as far back as A.D. 1400, but not earlier than that.

Onto the World Stage

When we initially wrote up our results for publication, we focused on what we felt was most scientifically interesting, for example, that we recovered an unusual pattern for the 1816 "year without a summer" that indicated a very cold Eurasia and lower than average temperatures in North America (observations that are independently confirmed by historical accounts), but a warmer than usual Middle East and Labrador (who knew?). Or that we had independently affirmed anecdotal accounts that there was a whopper of an El Niño event in 1791—a year that, according to our reconstruction, also happened to be a comparative scorcher for Europe and a large part of North America. Then we did the least scientifically interesting thing one could possibly do with these rich spatial patterns: We averaged them to obtain a single number for each year: the Northern Hemisphere average temperature.

That single aspect of our work got all the attention. In truth, it did also represent an advance. Compared to previous Northern Hemisphere temperature reconstructions, ours went further back, it resolved individual years rather than just decades, and it had error bars associated with it. Thus, we were able to draw the specific conclusion that, even taking into account the margin for error, three recent years—1990, 1995, and 1997—all appeared warmer than any other in the past six centuries.

Our study, which has come to be known as "MBH98" for the authors—Mann, Bradley, Hughes—was published in *Nature* on April

22, 1998—Earth Day.[17] I was caught completely off-guard by the amount of media attention the article received. Generally, one is lucky to get a nibble or two from the local media in response to press releases on a published scientific paper. This time was different. No sooner had the press releases gone out (one from U. Mass, another from *Nature*, and a third from the National Science Foundation) than the phone calls began coming in nonstop. Our study was written up in the *New York Times, USA Today, Boston Globe*, and a host of other major U.S. newspapers. Articles soon appeared in *Time* magazine and *U.S. News and World Report*. We even made it into *Rolling Stone* (though not the cover). I was asked in one afternoon to do television interviews with CNN, CBS, and NBC. In the CBS interview, John Roberts put it to me bluntly: "So does this prove humans are responsible for global warming?" He repeated the question at least three times during the interview, clearly not having gotten the money quote he was fishing for. I wouldn't take the bait. I repeated that our results were "highly suggestive" of that conclusion, but I wouldn't go further than that. I well knew that establishing that recent warming is anomalous in a long-term context alone did not establish that human factors were responsible for it. Any conclusion about *causality* required the use of climate models to estimate the relative contributions of the various factors, including human increases in greenhouse gas concentrations, hypothesized to be responsible for the observed changes.

There are several reasons that our paper might have received more than the usual expected attention. The globe had experienced record-breaking warmth that winter. The first three months of 1998 were the warmest on record (in western Massachusetts, where I was living, it barely felt like we'd had a winter). Temperatures had likely been spiked, to some extent, by a fluke of nature, a particularly large El Niño event. It was in part due to this fluke that the 1998 record for the hottest year in the instrumental record had still not unambiguously been broken by 2011 (one group has 2005 and 2010—in a statistical tie—narrowly beating it out, but another group has 1998 still holding the title).[18] In any case, 1998 was as of that date the warmest year in the instrumental record; with the advent of our study, "warmest on record" meant not just in 150 years but in at least 600!

That our paper coincidentally happened to be published on Earth Day no doubt gave journalists an extra news hook to cover the study.

Some commentators attached an almost diabolical significance to the timing, as if *Nature* was somehow conspiring with the world's environmental activists. The truth is much less interesting; the publication date at *Nature* is determined by the date of a paper's final acceptance and placement in the journal's publication queue.

Extending the Handle of the Stick

The original MBH98 hockey stick had a comparatively short "handle,"[19] extending back six centuries, and did not reach back far enough to establish whether a medieval warm period actually existed, let alone when it might have ended or begun. Some went so far as to attack us for only going back six centuries, claiming we had intentionally stopped short to avoid running into the putative medieval warm period.[20] We were guided, of course, only by what the validation tests had objectively indicated: that with the data we had, we couldn't go back any further and still obtain meaningful estimates. Why would anyone impugn our integrity so brazenly and make an allegation so false? Boy, would I later be in for a lesson.

Shortly after our study was published, Phil Jones and collaborators published another attempt to trace the Northern Hemisphere's mean temperature. They used the same composite approach as had Bradley and Jones in 1993, but this time they extended the estimate back over the entire past millennium.[21] We were a bit skeptical. Our own statistical tests, after all, had told us that it was not possible to obtain a meaningful reconstruction further back than six centuries, using more or less the same proxy data.[22] Nonetheless, my colleagues and I decided it was worth taking a closer look.

We decided to examine more closely the two dozen proxy records we had that extended back six centuries or more, which included those used by Jones and colleagues, and several others. I performed a series of so-called sensitivity tests, in which various proxy records are removed or—to use standard statistical terminology, "censored"—from the network, and the sensitivity of the results to those records is gauged by noting how much of an effect their removal has on the result. I titled the computer directory "censored" accordingly—a choice I would later regret.

The tests revealed that not all of the records were playing an equal role in our reconstructions. Certain proxy data appeared to be of critical importance in establishing the reliability of the reconstruction—in particular, one set of tree ring records spanning the boreal tree line of North America published by dendroclimatologists Gordon Jacoby and Rosanne D'Arrigo.[23] These records didn't extend any further back than A.D. 1400, however, and that's when we could no longer achieve a reliable reconstruction. This result actually made perfect sense since the pattern of twentieth-century warming in the instrumental record showed particularly strong warming in western and northern North America—precisely where these tree ring data were located. Moreover, a previous analysis of long climate model simulations had suggested that western North America was a "sweet spot" for estimating the average temperature of the Northern Hemisphere.[24]

We had other proxy data from this region, and they extended considerably farther back in time, in many cases more than a thousand years. There was a group of tree ring records from high-elevation sites in the western United States (primarily California, Nevada, and Arizona) that should in principle have reflected temperature variations in the region. Yet, despite the availability of these data, our validation tests were telling us that we couldn't go further back than six centuries. Intrigued by this apparent inconsistency, I undertook a closer comparison of the two different North American datasets.

Something rather remarkable emerged when the two datasets were laid on top of each other. The two tracked each other almost perfectly from the beginning of their period of overlap (A.D. 1400) until the early nineteenth century. Then the two series began to diverge, with the western U.S. series showing an almost exponential increase in tree ring growth relative to D'Arrigo and Jacoby's boreal tree line series. Then, at the beginning of the twentieth century, the two series began tracking each other again. As it turns out, the likely reason for this enigmatic observation had already been discussed in the scientific literature.

It was known that high-elevation trees are often not limited in growth simply by climate conditions such as growing season warmth, but also by carbon dioxide levels. Trees need carbon dioxide for photosynthesis, but immersed in the thinner atmosphere of high elevations they may be somewhat starved of this resource. In a 1993 article, Graybill and Idso had shown that these very trees might be expected to exhibit a positive growth response to increasing atmospheric carbon dioxide levels.[25] This so-called CO_2 fertilization mechanism could explain the divergence between the growth rates of the high-elevation western U.S. trees and the low-elevation boreal tree line stands: The timing of the divergence was almost perfectly correlated with the exponential rise of atmospheric carbon dioxide since the early nineteenth century associated with the Industrial Revolution. The disappearance of the divergence in the twentieth century was consistent with warmth once again returning as the key factor controlling the growth of the high-elevation trees in the presence of adequate carbon dioxide.

By correcting for that carbon dioxide effect through comparing otherwise similar trends in low- and high-elevation temperature-sensitive North American trees,[26] it seemed we might now be able to make use of the far-longer-term western U.S. data. Indeed, when we used the corrected version of the western U.S. tree ring data in our analysis, our validation tests gave us the green light; we could indeed now meaningfully reconstruct Northern Hemisphere average temperatures over the entire past millennium.

Despite this success, we were rather guarded in our conclusions, recognizing that our ability to obtain a reliable millennial Northern Hemisphere temperature reconstruction relied heavily on those western U.S. tree ring data as well as our somewhat *ad hoc* correction for potential CO_2 fertilization effects. The abstract of our article, entitled "Northern Hemisphere Temperatures During the Past Millennium: Inferences, Uncertainties, and Limitations," stated: "We focus not just on the reconstructions, but the *uncertainties therein*, and *important caveats*," and "Though *expanded uncertainties* prevent decisive conclusions for the period prior to A.D. 1400, our results *suggest* that the latter 20th century is anomalous in the context of at least the past millennium" (emphasis added). The expanded uncertainties were a result of additional cross-checks that indicated that the reconstruction was less effective at capturing century-scale and longer variations in earlier centuries, which led us to increase the error bars relative to MBH98. Even with these larger uncertainties taken into consideration, the reconstruction indicated that the 1990s were likely the warmest decade and 1998 likely the warmest year of the millennium.[27]

Our new paper[28] (henceforth "MBH99"), published just under a year after our original *Nature* article, once again garnered a fair amount of media attention, including a major spread in the Tuesday science section of the *New York Times*. But perhaps because it represented a cautious and incremental development relative to our earlier work and appeared in the lower-profile journal *Geophysical Research Letters* (and wasn't published on Earth Day), it didn't make quite the same splash, at least at the time. The article, however, would soon gain the attention of climate change contrarians, who perceived that our findings undermined one of their primary arguments against human-caused global warming: that a period of warmth comparable to that of the present day had existed in the relatively recent past, prior to any appreciable increase in greenhouse gas concentrations.

The Third Assessment Report

In order to understand the rise to prominence of the hockey stick, it is necessary to delve into the history behind the IPCC Third Assessment Report published in 2001. It was in that publication that our work truly entered onto the world stage. There is a pyramid-like hierarchy of IPCC report authorship. For any given chapter, there are dozens of contributing authors, fewer than a dozen lead authors, and, among the latter, two convening authors who have primary responsibility for the chapter. While I had been nominated to be an author of some sort prior to the publication of our work on paleoclimate reconstructions, I was still relatively fresh out of graduate school, and I was surprised when I learned in late 1998 that I had been selected as a lead author for the new report—a choice presumably related to the scientific attention our recent work had received.

My task as a lead author was to work with the numerous contributing authors in assessing the state of knowledge regarding evidence from the paleoclimate record, which—as in the previous IPCC reports—was part of the "observations" chapter. The first job was to solicit input from the leading experts in the field, including Keith Briffa and Phil Jones of the University of East Anglia's Climate Research Unit, who had performed their own proxy reconstructions of past climate, and Henry Pollack, a researcher at the University of Michigan who had derived an independent assessment of past temperatures from boreholes.[29] Boreholes are deep holes in the ground generated for purposes of geophysical exploration. Under the right circumstances (when there is minimal intrusion of fluid flow, the properties of the bedrock are reasonably uniform, changes in seasonal snow cover are minimal, etc.), the penetration of heat from the surface down into the upper layer of Earth can reasonably be assumed to follow a simple diffusion process. The physics of diffusion dictates that the vertical temperature profile down the hole can then be used to provide an estimate of how temperatures at the surface changed back in time.

To be credible as a true assessment of prevailing scientific understanding, IPCC reports must accurately reflect the diverse views within the scientific community on any particular issue—no simple task when it comes to contentious disciplines such as climate change. In addition, since the IPCC report is an assessment report, not simply a literature review, it is necessary that the IPCC evaluate the collective work of the scientific community in such a way that it is clear to readers on what points scientists agree, and where there is still active debate.

There were several rounds of discussion among the lead authors regarding each of the various drafts of our chapter. The lead authors included scientists with a wide range of views such as John Christy, a scientist with a somewhat contrarian outlook on climate change. In the end, whatever was said in the chapter would have to meet with John's approval. Moreover, whatever would be concluded in our chapter would have to withstand the scrutiny of the rigorous IPCC review process.[30]

My primary responsibilities involved the assessment of the paleoclimate records of past centuries.[31] It would be essential in this work to reflect the diverse range of views in the recent scientific literature and to present the range of estimates that had been published. Any key conclusions arising from the chapter could not rely on one study or one group's findings. They had to reflect a consensus of recent studies, if indeed that existed. The MBH99 hockey stick was shown in a plot by itself for two important reasons: (1) It was the only reconstruction done at the level of individual years rather than decadal or longer-term averages, and (2) it came with error bars, which the other reconstructions didn't. Thus, unlike other studies, it spoke to whether recent years, such as 1998, stood out as unusual against the backdrop of the longer-term reconstruction and its uncertainties.

An additional plot in the chapter compared three different independent or largely independent quantitative reconstructions of Northern Hemisphere temperatures over the past millennium. One was an estimate by Keith Briffa and collaborators based on tree ring density (as opposed to ring width) measurements that extended back six centuries,[32] while the other two—each extending back the full millennium— were the Jones et al. multi-proxy estimate discussed earlier in the chapter and the MBH99 hockey stick. Each showed recent warming to be anomalous in the context of the longer-term temperature history of the record. Another reconstruction of Northern Hemisphere temperatures over the past millennium by Crowley and Lowery was published too late to be included in the comparison figure, but its conclusions, similar to those of the other three reconstructions, were summarized in the IPCC chapter.[33]

Yet another figure in the chapter depicted the Pollack et al. estimate of past temperature change from borehole records, an estimate that was entirely independent of the other proxy estimates. Though the borehole temperature reconstructions came with their own caveats, they provided independent evidence that recent warming was unusual in at least a five-hundred-year time frame—as far back as reliable borehole temperature estimates went. A separate plot in the chapter showing glacier mass balance records spanning the past five hundred years indicated that the melting of mountain glaciers worldwide too was unprecedented over at least this time frame.

Collectively, the data clearly indicated that the modern warming was unprecedented in a long time frame. But only two estimates (the Jones estimate and the MBH99 estimate) gave a quantitative depiction of hemispheric-scale temperature changes for the full millennium, and only one (the MBH99) had error bars attached to it. After much discussion among all the lead authors, a consensus was reached on a tentative conclusion. The word *likely*, the group decided, would be attached to the conclusion that recent warming for the Northern Hemisphere on the whole was anomalous in a millennial context. In the parlance of the IPCC, this careful phrasing indicated confidence of about 67 percent, that is, a two-out-of-three chance that the conclusion was correct. That is a far cry from the 90 percent threshold required for "confident" inferences—what in IPCC parlance was referred to as "very likely." Like MBH99 itself, the IPCC Third Assessment Report was extremely cautious in the level of confidence it attached to its conclusion that recent warming was anomalous in a millennial context.

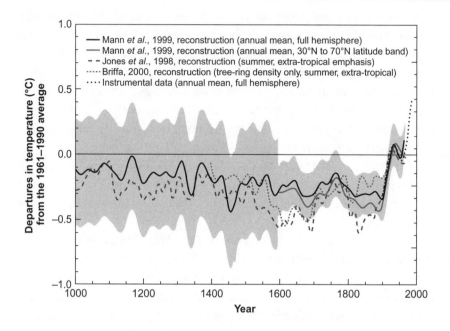

Figure 4.2: Competing Estimates of Millennial Temperature Trends
The graph compares different reconstructions of Northern Hemisphere temperature trends over the past thousand years as shown in the 2001 IPCC *Third Assessment Report*. Shown for comparison is the instrumental record from the midnineteenth to late twentieth centuries. The data have been smoothed to emphasize multidecadal and longer timescale variations.

The Implications

Though we didn't quite realize it at the time, the gauntlet had been laid down with the initial publication of the extended MBH99 millennial hockey stick and, especially, its subsequent prominence in the 2001 IPCC report. We had taken on a sacred cow of climate change contrarianism: the medieval warm period (MWP). Our reconstruction did not "eliminate the MWP," as our detractors liked to claim.[34] It did in fact include a period of relative warmth during the medieval era of the eleventh to fourteenth centuries. While medieval conditions were relatively warm, however, the modern tip of the blade—as can be seen in the graphic featured in the prologue—was warmer than the peak reconstructed medieval warmth.

One pillar of the contrarian case against human-caused climate change was that the mere existence of a warmer period centuries earlier somehow disproved any human influence on modern warmth. In reality, this was not true. Scientists had known for some time, for example, that there were periods in the deep geological past during which temperatures were warmer than today, such as in the mid-Cretaceous period 100 million years ago. The reason was atmospheric CO_2 concentrations that were several times higher than today owing to the slow geological processes that modify atmospheric composition on very long timescales. It was indeed possible that other natural factors, be they changes in solar output or volcanic activity, could have led to conditions that were as warm as today.

Whether conditions in past centuries might have been warmer than today, then, would not have a scientific bearing on the case for the reality of human-caused climate change. That case, as we've seen, rests on multiple independent lines of evidence: the basic physics and chemistry of the greenhouse effect, for example; the relationship between greenhouse gas concentrations and temperatures over geological time; and a pattern of observed climate change that can only be explained by climate models when human-produced greenhouse gases are included in the calculations. While the presence of a medieval warm period warmer than today would not negate the reality of modern human-caused climate change, evidence of its absence nonetheless would take away an important (if misguided) talking point of contrarians—something of which they were well aware.

Our finding that recent warming was anomalous in a long-term (now, apparently, millennial) context was suggestive of the possibility that human activity was implicated in the warming. I was always very careful not to claim that our work could firmly establish a human role in the warming. To draw such a conclusion based on our work alone would necessarily buy into the classic logical fallacy of "correlation without causation." We had established correlation—the anomalous warming that we documented coincided with the human-caused ramp-up in greenhouse gas concentrations—but we hadn't established causality.

A little more than a year after we had published our millennial hockey stick reconstruction, paleoclimatologist Thomas Crowley of Texas A&M University (and coauthor of the Crowley and Lowery reconstruction discussed earlier) published findings based on the use of a theoretical climate model simulation designed to investigate causes of past temperature change.[35] Crowley subjected the model to estimated changes in natural factors over the past thousand years using indirect measures of changes in solar output and explosive volcanic activity, information on both of which can be recovered from atmospheric deposits in polar ice cores. These simulations revealed that the natural factors could explain the extent of medieval warmth in our reconstruction; in the model, this warmth arose from a relative lack of cooling volcanic eruptions combined with relatively high levels of solar output. The natural factors could also explain the cooler conditions of the ensuing Little Ice Age, which resulted from relatively low levels of solar output and more frequent explosive volcanic eruptions. Fed the natural factors only, the model could not, however, reproduce the abrupt twentieth-century warming. In fact, the model predicted that the climate should have cooled in recent decades, rather than warmed, if only natural factors had been at play. It was only when Crowley added the modern human influences—increasing greenhouse gas concentrations primarily from fossil fuel burning and the regional cooling effect of industrial sulfate aerosols emissions—to the model simulation that it was able to track the hockey stick all the way through to the present. The conclusion was clear: Natural factors could explain the temperature changes of the past millennium through the dawn of the Industrial Revolution, but only human influences could explain the unusual recent warming. Causality was at least tentatively established now.

Chapter 5

The Origins of Denial

Doubt is our product since it is the best means of
competing with the "body of fact" that exists in the
mind of the general public.

—Unnamed tobacco executive, Brown & Williamson (1969)

In the 1990s, as the scientific evidence for human-caused global warming grew stronger and calls for action to curtail greenhouse gas emissions grew louder, fossil fuel industry executives made a critical decision. Rather than concede the potential threat climate change posed and the necessity of ultimately reducing fossil fuel use, they would instead engage in a massive, media-savvy public relations campaign. The strategy was simple: While presenting a seemingly forward-thinking, pro-environmental public face, oil companies and allied economic and political interests would, behind the scenes, use various means to sow doubt about the validity of the underlying science on climate change. It was a finely tuned balancing act intended to forestall any governmental policy action to regulate greenhouse gas emissions while seeking to maintain a positive corporate image.

Doubt Is Their Product

The source of the chapter's opening quote is David Michaels's *Doubt Is Their Product*.[1] The book describes the corporate public relations campaigns that the tobacco and other industries used for decades to discredit research demonstrating adverse health impacts of their products—a campaign that was successfully satirized in the 2005 movie *Thank You for Smoking*. The striking similarity with the tactics of climate change denial did not go unnoticed by former *Science* magazine editor-in-chief Donald Kennedy, who commented on the book's jacket, "if you're worried about climate change, keep worrying, because the same program is underway there."

In *The Republican War on Science*, Chris Mooney argues that the corporate-funded public relations campaigns of recent decades aimed at discrediting the science behind policies designed to protect our environment and health arose from conservative distaste of governmental regulation.[2] Those campaigns came to a head, he suggests, in the extreme antiregulatory atmosphere of the George W. Bush administration of 2000–2008. Legislation such as the Data Quality Act of 2001 saddled government agencies with onerous requirements on how they must respond to demands and complaints from industry groups regarding any data or scientific studies used in establishing government policy. It was, in Mooney's words, "a science abuser's dream come true," and it signaled the increasing politicization of science at the very time the hockey stick was coming to prominence.

Industry groups sought to frame the public discourse by constructing, to use the characterization of Naomi Oreskes and Erik Conway in *Merchants of Doubt*,[3] a virtual Potemkin village of pseudoscience institutions—think tanks, journals, news sites, and even a cadre of supposed experts, ideally with prestigious affiliations—to promote their own scientific (or, more aptly, antiscientific) messaging. These professed experts were used to promote industry-favorable views in the framing of policy-relevant matters of science, to manufacture doubt about mainstream scientific findings disadvantageous to their client, and to generate pseudoscientific sound bites that could be presented to the public under the auspices of neutral-sounding groups.[4] Using this tactic, industry advocates have, in the words of famed Stanford environmental scientists Paul and Anne Ehrlich, "sowed doubt among journalists, policymakers, and the public at large about the reality and importance" of an array of societal and environmental threats.[5] The Ehrlichs coined the term *brownlash* to characterize this orchestrated backlash against "green" policies.

The choice of language employed in antiscientific attacks is worthy of particular attention, as it has been exploited by purveyors of disinformation in a distinctly Orwellian manner, often to great effect in the public discourse. Their lexicon features simple, pithy terms like "sound science" that are repeated as mantras. Who, after all, could be against sound science? Implicit in this motif, however, is the notion—wholly inconsistent with the way science actually works—that the scientific enterprise must offer absolute proof if it is to be used to inform policy. Like a defense lawyer, industry special interests seek to introduce some measure of doubt into the public mindset. The demand for sound science is made for a curiously selective array of findings, be they the ozone-depleting properties of the chlorofluorocarbons (CFCs) once used in spray cans, or the adverse health effects of industrial mercury pollution. The true metric applied by industry special interests is, of course, not the actual quality of the underlying science, but simply this: Are the scientific findings in some way inconvenient to their clients (the health insurance industry, the pharmaceutical industry, the chemical industry, the fast food industry, or, of course, the fossil fuel industry)?

If so, those findings are quickly labeled "junk science" and are purported to represent the flawed or even fraudulent claims advanced by a cabal of ostensibly corrupt university professors, scientists, journal editors, and governmental science funding agencies. Mooney notes that, from the language being used, "you would think that environmental science, as conducted by America's leading universities, suffers from endemic corruption on a scale reminiscent of Tammany Hall."[6] The effort seeks simultaneously to paint scientists as enemies of the people and to spread doubt and confusion about established scientific findings. Meanwhile, it encourages educators to "teach the controversy"[7] when—scientifically speaking—there is none.

There is rich irony here, as the clearest cases of true junk science seem to have resulted from the corruptive influence of industry itself. Particularly striking examples can be found in the area of biomedical science and pharmaceuticals, where there have been numerous high-profile scandals involving companies that either ghostwrote articles for scientific journals singing the praises of their particular pharmaceutical product[8] or suppressed, through threat of litigation, scientific publications damaging to the credibility of their advertised claims.[9]

The attacks are typically carried out by organizations and groups with names like "Citizens for a Sound Economy" that masquerade as grassroots entities but in reality represent powerful industries, and have hence been termed "Astroturf" organizations. These groups employ ideologically aligned media outlets and a network of lawyers, lobbyists, and politicians to advance their message. Their efforts are aided

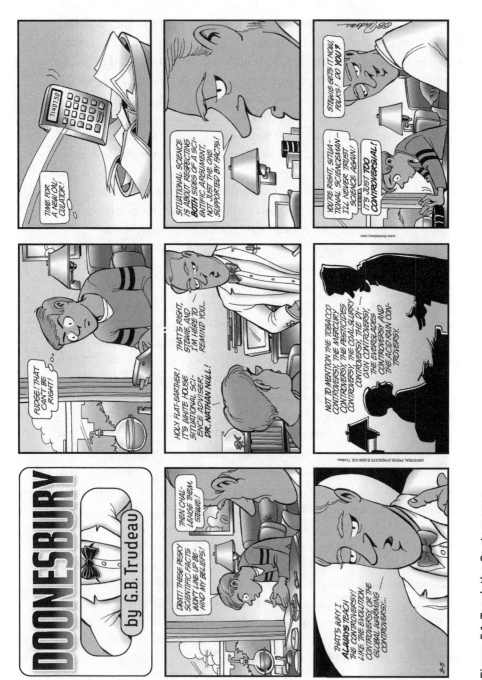

Figure 5.1: Teach the Controversy

A Doonesbury cartoon encapsulates the spirit of attacks on the science relevant to environmental and consumer protections. [DOONESBURY © G. B. Trudeau. Reprinted with permission of UNIVERSAL UCLICK. All rights reserved.]

by honest citizens, and sometimes even by mainstream media outlets, who are taken in and exploited, often unwittingly, to create an echo chamber of mass disinformation that permeates our airwaves and television screens and the Internet.

A central focus of many of these campaigns in recent years has, of course, been the discourse over global warming and climate change. For more than a decade, the scientific community, in its effort to communicate the threat of climate change, has had to fight against the headwind of this industry-funded disinformation effort. The collective battles are what I term the "climate wars."

The Climate Wars

The evidence for a well-organized, well-funded, and orchestrated climate change disinformation campaign has been laid out in detail on public interest group Web sites,[10] in articles in popular magazines,[11] and by an increasingly rich array of scrupulously researched books on the topic.[12] The campaign has its roots in the larger industry-funded public relations efforts that emerged during the 1970s and 1980s over acid rain, ozone depletion, missile defense, stem cell research, biodiversity loss, and a host of other issues.[13] Foreshadowing the climate change denying tactics outlined in the 2002 Luntz memo were the activities of the Global Climate Coalition (GCC). Formed in the late 1980s, the GCC was a consortium of more than fifty companies and trade associations representing chemical, mining, automotive, transportation, fossil fuel, shipping, farming, power, defense, pharmaceutical, and manufacturing industries with the purpose of funding and organizing opposition to emerging policy efforts aimed at greenhouse gas emission reductions. They played a critical role, it may be recalled, in the attacks on Ben Santer, accusing him of "political tampering" and "scientific cleansing" following publication of the IPCC Second Assessment Report.

In April 1998, just days after the publication of our original hockey stick article, new revelations surfaced about a prominent GCC member, the American Petroleum Institute (API). Internal documents leaked to the *New York Times* showed it was hatching a plan to "recruit a cadre of scientists who share the industry's views of climate science and to train them in public relations so they can help convince journalists, politicians and the public that the risk of global warming is too uncertain to justify controls on greenhouse gases."[14] The GCC itself was disbanded in 2001 following the defection of prominent members such as British Petroleum that—with some irony, in retrospect—were concerned about the negative public relations of being associated with an anti-environmental agenda.[15] While the GCC no longer itself exists, the denialist campaigns continued unabated. Other fossil fuel interests—oil giant ExxonMobil being a big player among them—have continued to fund groups spreading climate change disinformation for years.

Wealthy privately held corporations and foundations with close interests in, or ties to, the fossil fuel industry, such as Koch Industries[16] and the Scaife Foundations,[17] have become increasingly active funders of the climate change denial campaign in recent years. Unlike publicly traded companies such as ExxonMobil, these private outfits can hide their finances from public view, and they remain largely invulnerable to outside pressure. In recent years, as ExxonMobil has been pressured by politicians on both sides of the aisle to withdraw from funding the climate change denial movement,[18] Koch and Scaife have stepped up, contributing millions of dollars to the effort.

Many organizations have settled in the Potemkin village of climate change denial. Among them are the American Enterprise Institute, Americans for Prosperity, Advancement of Sound Science Center, Competitive Enterprise Institute, Cato Institute, Hudson Institute, George C. Marshall Institute, Fraser Institute, Heartland Institute, Alexis de Tocqueville Institution, Media Research Center, National Center for Policy Analysis, and Citizens for a Sound Economy (better known now as Freedomworks). There are literally dozens of others.[19]

Among the willing accomplices in the campaign of deceit are the various media outlets that often propagate climate change disinformation in their editorial and opinion pages. These venues include newspapers such as the *National Post* and *Financial Post* in Canada; the *Daily Telegraph, Times*, and *Spectator* in the United Kingdom; and U.S. newspapers such as the *Washington Times* and the various outlets of the Murdoch, Scaife, and Anschutz conservative media empires, which include not only prominent outlets such as Fox News and the *Wall Street Journal*, but syndicates such as the regional Examiner.com network and Web sites like Newsbusters.

Agents of Denial

Not only are there connections between the current campaign to attack the science of climate change and past industry-funded campaigns to deny other industrial health and environmental threats such as the dangers of smoking tobacco and of acid rain, environmental mercury contamination, and ozone depletion. Some of the very same scientists have been employed as advocates for not just one or two, but many of these issues. Think of them as all-purpose deniers.

The grandfather of all-purpose denial was Frederick Seitz, a solid-state physicist possessing impressive scientific credentials. Seitz was a former head of the U.S. National Academy of Sciences and in 1973 was awarded the prestigious Presidential Medal of Science. Seitz found common cause with two other similarly minded physicists—Robert Jastrow, founder of the NASA Goddard Institute for Space Studies (GISS) laboratory now directed by James Hansen, and Nicholas Nierenberg, one-time director of the Scripps Institution for Oceanography—in supporting and advocating for President Ronald Reagan's 1980s missile defense program.[20] The Strategic Defense Initiative was controversial enough that the issue of whether it was wise, let alone efficacious, divided the physics department faculty at UC Berkeley where I was doing my degree at the time.[21]

In 1984, the three scientists joined together to form the George C. Marshall Institute—a conservative think tank that *Newsweek* magazine called a "central cog in the denial machine."[22] Their chief mission was to combat efforts by Cornell University planetary scientist Carl Sagan and others who sought to raise awareness about the potential threat of "nuclear winter." The massive detonation of nuclear warheads during a thermonuclear war, Sagan and others hypothesized, might produce a global dust cloud as devastating for humanity as the asteroid-induced global dust storm that ended the reign of the dinosaurs. The concept had even penetrated into popular culture with the 1983 song "Walking in Your Footsteps" by the Police.[23] That nuclear winter projections were based on climate models brought climate modeling onto the radar screen of Seitz, Jastrow, and Nierenberg, and it set the stage for their later role as key climate change deniers.

Upon retirement from academia in the late 1970s, Seitz worked for the tobacco giant R.J. Reynolds for roughly a decade. In this capacity, he accepted more than half a million dollars while lending his scientific credibility to advocacy efforts aimed at downplaying the health threats posed by the smoking of tobacco.[24] In the early 1990s, Seitz went on to chair the George C. Marshall Institute full time, where he campaigned against the reality of global warming and the threat CFCs posed to the ozone layer.[25]

In 1998, in conjunction with yet another climate change denial group, the Oregon Institute of Science and Medicine, Seitz spearheaded a petition drive opposing the Kyoto Protocol to limit greenhouse emissions, mailing the petition with a cover letter and an article attacking the science of climate change to a broad list of recipients. He portrayed these materials as having the imprimatur of the National Academy of Science (NAS) by formatting the article—"Environmental Effects of Increased Atmospheric Carbon Dioxide" by Arthur B. Robinson, Noah E. Robinson, and Willie Soon—as if it had been published in the prestigious *Proceedings of the National Academy of Science* (*PNAS*), which it definitely had not been. Seitz even signed the enclosed letter using his past affiliation as NAS president. The NAS took the extraordinary step of publicly denouncing Seitz's efforts as a deliberate deception, noting that its official position on the science was the opposite of that expressed in Seitz's letter. The matter, coincidentally enough, played out just days before the publication of our 1998 hockey stick article in *Nature*.[26]

The "Oregon petition," with thirty-one thousand nominal "scientist" signatories, has often been touted as evidence of widespread scientific opposition to the science underlying human-caused climate change. However, a subsequent analysis by *Scientific American* found that few of the signatories were even scientists (the list included the names Geri Halliwell, one of the Spice Girls; and B. J. Hunnicutt, a character from the TV series MASH).[27]

Questionable petitions, misleading articles, and, as we'll see, even one-sided conferences constitute key *modi operandi* in the world of climate change denial. There was indeed a distinct feeling of déjà vu in fall 2007, when I, and many other scientists and engineers, received a packet in the mail consisting of an updated "article" by several of the same authors promoting the same myths and half-truths (e.g., the medieval warm period was warmer than today, the Sun is driving observed temperature changes, and so on). This article, too, was formatted to look as if it had been published as a peer-reviewed journal article,[28] and yet again was accompanied by a petition demanding the United States not sign the Kyoto Protocol. The origin of these materials was, once again, the Oregon Institute of Science and Medicine.

That group's activities seemed to be part of a coordinated effort. One year earlier, Kenneth Green of the American Enterprise Institute (AEI) was implicated in what at least appeared to be an attempt to solicit pieces from climate scientists critical of a recently published IPCC report in return for a cash award of $10,000.[29] In addition, in recent years, the Heartland Institute, a group that has been funded by both tobacco (Philip Morris) and fossil fuel (Exxon, Koch, Scaife) interests, has financed a series of one-sided conferences on climate change, featuring a slate of climate change deniers, many with no discernible scientific credentials, and most with financial connections of one sort or another to the fossil fuel industry or groups they fund.[30]

S. Fred Singer, whom we met in previous chapters, followed in Seitz's footsteps. Like Seitz, Singer's origins were as an academic and a scientist, and like Seitz, he left the academic world in the early 1990s[31] to advocate against what he called the "junk science" of ozone depletion, climate change, tobacco dangers, and a litany of other environmental and health threats.[32] He founded an entity in 1990 called the Science and Environmental Policy Project (SEPP)[33] that he used to launch his attacks and has also received considerable industry funding for his efforts.[34] Singer was the principal behind the denialist response to the IPCC Fourth Assessment Report, the so-called Nongovernmental International Panel on Climate Change (NIPCC) funded by the aforementioned Heartland Institute, and characterized by ABC News as "fabricated nonsense."[35]

Singer, like Seitz, has been accused of having engaged in serious misrepresentation, in this case involving the great scientist Roger Revelle.[36] Revelle was instrumental in our early understanding of human-caused global warming and the potential threat of continued fossil fuel burning. He is also credited with having inspired many of today's leading climate scientists and is cited by former U.S. vice president Al Gore as the origin for his concern about climate change. In 1991, shortly before Revelle's death, Singer added Revelle as a coauthor to a paper he published in the Cosmos Club journal *Cosmos*. The paper attacked the science of climate change and was nearly identical in both title and content to a paper that Singer had previously authored alone.[37] Reports from both Revelle's personal secretary and his former graduate student Justin Lancaster suggest that Revelle was deeply uncomfortable with the manuscript, and that the more dismissive statements in the paper were added after Revelle—who was gravely ill at the time and died just months after the paper's publication—had an opportunity to review it.[38]

While Seitz and Singer may have been the most prolific and versatile of the denialists, other scientists have served as specialists in the climate change denial movement. Frequently, though not always, they do so with either direct or indirect financial compensation and support from the fossil fuel industry.[39] Many write op-eds and opinion pieces for conservative-leaning newspapers or outlets supported by industry, such as TechCentralStation. Often they are sponsored to go out on the climate change denial lecture circuit, or they write books that are promoted, marketed, and even published by fossil-fuel friendly groups.

One of the more formidable among them is Richard Lindzen. His credentials, like those of Seitz, are impressive; he is a chaired professor at MIT and a member of the National Academy of Sciences. Lindzen—who also has received money from fossil fuel interests[40]—is perhaps best known for his controversial views that climate models grossly overestimate the warming effect of increasing greenhouse gas concentrations. It all has to do with the issue of climate feedbacks. Feedbacks, as we have seen, are mechanisms within the climate system that can act either to amplify (positive feedback) or diminish (negative feedback) the warming expected from increasing greenhouse gas concentrations. If a climate scientist has spent a career looking for missing feedbacks in climate models that are always of the same sign (positive for a "true believer" and negative for a "denier"), one might reasonably suspect that the endeavor has not been entirely objective. (Ironically, the one missing feedback I've argued for in the climate system is a negative one[41]—a rather inconvenient fact for those who would like to label me a "climate change alarmist.")

Lindzen has made a career of searching for missing feedbacks, but apparently only negative ones. Indeed, it seems as if he has never met a negative feedback he didn't like. And he has been quick to trumpet his claims of newly found negative feedbacks in op-eds, opinion pieces, and public testimony,[42] arguing time and again that his findings point to an overestimation of warming by models and are an indication that climate change is an overblown problem. Yet each of his past claims has evaporated under further scrutiny.

For years, Lindzen has argued that hypothesized but as yet unestablished negative feedbacks in the climate system will offset the very large positive feedbacks arising from increased evaporation of water into the atmosphere and melting of snow and ice associated with global warming. He has argued that a doubling of CO_2 concentrations will consequently only raise global average temperatures by roughly 1°C (and with zero uncertainty!). Yet the diversity of evidence from the paleoclimate and modern climate record suggests that less than 2°C warming for CO_2 doubling is highly unlikely.[43]

In 1990, Lindzen argued that a drying and cooling of the upper troposphere would mitigate global warming,[44] but later in effect conceded that further work had demonstrated that the mechanism he had proposed was not viable.[45] In 2001 he promoted a new hypothesis, the so-called "iris" effect,[46] in which warming ocean temperatures would supposedly lead to fewer high clouds, causing surface temperatures to cool down.[47] Once again, this hypothesis didn't hold up under scrutiny by other scientists.[48]

Undeterred, Lindzen claimed to find evidence for an additional, new negative cloud feedback, this time based on a putative statistical relationship between tropical sea surface temperatures and satellite measurements of the radiation escaping to space.[49] He claimed that when the tropics warm up, there are more low reflective clouds, causing more solar radiation to be returned to space, thus tending to cool the surface. When climate researcher Kevin Trenberth of the National Center for Atmospheric Research (NCAR) and his collaborators examined Lindzen's claims closely,[50] however, they found the data points Lindzen had chosen to be curiously selective, and the claimed relationship not supported when a more objectively chosen sample was used.[51] A subsequent analysis by other researchers concluded that the available data may actually support a positive overall cloud feedback, not a negative one.[52]

Pros and Amateurs

People like Seitz, Singer, and Lindzen have been in the front lines of professional climate change denial. But others have participated as well.

There is a whole corps of columnists and commentators who help promote climate change disinformation. In the United States, they include prominent radio and TV commentators such as Rush Limbaugh, Glenn Beck, and Sean Hannity, as well as many other lesser known, but similarly active and effective protagonists. Some, such as Bret Stephens of the *Wall Street Journal* and Debra Saunders of the *San Francisco Chronicle*, also operate with the imprimatur of ostensibly mainstream news organizations.

The boundaries between journalist, commentator, and paid industry advocate have become increasingly blurred with the development of the new media. Consider in the United States, for example, individuals such as Christopher Horner of the Competitive Enterprise Institute and James Taylor (no, not the singer-songwriter made famous by "Fire and Rain" and "Sweet Baby James") of the Heartland Institute. Though employed as lobbyists or lawyers by the industry-funded Competitive Enterprise Institute, they are regularly granted a forum by conservative news outlets to pen pieces attacking climate science and climate scientists. Tobacco and fossil fuel industry lobbyist Steven J. Milloy sometimes appears as a "junk science expert" on Fox News.[53] He runs a site called junkscience.org, billing himself, with no apparent sense of irony, as the "junk man." In the United Kingdom, Christopher Booker of the *Telegraph* has such a biased record of reporting on environmental issues that it has earned him the title of "patron saint of charlatans" from award-winning *Guardian* journalist George Monbiot.[54]

Video also has played an increasingly important role in climate change denial. Martin Durkin of the United Kingdom produced the ironically entitled documentary "The Great Global Warming Swindle." British media regulator Ofcom found that the film "did not fulfill obligations to be impartial and to reflect a range of views on controversial issues" and that it "treated interviewees unfairly."[55] This problem was particularly evident in Durkin's interview of MIT physical oceanographer Carl Wunsch, who was upset by the way his words were edited to imply a contrarian viewpoint very much at odds with his actual views.[56]

Then there is the recently deceased science fiction writer Michael Crichton. One of Crichton's last novels, *State of Fear,* was a thinly veiled climate change denialist polemic masquerading as an action adventure novel. Crichton even was invited as a witness in a U.S. Senate committee hearing held by Senator James Inhofe (R-OK) to sow doubt on the reality of climate change. It is telling that Inhofe had to turn to a science fiction novelist to make his case.

The United Kingdom has produced some of the more colorful climate change deniers. Christopher Monckton, the third viscount Monckton of Brenchley, has emerged on the denial scene in recent years. He claims to be an expert on climate change, though he has no formal scientific training. Richard Littlemore of the fossil fuel industry watchdog group DeSmogBlog tells us that Monckton has been caught on several occasions "indulging in deliberate manipulation of scientific data to arrive at misleading conclusions about climate science."[57] Monckton's assertions aren't confined to science; he has even claimed, falsely, to have won the Nobel Prize.[58] After he had repeatedly represented himself publicly as a member of the House of Lords, the clerk of Parliament took the unprecedented step of publicly demanding he cease and desist making this false claim.[59]

Then there are the amateurs down in the trenches who execute the ground game in the climate wars. Many of these individuals are simply ill informed, and are no doubt acting in good faith in expressing what they believe to be honest skepticism. But strident claims without substance abound, as do absurd accusations against others. Some of the amateurs are more than willing to engage in some degree of mischief, whether it be taking advantage of the IPCC open review process by flooding its authors with countless frivolous comments (each of which must be responded to, according to IPCC rules) or exploiting the Freedom of Information Act (FOIA) and related laws to launch frivolous requests for documents and private correspondence of scientists. A since deceased Tasmanian named John Daly, with his Web site "Still Waiting for Greenhouse," provided an early proof-of-concept for how a single individual with nothing more than a Web site could battle mainstream climate science by peddling contrarian views and maligning the work of dedicated scientists.

Today, much of the trench warfare takes place on the Internet. Former mining industry consultant Stephen McIntyre is especially well known for his broadsides against established climate science. McIntyre frequently uses his Web site climateaudit to launch attacks against climate scientists themselves, often leveling thinly veiled accusations of fraud and incompetence—once, for example, titling a post about a highly respected NASA climate scientist with the rhetorical question "Is Gavin Schmidt Honest?"[60]

Since then, a number of other amateur climate change denial bloggers have arrived on the scene. Most prominent among them is Anthony Watts, a meteorologist for a Fox News AM radio affiliate in Chico, California, and founder of the site "Watts Up with That?" which has overtaken climateaudit as the leading climate change denial blog. Watts also started the Web site SurfaceStations.org, which purported to identify poorly sited meteorological stations in the United States in an effort to demonstrate that the instrumental record of warming temperatures is hopelessly compromised by instrumental measurement biases. With the assistance of the Heartland Institute, Watts published a glossy, very official-looking report about the project, showing lots of photos of ostensibly badly sited meteorological stations, with plots of the supposedly compromised records.[61]

Curiously absent from that report, however, was any direct comparison showing what the surface temperature record looks like both with and without the sites that Watts deemed unworthy. Scientists at the National Oceanic and Atmospheric Administration (NOAA) went ahead and calculated it themselves, producing versions of the continental U.S. average temperature curve both with and without the records in question. You can probably anticipate the result: It was difficult to distinguish the "with" and "without" versions within the thickness of the plot curves. Eliminating the "suspect" data made virtually no difference at all; in fact, the small bias that was found was of the opposite sign. The "corrected" record showed slightly more warming![62] This is just one example of a favored *modus operandi* among climate change contrarians: hyping real or imagined errors that make no difference to any significant scientific conclusions—the scientific equivalent of a identifying a typo in a report.

Finally, there is the front line of the climate change denial ground attack. It consists of anti-climate-science activists and conspiracy theorists who operate largely under the radar screen but nonetheless play an essential role in the denial agenda. Their primary tool is the "cut-andpaste," the repetition of contrarian talking points in arguments with friends, neighbors, relatives, and coworkers; in letters to editors of local newspapers; in online newsgroups; in comments sections of Internet news articles; and on blogs. Their role is not to be underestimated, as false statements repeated often enough help create the echo chamber of climate change disinformation.

Not all "amateurs" are what they appear to be. A primary goal of the disinformation machine is to manufacture an illusion of grassroots support. This can be achieved by hiring ringers to pose as ordinary citizens, posting standard contrarian talking points and responses in online news threads, blogs, and the like. Prominent climate change deniers have occasionally been identified making use of a so-called sock-puppet (a "fake online identity to praise, defend or create the illusion of support for one's self, allies or company"[63]). Stephen McIntyre, for example, was found leveling online attacks hiding behind the sock-puppet "Nigel Persaud," while Michael Fumento of the Hudson Institute, perhaps best known for his attacks on environmental activist turned cinematic heroine Erin Brockovich, was once discovered posting self-supporting comments as "Tracy Spencer."[64]

Swiftboating Comes to Climate Change

One of the more unseemly features of the climate change denial campaign has been its use of character assassination as a tool for discrediting climate science itself. It is the art of the smear campaign that has come to be known as "swiftboating." The connection with the term is in fact remarkably direct.

Marc Morano got his start working for radio commentator Rush Limbaugh before moving on to work for the ExxonMobil and Scaife-financed Conservative News Service (now Cybercast News Service).[65] There, Morano was directly implicated in the original swiftboat attack on presidential candidate Senator John Kerry in the run-up to the 2004 presidential election.[66] That attack had taken one of Kerry's greatest strengths—he had been awarded three Purple Hearts for his service in Vietnam, while his opponent, George W. Bush, had avoided active duty—and, through a perversion of revisionist history, turned it instead into a perceived weakness.

Morano went on to become the pit bull of the climate change denial movement, launching swiftboat-like attacks as before, but this time directed against climate science and climate scientists. Among his many unsavory aspersions, he called NASA's James Hansen a "wannabe Unabomber" (suggesting it may be "time for meds").[67] I too have been at the receiving end of Morano's smears, having been called a "charlatan" responsible for "the best science that politics can manufacture."[68]

Beginning in 2006, Morano's efforts were funded on the taxpayer dime: He became a paid staff member on the Senate Environment and Public Works (EPW) Committee for Senate climate change denier Senator James Inhofe. From this perch, Morano promoted climate change denial talking points and launched attacks against climate scientists on the EPW Web site and through an e-mail listserv reaching large numbers of journalists and politicos. Undaunted after his position with Inhofe was terminated in 2009,[69] Morano headed back through the revolving door, this time hired by a Scaife- and ExxonMobil-funded[70] entity known as the Committee for a Constructive Tomorrow (CFACT) to run a new Web site called ClimateDepot.com. The site, which bills itself as "the Senate EPW website on steroids,"[71] provides Morano with a platform from which he can continue his barrage against the climate science community. In 2010, for example, he proclaimed that climate researchers "deserve to be publicly flogged" for speaking out on the threat of human-caused climate change.[72]

Shoot the Messenger

While the tactic of swiftboating or "shoot the messenger" may have been honed by people like Marc Morano, it has a deeper history when it comes to environmental science in America. Rachel Carson, whose book *Silent Spring*[73] in the early 1960s exposed the environmental threats from widespread use of the pesticide DDT, was the first to experience the wrath of industry-funded smear campaigns. The president of Monsanto Corporation, the largest producer of DDT, for example, called her "a fanatic defender of the cult of the balance of nature."[74] Despite the fact that her scientific findings have stood the test of time, attacks against Carson continue to this day. The Competitive Enterprise Institute boasts a Web site, rachelwaswrong.org, aimed solely at discrediting Carson's legacy. The thinking seems to be, if they can bring down Rachel Carson, they can bring down the entire environmental movement.

Then there was Paul Ehrlich, with his *The Population Bomb* in the late 1960s, which introduced the public to the notion that our patterns of consumption and population increase were on a collision course with environmental sustainability. Among the many others who denounced Ehrlich as an alarmist purveyor of doom and gloom was Julian Simon of the Cato Institute, who accused Ehrlich of having led a "juggernaut of environmentalist hysteria."[75] Yet Ehrlich's early warning has ultimately proven prophetic. In the 1990s, a group of more than fifteen hundred of the world's leading scientists, including half of the living Nobel Prize winners at the time, concluded that "Human beings and the natural world are on a collision course," inflicting "harsh and often irreversible damage on the environment and on critical resources."[76] The major national academies of the world have issued similar joint statements.[77]

A similar story holds for Herbert Needleman—like Rachel Carson, a fellow Pennsylvanian. Needleman's research in the 1970s identified a link between environmental lead contamination and the impairment of childhood brain development. Lead industry-funded scientists accused him of misconduct in his analysis of data.[78] He was ultimately exonerated after a thorough investigation by the National Institutes of Health, and his research findings have been validated by numerous independent studies over the decades.

Each of these scientists helped instill a wider recognition of the dangers posed by unprecedented, uncontrolled, and unchecked human alteration—be it biological, chemical, or physical—of our environment. Carson, Ehrlich, and Needleman were the forerunners of the climate scientists who would be similarly denounced for their inconvenient findings.

Stanford University's Stephen Schneider was among the most articulate scientific voices in the climate change debate from the 1970s through his untimely passing in 2010. He was particularly effective in the way he confronted specious claims by climate change deniers with humor and his own brand of pithy witticisms.[79] A respected scientist and member of the National Academy of Sciences, Schneider made seminal early contributions to the science of modeling Earth's climate system and performed some of the key early climate change experiments. Later in his career, he spearheaded efforts in interdisciplinary climate science, such as integrated assessment—coupling projections of climate change and its potential effects with economic models in order to inform real-world decision making. He was a leading voice in the public discourse over what actions we must take to mitigate potentially devastating future changes in our climate.

Needless to say, Schneider was a target. In the early 1970s, when it was still unclear[80] as to whether the warming effect of human-generated greenhouse gases or the cooling effect of sulfate aerosols would predominate, S. Ichtiaque Rasool and Schneider speculated, quite reasonably, that the latter might indeed win out if emissions of aerosols continued to accelerate.[81] As it turns out, the world's nations chose to follow a scenario in which the greenhouse warming would instead win out, an unintended consequence of the passage of clean air acts in the United States, Europe, and other industrial nations that required aerosols to be "scrubbed" from smokestacks prior to emission, primarily to solve the acid rain problem. But it easily could have turned out otherwise. The Rasool and Schneider paper nevertheless remains the source of the favorite contrarian talking point that goes something like: "Back in the 1970s, Steve Schneider was warning the world about global *cooling!*"

The attacks against Schneider didn't stop with the global cooling myth. One of the most persistent smears relates to a statement he gave in a 1989 interview with *Discover* magazine:

> On the one hand, as scientists we are ethically bound to the scientific method, in effect promising to tell the truth, the whole truth, and nothing but—which means that we must include all the doubts, the caveats, the ifs, ands, and buts. On the other hand, we are not just scientists but human beings as well. And like most people we'd like to see the world a better place, which in this context translates into our working to reduce the risk of potentially disastrous climatic change. To do that we need to get some broadbased support, to capture the public's imagination. That, of course, entails getting loads of media coverage. So we have to offer up scary scenarios, make simplified, dramatic statements, and make little mention of any doubts we might have. This "double ethical bind" we frequently find ourselves in cannot be solved by any formula. Each of us has to decide what the right balance is between being effective and being honest. I hope that means being both.[82]

Contrarians, like Martin Durkin in his "The Global Warming Swindle" polemic, are fond of editing Schneider down to the misleading snippet "we have to offer up scary scenarios, make simplified, dramatic statements, and make little mention of any doubts we might have" without the critical context, including the three sentences that followed it.

James Hansen was the first scientist to publicly testify to Congress that greenhouse warming was indeed upon us. In a sweltering Senate hall in the hot dry summer of 1988, Hansen asserted that "It is time to stop waffling. . . . the evidence is pretty strong that the [anthropogenic] greenhouse effect is here."[83] Though he has been criticized for that statement, in hindsight it appears that Hansen may have been correct that the signal of human-caused climate change had already emerged, albeit only weakly, by the late 1980s. The Reagan administration appeared to be unhappy with Hansen's public testimonies; as a NASA civil servant, he was not immune from their efforts to control his message. Representatives from the Office of Management and Budget repeatedly edited the drafts of his written congressional testimonies. Finally, in 1989, he'd had enough, and in bombshell testimony revealed that his words had been altered by the Bush administration.[84]

As Hansen has grown increasingly outspoken in recent years, the attacks against him by climate change deniers have grown more vicious. Critics have attempted to impugn his science by implying that he supplants objective scientific inquiry with political ideology. Among the baseless accusations have been that he received money from progressive activist George Soros and that he is secretly a Democratic Party operative because he received the Heinz Award in the Environment (in reality, Hansen has been a lifelong Republican, and the award was established to honor the memory of Republican politician John Heinz III, a Pennsylvania congressman who placed great value on environmental stewardship). The politically motivated attacks against Hansen over the years have been so extensive and profound that a separate book has been written on the topic.[85] But this is to get ahead of the story.

In chapter 1 we saw an early instance of an assault on climate scientists in the attack on Ben Santer for his groundbreaking work in the mid-1990s that helped establish a "discernible human influence on climate." He saw his integrity impugned as part of an industry-funded smear campaign, and his job and even his life were at times threatened.[86] The attacks against Santer were a sign of what was to come—for me. As Santer himself put it in an interview with the *New Scientist* a decade later, "There are people who believe that if they bring down Mike Mann, they can bring down the IPCC."[87]

Chapter 6

A Candle in the Dark

Skepticism is a lazy man's consolation, since it showed
the ignorant to be as wise as the reputed men of
learning.

—Bertrand Russell, *A History of Western Philosophy* (1945)

Skepticism plays an essential role in the progress of science. Properly employed, it is a key self-correcting mechanism that helps lead science inexorably, if erratically, toward a better understanding of the natural world.[1] Yet, as the philosopher Bertrand Russell's statement above reveals, skepticism in science can also be abused. His admonition has proven remarkably prophetic in the context of the climate change denial movement, wherein the term *skeptic* has often been co-opted to describe those who simply deny, rather than appraise critically.

Peer Review as Skepticism in Practice

Skepticism in the sense of critical consideration of evidence is intrinsic to the scientific enterprise. It is inherent in the challenges scientists make of each other to back up claims with logical reasoning and, where possible, hard data. The scientist must be willing to confront any holes in logic or flaws in reasoning noted by fellow scientists and, ultimately, the results must be subject to independent replication. This give-and-take occurs at scientific conferences, where scientists give presentations and can question each other on the details. It is exercised more formally through so-called peer review, a process that applies to articles describing original research, as well as to formal criticism of previously published work. Editors of scientific journals send papers out for formal evaluation of their intellectual merit by the authors' scientific peers—

other scientists, typically anonymous—who work in the same or a closely related area. Peer review is a kind of scientific "natural selection"; papers that can withstand the scrutiny of this process will find their way to publication and are often substantially stronger for it. Papers that cannot are rejected. Of course, the authors may have the opportunity to resubmit after making further revisions, or they may try their luck submitting to another journal.

Peer review does not necessarily determine whether the conclusions of a particular study are correct; that may ultimately require further work that either confirms or refutes the conclusions. Instead, the peer review process is designed to prevent the publication of papers so obviously flawed as to be clearly invalid with regard to the claims made or conclusions drawn, and unlikely to add usefully to the scientific discourse. This initial peer review process is hardly infallible, and in the pages that follow I provide examples where it has failed to catch fatal flaws in papers. In some subfields there are relatively few reviewers with the expertise necessary to evaluate a particular paper. Reviewers, moreover, are human beings subject to the same frailties that all people are. In some cases, they may be predisposed to assess unfavorably work that challenges their findings and theories, or assess favorably those papers that support them. The process of peer review is fallible, but nonetheless it is the best quality control mechanism we have in science. And it is scientific skepticism in action.

A paper may be in review and revision for a year or more before it appears in a scientific journal. And one might have to wait yet another year for any single authoritative (that is, peer reviewed) challenge of that paper to work its way through the process. The seeming disconnect between the slow pace of scientific discourse and the immediacy of nearly all other media in this new information age is striking. In these times of Twitter, blogging, and the twenty-four-hour news cycle, the old-fashioned, rather formal ways of science may seem archaic, if not outright prehistoric. Why can't we speed up the scientific process so that it meets our thirst for ready information and immediate response?

The process of publishing scientific work actually is more streamlined in some respects these days. Where once an author had to mail to the journal a cumbersome parcel containing a stack of duplicate hard copies of a manuscript, a paper can now be submitted online through a Web interface. And thanks to e-mail, no longer does an editor have to wait months to line up appropriate reviewers, get the paper off to them, and wait for return surface mail.

But this hasn't led to a dramatic decrease in the time from submission to publication during the Internet age. While the technology has changed, basic human nature has not—that's part of the answer at least. Modern scientists are no less susceptible to procrastination than scientists of yore, and they are every bit as busy. Furthermore, scientific work is no easier to evaluate today than it was in the past. A complicated theoretical derivation or sophisticated data analysis takes as long to review and assess today as it may have years ago. In short, scientific findings must still be vetted in the traditional manner: careful, deliberate review by other scientists with training and credentials that qualify them to evaluate the work. Scientific exchanges, critiques, and their replies must similarly work their way through that same process. Therein lies a conundrum.

Skepticism Meets the Internet Age

While publication of formal (peer reviewed) criticism of a peer reviewed scientific article may have to wait a year or more from the day the original article was published, a blog post expressing a critical opinion about the paper could appear on the Internet by nightfall of publication day and more soon thereafter. Now that opinions about science have become just another way to wage politics and scientists have become subject to the sorts of attacks once reserved for the political sphere, there are some lessons to be learned from politics. Chief among them is that a lie that is repeated often enough without refutation becomes perceived by many to be true.

When it comes to reporting findings in climate science and other fields of policy-relevant science, untruths and innuendo are readily propagated by a network of amateur pseudo-skeptics, specialty PR firms, and members of the corporate media who share their ideological views. Woe to the scientist who is slow to respond when subjected to such an attack, naively content to let the peer review process sort out the matter. By the time any formal response to potentially specious criticisms might be published, the damage to the credibility of the scientist, to the science, and even to society at large—which depends on access to unbiased assessments of policy-relevant science to make intelligent decisions about public policy—may have already been done. I should know—on one occasion, my work was denounced on the floor of the U.S. Senate by a prominent congressional climate change denier based on a deeply flawed critique of my work published just days earlier.[2]

Where do we draw the line between a politically motivated attack that needs to be responded to in real time through a blog post, a letter to the editor, or a news interview, and a legitimate criticism of scientific findings that reflects the natural, healthy, skeptical exchange of ideas of the scientific enterprise, and should be dealt with through the normal scientific channels? Legitimate, healthy scientific skepticism bears a certain hallmark. It does not, for example, focus on the character of the scientist or the policy implications of his or her findings. Instead, it follows a logical line of questioning. Do the results make sense? Do they contradict previous findings? Are the lines of evidence plausible? Do the data appear sound? One must approach all new claims and findings with appropriate skepticism, particularly when the conclusions appear to run rather dramatically against well-established thought. Carl Sagan's maxim that "Extraordinary claims require extraordinary evidence" reigns supreme.

When new findings are published that appear to challenge conventional wisdom, they should be carefully weighed against other currently available evidence either in support of or in contradiction to them. It is quite rare for a single new finding to overturn the weight of all previously assessed evidence. Generally, the most that happens is that existing theories are adjusted, refined, or expanded to incorporate new observations. Such was the case, for example, in physics with the twentieth-century developments of quantum mechanics and special relativity.[3] Occasionally, however, a study will suggest a fundamentally new way of looking at things.

Over the past decade, there have been numerous publications on climate science that were initially publicized as overturning prevailing thinking, only to have conventional wisdom win out in the end. In some of these cases, the authors and/or those promoting their work appear to have been driven by political agendas.[4] Other instances, however, simply bear witness to the erratic path of scientific progress.[5]

Consider the article by Israeli physicist Nir Shaviv and Canadian geologist Jan Veizer published in 2003 in the geological journal *GSA*

Today, which argued for a close relationship between long-term proxy records of cosmic ray activity and temperatures.[6] Seizing upon the authors' immoderate press release (entitled "Global Warming Not a Man-Made Phenomenon"),[7] contrarian outlets such as TechCentral-Station immediately trumpeted claims like "the warming we have experienced to date is entirely natural."[8] Just under a year later, a group of climate scientists demonstrated that Shaviv and Veizer's claim was a statistical artifact.[9] But the refutation appeared long after opponents of greenhouse gas regulation had a chance to get much mileage out of the paper—for example, in the form of credulous news coverage in mainstream venues such as the United Kingdom's *Independent*[10]—a reminder of the perils of science's measured pace when facing the media's eagerness to sell controversy.

The following year, Ross McKitrick—an economist with close ties to fossil fuel interests[11]—and Patrick Michaels of the Cato Institute teamed up to publish another article[12] that TechCentralStation claimed to be the death knell of human-caused climate change. In the TechCentral-Station piece,[13] Michaels himself characterized their article as a "bombshell" destined to "knock the stuffing out of" the mainstream view that human activity is causing the observed warming of the planet. McKitrick and Michaels purported to demonstrate that surface warming was an artifact of contaminated temperature records, based on a correlation they claimed to have found between economic activity and temperature change in various regions of the globe. Such a relationship, the pair argued, was unphysical and thus could only be explained by some sort of contamination of the underlying temperature data.

Though Michaels claimed the paper was subject to "four years of one of the most rigorous peer reviews ever," Australian computer science expert Tim Lambert discovered a fatal problem just weeks after its publication.[14] In attempting to account for the effect of latitude on temperature trends, McKitrick confused the mathematical entities of degrees and radians (two different ways to measure angles that differ by a factor of roughly fifty in magnitude!) and had in essence been feeding his analysis procedure random noise. This alone was enough to render the conclusions erroneous, but other equally significant problems in the paper were subsequently documented as well.[15]

Another article, this one published in *Nature* in 2005 by British physical oceanographer Harry Bryden and colleagues,[16] though its thesis did not pan out in the end, provides an example of the erratic path of scientific progress. The paper appeared to demonstrate a dramatic slowing over the previous several decades in the North Atlantic "thermohaline" or "conveyor belt" ocean circulation pattern—the scenario envisioned a year earlier in the movie *The Day After Tomorrow*. While nothing as dramatic as Hollywood had envisioned could conceivably befall us, credible climate model simulations indicated that the conveyor belt could indeed weaken or even collapse over several decades, with some regions in or neighboring the North Atlantic ocean paradoxically cooling, even as the globe continues to warm. Bryden and colleagues had claimed that such a thing was not only conceivable, but was possibly happening already—a 25 percent reduction in the strength of the conveyor belt had apparently taken place in just decades, they claimed. That no inkling of cooling in the North Atlantic had been witnessed and no state-of-the-art climate models had predicted any weakening of the conveyor belt to have taken place yet, however, suggested that considerable caution was needed before jumping to the conclusion that such a change in circulation was in fact underway.

More detailed measurements taken later showed that the *Nature* study likely suffered from a phenomenon known as "aliasing"— mathematically equivalent to the "wagon wheel effect" (the appearance that wagon wheels are spinning in the wrong direction in very old movies with limited frame rates). We know that the strength of the ocean current system varies on all timescales. The apparent thirty-year-long trend probably reflected sparse temporal sampling; there were only five measurements available over the entire time period. The early measurement appears to have caught the ocean current system, by chance, in a relatively strong phase, while the recent measurement, by contrast, happened to catch the ocean current system in a relatively weak phase. This chance set of measurements thus gave the false appearance of a negative long-term trend.

This episode reflects science working as it should. Experts analyzed a set of observations that motivated a rather unexpected hypothesis— that climate change was already leading to a dramatic weakening of the North Atlantic current system. This spurred scientists to make more refined and detailed measurements. Those measurements showed the original hypothesis to be wrong, but consider what happened in the process. We learned something new and important about the way a key ocean current system operates: that it may vary erratically in its properties on short timescales—a perfect example of science's own wonderfully erratic path toward greater understanding.

It's the Anomalies, Stupid

Science advances, then, not simply through the confirmation of previous findings, but by scientists trying to understand the data and observations that don't fit, that don't seem to make sense—the anomalies. That too is scientific skepticism in action.

It was an interest in understanding such anomalies that drove my own research into past climates. The Northern Hemisphere average temperature curve—the hockey stick—was in some sense the least interesting aspect to me of our temperature reconstruction work. What was most fascinating—and potentially instructive—wasn't the global trend, it was the features that bucked the trend. What was so special about Europe that caused it to cool more than other regions during the height of the Little Ice Age? Why did some proxy records appear to be pointing toward cold La Niña-like conditions in the tropical Pacific during the height of what was supposed to be the medieval warm period (or medieval climate anomaly), for example? These were the sorts of questions my colleagues and I were keen to pursue. The hockey stick itself was a by-product, indeed almost an afterthought.

I had always enjoyed an intriguing puzzle, especially one that called for novel and clever approaches. Here was one. How could we figure out what dynamic aspects of the climate system were leading to these perplexing regional features of the Little Ice Age and medieval climate anomaly? In pursuing such questions, I collaborated with scientists who had been using theoretical climate models to investigate the role of purely internal variability related, for example, to the AMO,[17] and—of increasing interest to me—forced variations driven by solar fluctuations and volcanoes.

Colleagues at the NASA Goddard Institute for Space Studies (GISS, located on Manhattan's Upper West Side, literally sitting above Tom's Diner of *Seinfeld* fame) had been using their climate model to investigate how the Northern Hemisphere jet stream responds to solar cycles.[18] Meanwhile, my U. Mass. collaborators and I had been looking at the pattern of surface temperature changes in our reconstructions associated with variations in solar output over time.[19] We found that the decrease in solar output during the Maunder minimum of the seventeenth century, while leading to global cooling of only a few tenths of a degree Celsius, gave rise to much greater regional cooling (nearly a degree in some places) over much of Eurasia and North America. Some regions, like western Greenland and the Middle East, though, actually showed a temperature increase! Working with our GISS collaborators, we were able to demonstrate that this pattern was consistent with the way a lowering of solar output weakens the Northern Hemisphere jet stream, altering its north-south undulations, so that some regions cool more than the global mean, while other regions cool less, or even warm.[20]

Other anomalies appeared to be related to the impact of volcanic eruptions on climate. In the mid-1980s, a scientist named Paul Handler from the University of Illinois had suggested a relationship between explosive tropical volcanic eruptions and the timing of major El Niño events,[21] but his findings were based on only a short interval of time: the instrumental record available back through the late nineteenth century. Other properly skeptical scientists argued that the statistical relationship was not very robust—remove one major volcanic eruption, for example, and the relationship is no longer statistically significant—and there was no convincing physical explanation given for why this relationship between volcanic eruptions and El Niño events should exist in the first place.[22]

With the longer-term record of El Niño events that we now had from our proxy temperature reconstructions (we could reliably reconstruct the history of El Niño back through the early seventeenth century), we could see whether the relationship held up over a time period that was both longer than, and completely independent of, the modern period Handler had analyzed. What's more, Amy Clement and Mark Cane of Columbia's Lamont Doherty Earth Observatory had provided an important link in establishing a plausible physical mechanism for the purported relationship. In a provocative 1997 article, they used the so-called Cane-Zebiak model of the El Niño phenomenon to demonstrate that the very same mechanisms responsible for El Niño held some counterintuitive implications for how the tropical Pacific Ocean and atmosphere system might respond to an external heating.[23] The mechanism they identified, known as the "tropical Pacific Ocean thermostat," implied that the eastern and central tropical Pacific might actually cool down! The pattern of cooling resembled the opposite of El Niño—La Niña—and it resulted from subtle ways that wind patterns and ocean currents influence each other in the equatorial Pacific. Clement, Cane, and collaborators argued that this mechanism could paradoxically cause the climate to look more like the cold La Niña state even as global warming proceeds.

A pseudo-skeptic who denies the reality of climate change might like this result simply because it provides an argument for why fossil fuel burning could lead to less global warming than the models predict. But as true skeptics, we liked the result for an entirely different reason; it questioned the simplistic notion that the globe should simply warm or cool in unison in response to external influences (be they natural or human and nature), and it challenged us to think about the subtleties of how the climate system responds to such influences—the anomalies.

If the thermostat mechanism caused the climate to look more like La Niña in response to heating (by either increased greenhouse gas concentrations or an upturn in solar output), it ought—my colleagues and I reasoned—to exhibit an El Niño-like pattern in response to the cooling influence of an explosive volcanic eruption. Not any old volcanic eruption would do; the volcanic aerosols would have to block out sunlight from reaching the surface of the tropical Pacific Ocean, something that—because of large-scale wind patterns—only a tropical eruption will do.[24] It made perfect sense to us, now, why Handler had seen a clear relationship only with tropical volcanic eruptions.

We confirmed Handler's original findings in our own independent analysis of the relationships between volcanic activity and El Niño over the past several centuries.[25] Even scientists who were skeptical of the original Handler work, such as climate researcher Alan Robock of Rutgers University, appeared to view our findings as credible. We had established the same relationship as Handler using a completely independent and longer time interval. Moreover, a reasonably compelling physical mechanism could now be articulated for why tropical volcanic eruptions would lead to an El Niño-like response of the climate system. We weren't claiming that volcanoes were the trigger for El Niño events. Our results, instead, simply suggested that tropical volcanic eruptions made the environment somewhat more favorable, roughly doubling the probability of an El Niño event occurring. I would subsequently collaborate with Cane and coworkers to show that the Cane-Zebiak model of El Niño could indeed reproduce the relationships we found in the proxy data.[26] Sadly, Handler had passed away in 1998 and hadn't lived to see the vindication of his original hypothesis.[27]

In the years since the original 1998 hockey stick article, my collaborators and I continued to investigate such puzzles of past climate change in search of possible insights about the present and future. We showed in recent work,[28] for example, that the same mechanisms described above may help to explain many of the now-better-established features of the medieval climate anomaly.[29] For example, the La Niña-like temperature pattern in the tropical Pacific we have discerned for the MCA, when solar output was high and volcanic eruptions were few, seems to be consistent with the tropical Pacific thermostat mechanism.

The implications of this seemingly innocuous finding are not trivial. It suggests the possibility that heating by increased greenhouse gas concentrations could lead to a more La Niña-like state of the climate, associated, for example, with intensified drought in the desert Southwest and increased Atlantic hurricane activity. If the minority of climate models that produce that response are correct, we might see a greater exacerbation of these effects than the IPCC currently projects. We have then a genuine area of uncertainty in the science with significant potential societal implications, one where a healthy dose of skepticism is warranted in interpreting the predictions of state-of-the-art models. Climate scientists continue to seek more data and make further refinements in the models as they strive to resolve this issue. Here we are very much in the midst of science's erratic path forward.

Science and the Media Collide

There is a basic inconsistency between the here-and-now incentive that drives our popular media, and the slower, more deliberate manner in which science advances. While dramatic (and, unfortunately, many times misleading) headlines may help sell newspapers and get us to view evening news and commentary shows, they are not faithful to the way scientific progress generally occurs—or indeed to how any complex situation unfolds. Individual scientific studies, as discussed earlier, rarely change our basic scientific understanding. Rather, the accumulation of evidence from many studies and the process by which some findings are reinforced and validated, while others fall to the wayside, typically lead to slow but steady scientific advancement.

Newspaper articles and television or radio broadcast news segments (sometimes aided by carelessly written press releases) tend to play up controversy and minimize uncertainties and caveats. Yet it is those caveats—the "error bars," the conditional and tentative assertions, the qualifications—that scientists often emphasize (frequently to a fault) in their efforts to prevent their findings from being misinterpreted, overinterpreted, or generalized beyond their range of applicability. Evidently, scientists and journalists (or at least their editors), in this respect, tend to work at cross-purposes.

Who can blame press officers, journalists, and their editors for emphasizing that which appears most novel, unusual, or surprising about a breaking science story? They must find a "hook" to sell their story if it is to compete effectively with the numerous other stories seeking a place in the ever shrinking "news hole." Incremental refinements may seem dull and uninspiring to the lay public, but controversy sells, and conflict, if a reporter can find it in a story—well, that's the mother lode.

As a result of this dynamic, the public is subjected to extreme viewpoints, alarmist headlines, and a barrage of seemingly contradictory findings—the "whiplash" effect that has been noted by the long-time *New York Times* science writer Andrew Revkin.[30] The effect is, of course, worsened by the existence today of agenda-driven news outlets that seek to actively discredit scientific findings that conflict with their ideological views. It is not difficult to see why confused observers attempting to follow scientific developments would throw up their hands, resigned to the notion that all we can safely conclude is that "the scientists don't agree."

Such a scenario plays right into the hands of vested interests. If one's agenda is, for example, to impede efforts to combat the climate change threat, it is not necessary to convince the public that the science of climate change is wrong, simply that it is grossly uncertain. It is hardly coincidental, then, that climate change contrarians have increasingly focused on manufacturing controversy where there essentially is none among those expert in the field (the overwhelming majority of practicing climate scientists concur that human activity is warming the planet and changing our climate[31]), rather than engaging in good-faith debate over the actual remaining uncertainties in the science or what implications they might have for public policy.

This is not to say that the media in some monolithic way are acting as willing accomplices to the climate change denial campaign. Nonetheless, emphasizing controversy over substance, and retreating all too often to the rather uncritical notion of "journalistic balance" when covering "debates" between legitimate scientists and antiscientific advocates, play unwittingly to that agenda.

The IPCC and the Path of Science

An essential means of evaluating science's path in a given area is not to emphasize individual papers, but instead to focus on the larger-scale evolution of scientific thinking over time. Such is the role of so-called scientific assessments, which involve extensive reviews of the available science in a particular research area by a group of leading scientists in the field. Scientific assessments are conducted at the national level in the United States, for example, by the National Academy of Sciences (NAS), an organization established by Abraham Lincoln in 1863 to "investigate, examine, experiment, and report upon any subject of science or art" for the purpose of informing governmental policy.

At the international level, there are no assessments in the world of climate science as important as those of the Intergovernmental Panel on Climate Change (IPCC), which in 2007 was awarded the Nobel Peace Prize "for their efforts to build up and disseminate greater knowledge about man-made climate change, and to lay the foundations for the measures that are needed to counteract such change." The IPCC was established in 1988 by the World Meteorological Organization (WMO) and the United Nations Environment Program (UNEP), and was tasked with assessing the evolving state of scientific understanding of climate change. Every five to seven years beginning in 1990, the IPCC has published a weighty three-volume set of reports addressing the basic science of climate change (working group 1), the projected impacts (working group 2), and the potential for mitigation of climate change (working group 3).

The report of each of the three working groups consists of detailed individual chapters assessing developments in the various relevant subdisciplines of climate science. Each IPCC report chapter is fifty to one hundred pages long, written by about a dozen scientific experts in that subdiscipline who serve as lead authors (I was one in the 2001 Third Assessment). Along with fifty or more additional contributing authors for each chapter, they collectively review the developments detailed in hundreds of peer reviewed research articles relevant to the topic at hand. Each of the three working groups also produces a summary for policy makers, which provides a brief synopsis of the working group's key findings written in language accessible to a lay audience and policy makers.

The IPCC review process remains the most rigorous, comprehensive, and transparent of any major scientific assessment.[32] IPCC reports are subject to three distinct rounds of peer review, each of which takes place over roughly two months. First, there is an initial round of expert review, wherein several thousand scientists from all disciplines and with a wide range of backgrounds and perspectives, drawn from academic, government, and nongovernmental organizations and industry, are called upon to provide detailed comments on the content of the report. Lead authors are required to consider and respond to all comments and make appropriate revisions, all of which are documented and available online. The revision process is overseen by two independent review editors with expertise in the specific subject areas of the chapter, to ensure that any legitimate issues reviewers raise are dealt with in a satisfactory manner.

The revised draft report is then subject to the next round of review, the so-called government review, which includes rereview by the original expert reviewers and additional review by government representatives from all participating United Nations member nations. Each government may choose how it implements its review. The U.S. government, for example, solicits public comments through a notice in the *Federal Register*, in essence allowing anyone at all to serve as a reviewer of the IPCC report.[33] The revision process is again repeated. Finally, national governments are again invited to comment on the report.

The final wording of the all-important summary for policy makers for each of the three working groups is agreed upon word-by-word in a final plenary meeting in which government delegations are present and can propose variant terms or make objections to specific wording, but the scientists are in charge. The final version reflects a consensus on the precise wording of the report between the scientists involved in the writing of the report and the representatives of the government delegations. It is difficult to imagine a more open, inclusive, and responsible assessment process than that which the IPCC follows.

The IPCC reports remain the gold standard for evaluating the state of scientific understanding of climate change. They are intended to inform but, importantly, not prescribe policies for avoiding so-called *dangerous anthropogenic interference* with the climate system.[34] The actual policy prescriptions, as they should be, are left for policy makers and their constituents to decide.

These reports are consensus documents in every respect. They emphasize conclusions that are based on results that have been replicated by independent studies, using differing approaches and assumptions. They highlight where there is true uncertainty. They put greatest weight on findings that are widely accepted by the scientific community, and downplay the more tentative findings of individual articles departing from conventional wisdom that, as we have seen, frequently—but not always—do not hold up when subject to further scrutiny by subsequent independent efforts.

Given the importance placed on the IPCC reports, there should be little surprise that a prime tactic of climate change deniers has been to attempt to discredit the IPCC and its assessment process. We saw such attacks in the assaults in 1995 on Ben Santer and the Second Assessment Report, and they have escalated in more recent times.

Some of the attacks against the IPCC appeal to the belief that the organization is somehow beholden to an activist UN agenda aimed at establishing a "new world order."[35] This is a curious notion. In 2001, the NAS reviewed the findings of the IPCC at the request of Republican president George W. Bush.[36] It concluded that "changes observed over the last several decades are likely mostly due to human activities"— almost word-for-word the conclusion of the IPCC. The NAS conclusions were in turn endorsed by the national academies of science of all of the major industrialized countries.[37] In order to believe that climate change science is part of some new world order plot, one would therefore need to imagine a rather implausible conspiracy among the world's leading capitalist nations to undermine their own power and authority through the creation of a world government. Ironically, while the Bush administration was casting doubt on the science of global warming, the

Pentagon was already getting busy planning for it, with its primary concerns not about international regulation, but about the national security threats posed by a changing climate.[38]

Despite the multiple layers of review, errors sometime occur, not surprising given the huge size of the multivolume reports. Detractors of the IPCC, for example, seized upon two minor errors in the impacts (working group 2) part of the report in an attempt to undermine confidence in the 2007 IPCC fourth assessment, as is discussed in a later chapter. One instance involved the year projected for the disappearance of Himalayan glaciers, which, thorough a transposition error, turned 2350 into 2035. In the other error, the report mistakenly cited the area of the Netherlands currently lying below sea level as 55 percent rather than the true value of 26 percent. Neither of these two minor errors in the several-thousand-page-long report made it into the technical summary report or the summary for policy makers.

Ironically, a more defensible criticism was that the IPCC underestimated prospects for future sea level rise. The report estimated the maximum sea level rise by the end of this century at less than a foot and a half. However, that estimate—as the IPCC itself readily acknowledged—did not include the effects of melting ice sheets, which are already known to be contributing to sea level rise. More recent alternative estimates that attempt to account for such contributions estimate the maximum sea level rise at about four feet.[39] So there was reason to be skeptical of the IPCC sea level rise estimates, but for a reason opposite to what climate change deniers argue.

Breaking of Paradigms

With all of this emphasis on consensus in science from initial paper publication to the elaborate procedures of the IPCC, where is the room for outside-the-box thinking? Certainly there must be a role for "pushing the envelope" in the progress of science? A healthy, functioning scientific skepticism requires that we question, in good faith, any scientific finding, but particularly, as Carl Sagan suggested, findings that seem most surprising or whose implications are potentially most far-reaching. The underlying sentiment was perhaps best expressed in a quotation commonly attributed to Mark Twain: "It ain't what you don't know that gets you into trouble. It's what you know for sure that just ain't so." At times, dramatic and even fundamental revision in scientific thinking is required, growing out of findings that cannot be accommodated by prevailing scientific theories. Such was the viewpoint popularized in 1962 by Thomas Kuhn's *The Structure of Scientific Revolutions.*[40]

Some of the greatest advances in modern scientific thinking have occurred through such a breaking of scientific paradigms, with individual scientists bucking conventional wisdom. The poster child is Galileo, who in the early seventeenth century championed the view of heliocentrism earlier advanced by Copernicus.[41] Galileo insisted that Earth orbits the Sun, rather than the other way around, against the prevailing religious teachings of his day.[42] Charles Darwin's theory of evolution is another example.

In the Earth sciences, the archetypal example is the theory of continental drift advanced by Alfred Wegener in the early twentieth century. Wegener defied the conventional geological thinking of his day, arguing that Earth's continents slowly move relative to each other to yield the dramatic changes observed in the distribution of continents over geologic time. The theory was not accepted until the 1960s, by which time enough evidence had amassed—in particular, geological measurements indicating the slow spreading of the seafloor—to support a break with the old paradigm, giving way to the modern theory of plate tectonics.

Another example still closer to the field of climate research is what is known as chaos theory or, in more common parlance, the proverbial "butterfly effect"—the idea that the behavior of certain systems is so sensitive to small perturbations that an effect as small as the flapping of a butterfly's wings in one part of the world can affect what transpires on the other side of the world. That natural systems could, in principle, exhibit a dramatic sensitivity to the tiniest changes in their initial state was first suggested by the French mathematician Henri Poincaré in the late nineteenth century.

It was not until the discovery in the early 1960s that such behavior could be seen in real-world systems such as the weather, however, that chaos theory became viewed as a revolution in science. We now know that it applies to diverse physical phenomena ranging from certain types of chemical reactions to the behavior of pendulums. Chaos theory even applies to the way pinball machines and amusement park Tilt A

Whirl rides behave. It also applies to the climate oscillations associated with the El Niño phenomenon. The influence of chaos, of course, has its limits. Just because certain aspects of the weather and climate are chaotic in character, for example, does not mean that climate changes are unpredictable, as anyone familiar with the rather easily forecasted climate phenomena known as *winter* and *summer* is no doubt aware.

This scientific advance is generally attributed to Edward Lorenz, an atmospheric scientist at MIT. It was nonetheless a 1963 paper on the problem of thermal convection published by my Ph.D. adviser at Yale University, Barry Saltzman, that—as acknowledged by Lorenz in his original publication[43]—led to Lorenz's identification of chaos in real-world behavior. Barry had noticed that the solutions to this atmospheric physics problem displayed unstable behavior. The behavior wasn't simply some run-of-the-mill numerical instability, however. Though Barry didn't realize it at the time, he had stumbled onto the next major scientific revolution.

The Galileo Gambit

A favorite tactic of contrarians is to attempt to usurp the mantle of the great paradigm breakers of scientific history such as Wegener or even that patron saint of paradigm breakers, Galileo himself. Indeed, there is even a term for the phenomenon: the "Galileo gambit."[44] Yet for every Wegener, Galileo, or Copernicus, there are a thousand charlatans, pretenders, and false prophets. Almost invariably, the culture of science allows them a chance to make their case. For example, at the fall 2008 meeting of the American Geophysical Union (AGU), the largest scientific society in America in the field of Earth sciences, most scientists were discussing incremental advances in, say, our understanding of the dynamics of ice sheets or the physics of auroras. One individual, however, presented what he termed a "new cosmological concept called Ac-creation," which he asserted would overthrow the conventional understanding of how our solar system was formed, and which he represented as "a paradigm shift equal to that wrought by Copernicus."[45] There was, as it turned out, no paradigm shift on that day.

Climate change contrarians frequently present themselves as the "plucky iconoclast" repressed by the scientific establishment.[46] Patrick

Michaels, as previously noted, claimed to have overturned conventional wisdom of a human role in the warming of the globe with a single (rather flawed) study. Assessing the quality of Michaels's overall scientific work, John Holdren, Harvard professor and presidential science adviser under Barack Obama, provided this blunt characterization: "He has published little if anything of distinction in the professional literature, being noted rather for his shrill op-ed pieces and indiscriminate denunciations of virtually every finding of mainstream climate science."[47]

Tim Ball is perhaps the most prominent climate change denier in Canada and the public face of a group that is called (no joke) Friends of Science.[48] Ball likes to repeat the standard contrarian talking points, for example, that it is impossible to predict a human impact on climate because we cannot predict the weather a week in advance[49]—a claim that confuses weather (e.g., will it rain a week from today?) with climate (e.g., will it be cold next winter and warm next summer?). In an unabashed play of the gambit, Ball (along with a number of other climate change contrarians we've encountered[50]) serves as an adviser to a group that actually bills itself the "Galileo Movement." The group's Web site offers this raison d'être for its founders: "At first they simply accepted politicians' claims of global warming blamed on human production of carbon dioxide (CO_2). When things didn't add up, they each separately investigated. Stunned, they discovered what many people are now discovering: climate claims by some scientists and politicians contradict observed facts."[51]

Back to Ball: In an opinion piece in the *Calgary Herald*, he claimed, among other things, to have been the "first climatology Ph.D. in Canada." When University of Lethbridge environmental science professor Dan Johnson pointed out in a letter to the editor in the *Herald* that not only is that claim untrue, but that Ball's record "does not show any evidence of research regarding climate and atmosphere,"[52] Ball sued for defamation, claiming that Johnson had impugned his reputation. The contrarian-leaning *Herald*, in a statement of defense, asserted that Ball "never had a reputation in the scientific community as a noted climatologist and authority on global warming" and "is viewed as a paid promoter of the agenda of the oil and gas industry, rather than a practicing scientist." Ball quietly dropped the lawsuit.[53]

Other contrarians, like Christopher Monckton and Stephen McIntyre, are amateurs with no formal background or training in the

Figure 6.1: Galileo
Portrait of Galileo Galilei (1564–1652). [Hulton Archive, Getty Images.]

key scientific areas (climate science, geophysical fluid dynamics, atmospheric physics, oceanography, and geosciences). With rare exception, they do not publish their claims and assertions in peer reviewed journals or attend and participate in panels, workshops, or scientific conferences. In short, they operate almost exclusively outside the normal scientific process.

What is often forgotten is that Galileo, Darwin, Wegener, Lorenz, and their fellow scientific revolutionaries were all respected, serious scientists or mathematicians of their day. Galileo, chair of the University of Pisa's math department, invented the telescope and discovered sunspots (a contribution that today allows researchers in climate science to infer past changes in solar activity and their relationship with climate). Theoretical physicist Stephen Hawking has asserted that Galileo "perhaps more than any other single person, was responsible for the birth of modern science."[54] Wegener not only gave us continental drift, but was also a leading atmospheric scientist and is credited, along with two other scientists, with having discovered the main mechanism responsible for raindrop formation and rainfall (known as the Wegener-Bergeron-Findeisen mechanism). His original lectures on the thermodynamics of the atmosphere are still regarded as seminal texts.

There could be no sharper contrast between modern-day climate change deniers and the true paradigm breakers of scientific history. The very notion of such a comparison brings us back to the chapter's epigram from Bertrand Russell. For it would be both misguided and dangerous to mistake the "lazy man's consolation" of climate change denialism for the legitimate skepticism practiced by the "reputed men [and women] of learning" working to advance the science.

Chapter 7

In the Line of Fire

I went to a fight the other night and a hockey game
broke out.

—Rodney Dangerfield

The most critical development in the tale of the hockey stick—the prominent display of the MBH99 graph in the IPCC Third Assessment Report's Summary for Policy Makers in 2001—was beyond my control. In hindsight, it would have been wiser for the authors of the summary to instead have shown the other figure from our chapter which compared three different reconstructions that collectively pointed to the same conclusion: that recent warmth was anomalous in a millennial context. The fact that the MBH99 hockey stick alone had been featured in the policy makers' document would establish it as a fundamental icon in the climate change debate. That alone was not a problem. However, sooner or later such a prominent icon would find itself in the crosshairs of the assault against the scientific case for human-caused climate change.

The MBH99 hockey stick was now thrust into the limelight. It was not only featured prominently in the 2001 Summary for Policy Makers, but it would be displayed in public lectures by IPCC chairman Sir John Houghton of the United Kingdom. It would appear in books written about climate change. There was even a campaign, at one point, to plaster the hockey stick on billboards around the world—an idea that fortunately did not come to fruition. Vice President Al Gore cited our findings in a number of his speeches on climate change during 1999. In late 1999, I was contacted by staff of the White House Office of Science and Technology Policy who wanted to make sure they understood key details of MBH99. Bill Clinton, I was told, would be mentioning our study in his January 20, 2000, State of the Union address.

There is a potential downside to one's scientific study becoming the focus of so much attention. Many scientists enjoy almost nothing more than proving the other guy or gal wrong. The more prominence a particular study or idea gets, the more tantalizing it is to best it in some way. Scientists, of course, should be skeptical of any new claims, hypotheses, or theories and study them carefully. Such scrutiny is an essential part of the self-correcting machinery of the scientific enterprise, as we saw in the preceding chapter. Scientists are passionate about their work, and debates among them can grow quite heated. However, in the case of controversial areas of science such as climate change (or evolution, stem cell research, environmental contaminants, and so on), there is additional fuel that is often thrown on the fire.

In such fields of science, scientific claims are subject not only to the potentially legitimate criticisms of one's scientific colleagues. There are, in addition, attacks by those, whether in science or (more likely) out, who are more interested in a political agenda than in scientific truth. As we have seen, a whole disinformation industry has developed in recent decades with the purpose of discrediting the science underlying climate change. Any scientific work perceived to bolster the case for the reality and threat of human-caused climate change is likely to receive the unwanted attention of the climate change denial machine. This applied to the early work of climate scientists Stephen Schneider and James Hansen. It applied to Benjamin Santer and his important contributions to the detection of human-caused climate change in the mid-1990s. And it now had come to apply to my coauthors and me, and the hockey stick.

Scientific Give-and-Take

In the years following publication of the hockey stick, a healthy scientific debate did develop within the paleoclimate research community. That debate played out in the peer reviewed scientific literature. A number of different reconstructions were produced, and while they all appeared to come to the same overarching conclusion—that recent warming was unusual in at least a millennial context—they didn't agree on all the details: How cold was the Northern Hemisphere on the whole during the Little Ice Age? How warm was it at the height of the medieval climate anomaly/medieval warm period? To the extent that various reconstructions[1] differed on these points, were the differences due to the differing mix of proxy data being used, the differing statistical methodologies, or both?

Some of the differences, my coauthors and I argued, could be attributed to the differing seasonality or differing regional emphasis reflected in the various competing estimates. Our own reconstruction was based on an averaging of temperatures over the full Northern Hemisphere—tropics and extratropics, land and ocean—and over the entire year (winter and summer). Other estimates, such as those of Jones et al. and Briffa et al., primarily reflected summer only and were weighted more toward the extratropics and the continents only. Fluctuations in the pattern of the jet stream could lead to temperature shifts over regions such as North America and Eurasia much larger than those observed in other regions, for example. And, as noted earlier, opposing temperature shifts in the tropical Pacific might tend to cancel out temperature changes in extratropical regions. Such factors complicated comparisons of different proxy reconstructions that emphasized different regions. We debated the implications of these various considerations with these other researchers in numerous exchanges in the scientific literature.[2] These differences were about the details; none of this work disputed the basic hockey stick finding that recent warmth is anomalous in a millennial context.

Other challenges to our work came from scientists reconstructing past temperatures from borehole measurements. Henry Pollack and his team published a revised temperature reconstruction in *Nature* in 2000.[3] They reiterated their original conclusion that recent warming was anomalous in a long-term context (five hundred years), but they argued that the Little Ice Age during its peak was roughly 0.5°C cooler than reconstructions from "traditional proxies" (tree rings, ice cores, corals, etc.) indicated. Pollack and colleagues asserted that the discrepancy might be due to the inability of data from those proxies (in particular, tree ring data) to resolve very long-term variations. This claim was certainly plausible. It was well established that there were potential limitations in reconstructing long-term trends from tree ring data owing to something known as the "segment length curse": When joining together data from trees that lived at different times to form a single, long-term composite measure, it can be challenging to retain climate information on timescales longer than the lengths of the individual segments (i.e., the lifetime of the individual trees).[4] We were careful to acknowledge such caveats in our own work, and indeed sought to use tree ring data that were processed to retain long-term trends.

While we did agree with the primary conclusion of Pollack et al. that "to obtain a complete picture of past warming, the differences between the approaches need to be investigated in detail," we were skeptical of the very large Little Ice Age cooling indicated by the borehole data. It was inconsistent, we argued, with the few long instrumental records that were available,[5] and we showed that certain influences, such as changes over time in the extent of insulating winter snow cover, could adversely impact the borehole estimates.[6] In the years ahead, we would maintain a healthy debate in the scientific literature with these researchers on the relative merits of the traditional proxy and borehole estimates.[7] But once again, none of this called into question the basic hockey stick conclusions regarding anomalous recent warmth.

Taking another tack, prominent paleoclimatologist Wallace Broecker of the Lamont Doherty Earth Observatory contended that there was a more pronounced medieval warm period than was indicated by either our work or the other studies cited in the IPCC Third Assessment Report.[8] He argued that the type of data derived from recent multi-proxy studies (of tree rings, corals, ice cores, etc.) lacked the precision needed to distinguish the small (roughly 0.5°C) apparent difference between medieval and modern hemispheric warmth. Moreover, he argued that proxy reconstructions of Northern Hemisphere temperatures were contradicted by other regional climate evidence (e.g., changes in glacial extent or boreholes drilled in polar ice, and historical accounts of the Viking colonization of Greenland during medieval times). However, we felt that Broecker hadn't backed up his claims, in particular his claim that proxy reconstructions lacked the precision to distinguish temperature differences of 0.5°C.[9] Much of the supposedly contradictory data he cited conflated regional temperature changes—which often cancel out or differ in timing between regions—with truly global or hemispheric-scale periods of warming and cooling. Indeed, later work of our own would provide both evidence and a physical mechanism for why the Viking-colonized regions of southern Greenland might have been warmer during medieval times than today, even while most other regions were substantially cooler.[10]

The larger context for Broecker's argument, however, was his belief that the medieval warm period represented the warm phase of an internal millennial climate oscillation. Broecker and his colleague Gerard Bond had articulated the view that the climate exhibits pronounced natural oscillations with a periodicity of roughly 1,500 years.[11] Broecker believed that the oscillations were intrinsic to the climate system, related to the so-called thermohaline or conveyor belt pattern of ocean circulation that warms the North Atlantic ocean and parts of coastal Europe. Broecker argued that the action of this climate system component was rather unpredictable and could either oscillate with a characteristic millennial periodicity or react erratically and abruptly to external influences, such as changes in solar output, or perhaps human-caused global warming.

Broecker had hinted at the belief that current warming could in substantial part be simply the most recent warming period in the natural millennial cycle. Even so, human activity was, in Broecker's view, "poking" an "angry beast" with "sticks." We had no way of knowing when the beast might react, and when it did, he suggested, the response might be both extreme and abrupt—the conveyor belt ocean circulation pattern could suddenly collapse.

The premise appeared to have at least a grain of truth to it. The sinking of surface waters in the high latitudes of the North Atlantic, by drawing in the warmer surface waters from the south to replace them, plays a key role in maintaining the conveyor belt pattern of poleward traveling surface currents in the North Atlantic ocean. The sinking is due in part to the high salinity of the surface waters, as saltwater is denser than freshwater. The large amount of freshwater released into the North Atlantic when the ice sheets were melting toward the end of the last Ice Age, roughly twelve thousand years ago, is believed to have led to a freshening of North Atlantic surface waters, inhibiting their sinking and thereby shutting down this ocean circulation system. The shutdown is believed to have induced a sudden return to glacial conditions in regions neighboring the North Atlantic before the ultimate termination of the Ice Age. Current climate models predict an analogous, though far weaker, response to the rapid melting of remaining continental ice associated with global warming. Indeed, it is this thinking that became popularized, albeit in a caricatured manner, by the movie *The*

Day After Tomorrow, where global warming is depicted as paradoxically leading to a sudden new Ice Age in the Northern Hemisphere.

The evidence supporting a prominent role for the conveyor belt in generating millennial oscillations was highly controversial however. Though Bond's team had offered some suggestive evidence, there seemed to be scant additional support that the putative millennial cycle had been operating during the modern interglacial period (the Holocene period of the past twelve thousand years). Broecker's hypothesis that the medieval warm period and Little Ice Age were part of such a natural millennial cycle seemed more tenuous still.

The "segment length curse" issue would come again to the fore the following year. Ed Cook of Lamont Doherty was one of the seminal contributors to the science of dendroclimatology and had brought considerable statistical rigor and insight to this field. Along with Jan Esper of Switzerland and others, he had been working on a new tree ring-based reconstruction of Northern Hemisphere temperatures. Esper et al.[12] attempted to deal with the segment length problem using a relatively novel method that was rather aggressive in its effort to retain long-term trends, but at the price of potential pitfalls.[13]

The Esper et al. reconstruction appeared to differ substantially from earlier reconstructions, including our own. In particular, it suggested larger temperature fluctuations from century to century and featured a warmer medieval warm period, closer to current warmth. The study got a fair amount of media attention, and I was asked to do an interview on NPR's *Science Friday* with Ira Flatow. At the time, I was on vacation on the island of Chincoteague, Virginia, with my fiancé and had to use a makeshift landline phone for the interview. Cook, in the magnanimous fashion anyone who knows him might expect, recommended that Flatow interview me about the paper, knowing full well that I had some misgivings about the new findings.

I felt that the study was a solid and useful contribution to the scientific discourse, but believed there were important caveats. First, the tree ring data used mainly represented summer temperatures in the continental interiors, and there were good reasons to believe that temperature variations in these regions are especially large (a point that the authors acknowledged). We know, for example, that explosive volcanic eruptions have the greatest cooling impact on continental interiors and in summer. This could plausibly explain why the greatest discrepancies between their estimate and ours were during periods of frequent volcanic activity. Second, their tree ring reconstruction didn't record any of the warming since the mid-twentieth century, so it could not be used to compare medieval and current warmth. Third, their temperature scaling was somewhat arbitrary, in my view, and arguably inflated the apparent amplitude of past temperature fluctuations.[14] Finally, I had some reservations about the earlier medieval portion of their reconstruction because of the relatively small sample size of their tree ring dataset during that time period—which could be especially problematic with their particular method.[15]

Cook and colleagues conceded some of these points in a follow-up paper.[16] When one accounted for the differing nature of the Esper et al. reconstruction and the hockey stick with regard to seasonal and spatial representativeness and simple issues such as the scaling of the records, many of the apparent discrepancies were resolved—though not all. For example, the thirteenth century (which falls within the medieval warm period of most reconstructions) remains quite a bit colder in the Esper et al. reconstruction than in other published reconstructions.[17]

The exchanges over the Esper et al. paper represent scientific give-and-take at its best: good faith challenges and responses between colleagues that took place both formally in the peer reviewed literature and informally in private correspondence among authors. We differed in our interpretations of the data, but we were respectful of one another's work. Our aim was the same: a better scientific understanding.

In 2002, I introduced the notion of "pseudoproxy" data as a means of testing climate reconstruction methods.[18] The idea was fairly simple: Introduce contaminating noise to actual temperature records (or model temperature output) to create synthetic proxy datasets that approximate the imperfect and noisy character of real-world proxy records. One can then test a climate reconstruction method using these synthetic proxy records and compare against the actual temperatures, which are known, to evaluate the performance of the method. In 2004, Hans Von Storch of Germany and his collaborators also used pseudoproxy tests of this sort,[19] employing a brand-new climate model simulation of the past thousand years—a potentially richer test bed for investigating climate reconstruction methods than those my colleagues and I had used in our past work. Based on these tests, Von Storch et al. argued that re-construction methods substantially underestimated the amount of long-term variability in past temperatures—in other words, they un-derestimated the amount of wiggle in the "handle" of the hockey stick, and other related climate reconstructions.

Their principal finding—that so-called "regression" methods underestimate the amplitude of variation—was already familiar. We had discussed the issue in 2002 in our earliest work with pseudoproxy networks,[20] and it was accounted for in the very wide error bars shown on the millennial MBH99 hockey stick. Though such criticisms applied equally to virtually all statistical climate reconstruction approaches, Von Storch et al. focused their attention on the hockey stick, perhaps because it was the most prominent of the published climate reconstruc-tions. The critique would have been perfectly fair and appropriate were it not for a number of significant complications to the story.

In a press release put out by his institution, Von Storch referred to the hockey stick as *quatsch*—the German equivalent of *nonsense* or *garbage*. This was shocking, arguably unprofessional language coming from a fellow scientist. But that was a matter of style. Perhaps more important, there were some problems of substance. There was a significant method-ological error in the Von Storch study that would largely undermine its conclusions. While it was well known that regression methods tend to underestimate the amplitude of variability, the real issue is by how much in the particular situation, and whether that underestimation is accounted for in the size of the error bars. The level of underestimation suggested in the Von Storch et al. analysis seemed implausibly large. After repeated inquiries from other researchers, it eventually came to light that there was a key undisclosed additional step in the Von Storch group's procedure that was not part of the MBH98/MBH99 approach or any other reconstruc-tion approach. That step introduced a dramatic but entirely spurious de-gree of apparent underestimation in the procedure.[21]

There would nonetheless be many constructive scientific develop-ments in the years to come. As we continued to collect further data and refine our own methods and analyses, many other researchers would introduce alternative approaches, challenge assumptions, and contrib-ute positively to the advance of the discipline. While none of this addi-tional work would call into question the overall hockey stick conclu-sions, the science undoubtedly would not have progressed nearly as much as it has over the past decade were it not for these collective efforts of scientists, all seeking a better understanding of the underlying scientific issues.

Contrarian Attacks

Not long after publication of our MBH98 hockey stick paper, many of the usual climate change deniers stepped out to attack our work, which was generally perceived as providing a compelling case for the reality of human-caused climate change. The criticisms were easily dismissed from a scientific standpoint, but their intent wasn't so much to contribute to the scientific conversation as it was to influence the public discourse.

Patrick Michaels was a prominent climate change contrarian at the University of Virginia primarily known for his advocacy for the fossil fuel industry.[22] Michaels, who would later become my colleague when I accepted a faculty position at the same institution in fall 1999, had put out two critiques of our work in the months following the first version of the hockey stick, MBH98. These critiques did not appear in the peer reviewed scientific literature, but on Michaels's Web site, the World Climate Report, a recipient of funding from the Western Fuels Association.[23] The first piece, "Science Pundits Miss Big Picture Again" (May 1998), was written by Michaels himself, while the second, "The Summer of Our Discontent" (August 1998), had been invited from Sally Baliunas and Willie Soon of the Harvard-Smithsonian Center for Astrophysics.

The pieces suggested that we had extended the MBH98 hockey stick no further back in time than A.D. 1400 for fear of encountering the warmer temperatures of the medieval warm period—a charge that, as noted earlier, is nonsensical, since the stopping point was entirely determined by objective statistical criteria. Second, they claimed that our reconstruction suffered from an issue known as the "divergence problem" that in fact was specific to an entirely different proxy reconstruction, not to ours.[24] Michaels was gracious enough to allow me to publish a response on his Web site in September 1998. This left me with the impression that climate change "skeptics," while tough, might be fair. I would later learn there was nothing fair in the way many contrarians treated either the science or the scientists.

MIT atmospheric scientist Richard Lindzen attacked our work in a *Scientific American* article penned by science writer Daniel Grossman.[25] Lindzen had claimed, quite incorrectly, that the hockey stick reconstruction of MBH99 for the years prior to 1400 was based only on tree rings and from only four locations. Grossman included my response, wherein I pointed out that the reconstruction for the first four centuries was based on ice cores as well as tree rings, and the tree ring data came from thirty-four independent sites in a dozen distinct regions around the globe. I was taken aback that a fellow researcher would attack our work based on such clearly erroneous claims. I was still relatively young and naive.

More specious critiques would follow. S. Fred Singer mounted a campaign against our work a year after the MBH99 millennial hockey stick was published. His organization (SEPP) organized a press event on Capitol Hill in May 2000, prior to the official publication of the Third Assessment Report. Singer used this event to assail the draft IPCC chapter on climate observations, referring specifically to the conclusion that recent warmth appears anomalous in the context of the past millennium. "We don't accept this. We challenge this," exclaimed Singer at the event. Fair enough, but the justification for this challenge was poor indeed. Singer invited Swedish paleoclimatologist Wibjorn Karlen to make his case. In a refrain that climate change deniers would repeat over and over as mantra, Karlen asserted that the hockey stick contained "neither a Medieval Warm Period nor a Little Ice Age."[26] Of course, the MBH99 hockey stick showed both; the conclusion that the contrarians really didn't like was that the medieval warm period, while rivaling mid-twentieth-century warmth in our reconstruction, simply did not reach subsequent levels of warming.

More carefully orchestrated and promoted contrarian attacks were to come. It would become increasingly clear to my colleagues and me that the intent of the attacks was to undermine not just the IPCC, but all of climate science. The attacks would make their way into White House policy and onto the floors of the U.S. Senate and House of Representatives. They would become increasingly personal, aimed at singling out specific individuals, me in particular, as if the entire weight of the scientific case for human-caused climate change rested on a handful of scientists.

Chapter 8

Hockey Stick Goes to Washington

Never argue with a fool, onlookers may not be able
to tell the difference.

—Mark Twain

As we entered the new millennium, the hockey stick emerged as a central object of attack in the broader battles waged against climate science, and the attacks became increasingly political. Among them was
an effort by the Bush administration to purge the hockey stick from a
government science report, a hostile hearing held by a prominent congressional climate change denier to attack our work, and a denunciation
of our work by that same congressman on the floor of the U.S. Senate on
the eve of a vote on a key climate bill.

The Scientization of Politics

Five years after the IPCC Third Assessment Report finding of a "discernible human influence" on the climate, during the run-up to the
2000 presidential election, one of the two candidates said: "As we promote electricity and renewable energy, we will work to make our air
cleaner. With the help of Congress, environmental groups and industry, we will require all power plants to meet clean air standards in order
to reduce emissions of sulfur dioxide, nitrogen oxide, mercury and carbon dioxide within a reasonable period of time. And we will provide
market-based incentives, such as emissions trading, to help industry
achieve the required reductions."[1]

If you guessed those were the words of Democratic candidate and
climate change crusader Al Gore, you guessed wrong. It was the vice

president's Republican opponent, George W. Bush. Given the reputation the Bush administration would eventually earn for being hostile to science[2] and the position it would eventually take, that there was still "a debate over whether [global warming] is manmade or naturally caused,"[3] it is easy to forget that Bush originally expressed a more enlightened view about climate change. He appointed a pro-environment northeastern Republican moderate, Christine Todd Whitman, to head the Environmental Protection Agency (EPA). One could be forgiven for thinking he might even have been on his way to fulfilling a promise that his father, former president George H. Bush, had made but, in the opinion of many, had failed to live up to: becoming the "environmental president."

In a Senate hearing in 2001, Whitman had stated that "there's no question but that global warming is a real phenomenon, that it is occurring." Increased flooding and drought "will occur" as a result, she had declared. "The science is strong there." She encouraged the president to support policies aimed at controlling greenhouse gas emissions, even stating in a confidential memo that doing so was a "credibility issue for the U.S. in the international community." She advised the administration "to appear engaged" in negotiations over the international treaty to limit greenhouse gas emissions known as the Kyoto Protocol.[4]

The forces of climate inaction quickly stepped in, and Whitman's advice was ignored. Vice President Dick Cheney formed an energy task force in early 2001 to draft a new national energy policy. The makeup of the task force was a secret, but the list was eventually leaked to the *Washington Post*. Among the most influential members, it turned out, were ExxonMobil vice president James J. Rouse and Enron head Kenneth L. Lay (who would later be convicted of securities fraud). Among the participants were numerous representatives of the country's leading electrical utilities and mining and fossil fuel interests, including the American Petroleum Institute, the largest fossil fuel industry trade group. According to the *Washington Post*, the list of participants bolstered "previous reports that the review leaned heavily on oil and gas companies and on trade groups—many of them big contributors to the Bush campaign and the Republican Party."[5] The task force in fact closely resembled a group we encountered in the first chapter, the Global Climate Coalition formed by the fossil fuel industry in the 1990s to oppose policies aimed at reducing greenhouse gas emissions.

Bush appointed Philip Cooney as chief of staff for the White House Council on Environmental Quality (CEQ) in 2001. A lawyer with a bachelor's degree in economics, Cooney had no formal scientific training. He had served as a climate change policy lobbyist for the American Petroleum Institute (API), an inauspicious background for someone who would be in charge of White House environmental policy. Cooney worked closely with Myron Ebell of the Competitive Enterprise Institute (CEI)—a lobbying group funded by the fossil fuel industry and conservative foundations[6] that advocates against governmental regulation in the areas of consumer and environmental policy—to undermine constructive climate change steps that the Bush administration EPA might be planning to take. Following the EPA's publication in June 2002 of the "Climate Action Report" originally commissioned by Whitman, which warned of the potential threats of climate change, Cooney and Ebell hatched a plan, it would later be revealed, to drive a wedge between the White House and the proactive EPA administrator.[7] A primary goal of the effort, according to Jake Tapper, writing for *Salon*, was to force Whitman, far too environmentally friendly for their purposes, to resign. She did a year later.[8]

Using the controversial Data Quality Act, CEI and CEQ worked together to invalidate a climate change report known as the National Assessment.[9] The National Assessment had been developed during the Clinton administration and published in November 2000 on the eve of the presidential election. It represented the most exhaustive assessment to date of the potential societal impacts of climate change. The fact that the 2002 "Climate Action Report" relied substantially upon this earlier assessment made it a highly attractive target to Cooney and the CEI. In August 2003, CEI filed a lawsuit against the Bush administration to have the report declared unlawful and to prevent its further distribution. The lawsuit was dismissed "with prejudice," meaning the suit could not be re-filed.[10] Senator Joe Lieberman of Connecticut was not impressed either, writing to the White House: "I hope that the lawsuit . . . is not the result of a collusive plan conceived by the CEI in concert with the Administration, itself. It would be wrong in any circumstance to reject the well-founded findings of the [Clinton-era report], but for the Administration to use an outside group to pursue such an ill-conceived goal would be doubly wrong, and could also be abuse of the courts at the expense of the taxpayers."[11]

As was later revealed, Cooney had tried to influence the administration's stance on the science behind climate change through unilateral editing of various government reports. Internal White House memos leaked to the *New York Times* in June 2005 indicated that Cooney "who once led the oil industry's fight against limits on greenhouse gases has repeatedly edited government climate reports in ways that play down links between such emissions and global warming."[12] The extent of this editing went well beyond an incident, previously leaked by the *New York Times* in 2003, in which Cooney had been caught editing the EPA's State of the Environment report to delete, for example, conclusions about the likely role of human activity in observed warming that had been documented in a 2001 National Academy of Sciences report—a report that the Bush administration, skeptical of the conclusions of the 2001 IPCC Third Assessment Report, had itself commissioned.

According to the documents obtained by the *New York Times*, Cooney—presumably with the tacit approval of the administration—had been making unilateral, undisclosed edits to numerous government climate change reports during 2002 and 2003, as revealed by his handwritten notes and edits of the original documents. His removal of important passages and addition of his own caveats had the net effect of weakening the conclusions expressed. These changes in many cases were made in the final versions of the reports, after supervisors and even senior Bush administration officials had already signed off. Andrew Revkin of the *Times* noted, for example, that Cooney crossed out a whole paragraph describing the effect climate change could have on the depletion of freshwater resources in the western United States, writing in the margins that the discussion was "straying from research strategy into speculative findings/musings"—as if he were somehow in a better position than the scientists who wrote the report to draw such fine distinctions. Other changes were more subtle, but they had the same net effect of watering down the conclusions, for example, inserting "significant and fundamental" before the word "uncertainties" in one report, and adding "extremely" before "difficult" in another.[13] Two days after the *New York Times* broke the story detailing these instances, Cooney announced his resignation. But no matter—in less than a week he would be welcomed back to the industry from whence he had come, courtesy of ExxonMobil.[14] Despite his hasty exit, Cooney had arguably already accomplished the task expected of him, to help execute from within the

White House the strategy that had been laid out in the 2002 Luntz memo: To sow doubt in the science of climate change.

The Hockey Stick Enters the Fray

The hockey stick did not escape Cooney's controversial edits. In attempting to water down the EPA's 2003 State of the Environment report, Cooney, among other things, had deleted a plot showing the hockey stick. In fact, he deleted all reference to our work. In its place, he added a reference to a controversial study by Willie Soon and Sallie Baliunas financed by Cooney's former employer, the American Petroleum Institute.[15] In an April 21, 2003, memo to Vice President Dick Cheney's office (eventually obtained in 2007 through a congressional investigation of administration misconduct[16]), Cooney readily betrayed his agenda:

> Soon-Baliunas contradicts a dogmatic view . . . in the climate science community that the past century was the warmest in the past millennium and signals human induced "global warming." . . . We plan to begin to refer to this study in Administration communications on the science of global climate change; in fact, CEQ just inserted a reference to it in the final draft chapter on global climate change contained in EPA's first "State of the Environment" report. . . .With both the National Academy and IPCC . . . holding that the 20th Century is the warmest of the past thousand years, this recent study . . . represents an opening to potentially invigorate debate on the actual climate history of the past 1000 years and whether that history reinforces or detracts from our level of confidence regarding the potential human influence on global climate change.

Cooney, in short, saw fit to reject the consensus of rigorous national and international scientific assessments by the IPCC and the U.S. National Academy of Sciences based on a single study funded by his former fossil fuel lobbying organization.

It was only later, after this information had made its way into the public domain in late 2007, that I was able to piece together certain odd and seemingly unconnected events from 2003. I'd had a series of exchanges in April 2003 with David Halpern, a senior policy analyst for the Office of Science and Technology Policy (OSTP), the unit within the White House that coordinates the president's science policy agenda. I had dealt with the OSTP before, in late 1999 in the lead-up to Bill Clinton's mention of our hockey stick conclusions in his January 2000 State of the Union address. But this wasn't the Clinton OSTP; it was the Bush OSTP. Halpern came right to the point, making no secret of the fact that Vice President Cheney's office was intrigued by the recent Soon and Baliunas study. Halpern was interested in what I thought of it and what my response was to the claims made in the study, especially the explicit and implicit criticisms of past paleoclimate research that included my own work. My exchanges with Halpern, it turns out, coincided in timing with Cooney's April 21 memo to Cheney's office.

I might have been more worried by ulterior motives in this inquiry, but Halpern was not just any political operative. He was a respected oceanographer and climate scientist from the NASA Jet Propulsion Laboratory who had published seminal research on the physics underlying the El Niño phenomenon and now had taken an advisory position within OSTP. He also happens to have been the editor who handled the very first article I submitted in the field of climate science, a paper in the early 1990s that, like much of my earlier work, dealt with natural climate variability.[17] I was relieved that Halpern was directly involved in this affair, not because I felt he would be somehow politically sympathetic, but simply because he was a member of the scientific community, and one who was quite familiar with climate science. We shared a common language, and this made it easier to communicate my views on the matters in question.

The Paper That Launched a Half-Dozen Resignations

So what was the Soon and Baliunas paper that was getting so much attention within the Bush administration? Sallie Baliunas, the more senior author, completed her Ph.D. in astrophysics at Harvard in 1980. She stayed on as a staff scientist at the Harvard-Smithsonian Center for Astrophysics, an institute loosely associated with Harvard University. Her best-known scientific contribution in the area of climate was her work with solar physicist, founder of NASA GISS, and cofounder of the conservative George C. Marshall Institute, Robert Jastrow.[18] The pair suggested that comparisons of the characteristics of our Sun with families of other stars could be used to relate past records of sunspot activity to past changes in solar irradiance (the solar quantity that directly influences climate). They argued for relatively large variations in solar irradiance in past centuries, such as during the seventeenth-century Maunder minimum of low sunspot activity. Recent assessments with a larger sample of stars have determined that Jastrow and Baliunas's findings were likely erroneous and point to substantially smaller variations in solar irradiance on centennial timescales.[19] Willie Soon completed his Ph.D. in aerospace engineering at the University of Southern California in 1991 and became a protégé of Baliunas at the Harvard-Smithsonian Center in 1992. The two went on to publish a number of articles analyzing the relationships between records of past solar variability and climate. Soon was also one of the coauthors of the controversial article that accompanied Frederick Seitz's Oregon Petition mass mailing in 1998 (see chapter 5).

Soon and Baliunas received considerably more attention for the new study than for their previous work. Actually, the Soon and Baliunas article took the form of two nearly identical papers published simultaneously in two different journals in spring 2003.[20] One version of the paper appeared in the journal *Climate Research* while the other (which, it turns out, was simply a longer, unedited version of the first, but with three more coauthors added) was published in the journal *Energy and Environment*. Duplicate publication of a paper is highly unusual, and in fact is strictly forbidden by most academic journals. That both the authors and the study had been supported by the American Petroleum Institute—each of the authors had a long history of fossil fuel industry funding[21]—combined with the highly unusual dual publication of the paper raised some eyebrows.

Questions had been raised, moreover, about the two journals that jointly published the paper. *Climate Research* had in the recent past published a spate of contrarian papers of questionable scientific merit. Some members of the editorial board[22] had already expressed concern that one editor at the journal known for his advocacy for the fossil fuel industry, Chris de Freitas, had been enabling publication of substandard papers with a contrarian bent.[23] De Freitas, it turns out, was also the editor who handled the Soon and Baliunas paper. The other journal, a social science periodical called *Energy and Environment*, is not recognized by the Institute for Scientific Information, the body responsible in essence for the accreditation of scientific journals. The journal's editor, Sonja Boehmer-Christiansen, is an outspoken critic of environmental regulation. In the wake of her publication of the Soon and Baliunas paper, she quite remarkably confessed in an interview with Richard Monastersky of the *Chronicle of Higher Education*,[24] "I'm following my political agenda—a bit, anyway. But isn't that the right of the editor?"[25]

The Soon and Baliunas study claimed to contradict previous work—including our own—that suggested that the average warmth of the Northern Hemisphere in recent decades was unprecedented over a time frame of at least the past millennium. It claimed to do so not by performing any quantitative analysis itself, but through what the authors referred to as a "meta-analysis"—that is to say, a review and characterization of other past published work.

A fundamental problem with the paper was that its authors' definition of a climatic event was so loose as to be meaningless. As Richard Monastersky summarized it in his article, "under their method, warmth in China in A.D. 850, drought in Africa in A.D. 1000, and wet conditions in England in A.D. 1200 all would qualify as part of the Medieval Warm Period, even though they happened centuries apart." In other words, their characterization didn't take into account whether climate trends in different regions were synchronous. The authors therefore hadn't accounted for likely offsetting fluctuations—the typical sort of seesaw patterns one often encounters with the climate, where certain regions warm while others cool.

An additional problem with the study is readily evident from Monastersky's characterization above. Rather than assessing whether there was overall evidence for widespread warmth, the authors were asking a completely different, practically tautological question: Was there evidence that a given region was either unusually warm, or wet, or dry? The addition of these two latter criteria undermined the credibility of the authors' claim of assessing the relative unusualness of warmth during the medieval period. These two criteria—were there regions that were either wet or dry—could just as easily be satisfied during a global cool period!

A third problem is that the authors used an inadequate definition of modern conditions. It is only for the past couple of decades that the hockey stick and other reconstructions showed warmth to be clearly anomalous. Many of the records included in the Soon and Baliunas meta-analysis either end in the mid-twentieth century or had such poor temporal resolution that they could not capture the trends over the key interval of the past few decades, and hence cannot, at least nominally, be used to compare past and present.

There was yet a fourth serious problem with the Soon and Baliunas study. The authors in many cases had mischaracterized or misrepresented the past studies they claimed to be assessing in their meta-analysis, according to Monastersky. Paleoclimatologist Peter de Menocal of Columbia University/LDEO, for example, who had developed a proxy record of ocean surface temperature from sediments off the coast of Africa, indicated that "Mr. Soon and his colleagues could not justify their conclusions that the African record showed the 20th century as being unexceptional," and told Monastersky, "My record has no business being used to address that question." To cite another instance, David Black of the University of Akron, a paleoclimatologist who had developed a proxy record of wind strength from sediments off the coast of Venezuela, indicated that "Mr. Soon's group did not use his data properly"; he told Monastersky pointedly: "I think they stretched the data to fit what they wanted to see."[26]

That such a deeply flawed paper not only was published in peer reviewed journals, but was then immediately promoted so widely and uncritically by those with a clear policy ax to grind, including people within the highest circles of our government, was cause for widespread concern. It was certainly a cause of concern at Harvard University, where a number of faculty members were uncomfortable with the way their institution's imprimatur had been used to serve what appeared to be a partisan political agenda. For example, John Holdren, the Heinz professor of environmental policy who went on to become president of the American Association for the Advancement of Science (AAAS) and presidential science adviser in the Obama administration, voiced the opinion[27] that "The critics are right. It's unfortunate that so much attention is paid to a flawed analysis, but that's what happens when something happens to support the political climate in Washington."[28]

There was little doubt that a response from the climate research community was needed, and various colleagues encouraged me to initiate the effort. A group of twelve leading climate scientists joined me in authoring a rebuttal to Soon and Baliunas in *Eos,* the official newsletter of the American Geophysical Union.[29] The rebuttal, in somewhat more technical terms and with specific reference to other recent work in the field, basically pointed out the flaws noted above. The American Geophysical Union considered our rejoinder important enough to issue a press release entitled "Leading Climate Scientists Reaffirm View That Late 20th Century Warming Was Unusual and Resulted from Human Activity" in early July 2003, just prior to the article's publication. Nevertheless, the Soon and Baliunas study was immediately taken up by the U.S. Senate's leading climate change denier, Republican James Inhofe of Oklahoma.

The Single Greatest Hoax Ever Perpetrated

Climate change is "the single greatest hoax ever perpetrated on the American public," exclaimed James Inhofe on the U.S. Senate floor in January 2005.[30] Personally, I thought Mr. Inhofe was paying my colleagues and me a tremendous compliment. After all, if thousands of highly opinionated and frequently cantankerous climate scientists had indeed conspired not only to coordinate such an elaborate hoax, but to get the ice sheets, sea level, and ocean temperatures to play along, we were certainly far more impressive a bunch than anyone had ever acknowledged.

The leading recipient of oil and gas money in the U.S. Senate at the time,[31] Inhofe now took the baton from Phil Cooney and the Bush White House CEQ. As head of the Senate Environment and Public Works (EPW) committee, he held a hearing in July 2003 aimed at calling the entire science of climate change into question based simply on the Soon and Baliunas study. As the Republicans held the majority, Inhofe largely controlled which witnesses were invited to testify. Of three witnesses, he would call two. The minority, led by Jim Jeffords of Vermont (an independent who caucused with Democrats), would get to call one—and they chose me. Inhofe chose Willie Soon as one of the two majority witnesses.[32] His second witness was a Soon and Baliunas coauthor on the *Energy and Environment* version of the paper, David Legates of the University of Delaware, yet another climate change contrarian with fossil fuel industry ties.[33]

The hearing took place on July 29, 2003, one day after Inhofe gave an extended speech about climate change and science on the Senate floor. In those remarks, Inhofe trotted out the "scientists were predicting cooling in the 1970s" canard we encountered in previous chapters. He misrepresented the work and views of several prominent climate scientists, including James Hansen, Stephen Schneider, and Tom Wigley.[34] He also unveiled his latest line of attack, the claim that the Soon and Baliunas study had discredited the hockey stick—the subject of the next day's hearing.

I dressed in my best suit for the hearing. I had been corresponding with members of the minority staff as I drafted my opening statement and gathered materials for my testimony, and I felt well prepared for what was to come. I was seated facing the EPW committee, which included chair Inhofe, ranking member Jeffords, Wayne Allard (R-CO), Tom Carper (D-DE), Hilary Clinton (D-NY), John Cornyn (R-TX), Craig Thomas (R-WY), and George Voinovich (R-OH). Soon and Legates sat on either side of me.

As expected, Soon and Legates used the hearing to promote the claims of their study. I, in turn, was more than happy to oblige in detailing the fundamental flaws in that study. When asked by Inhofe what I thought of a table he presented summarizing the Soon and Baliunas claims, my assessment was blunt: "just about everything there is incorrect." In response to a follow-up question from Jeffords, I elaborated: "They got just about everything wrong. They did not select the proxies properly. They did not actually analyze any data. They did not produce a reconstruction. They did not produce uncertainties in a reconstruction. They did not compare to the proper baseline of the late-20th century." While I recognized the importance of refuting the Soon and Baliunas study—and of defending our own and other independent confirmatory studies indicating the anomalous nature of recent warming—I also sought to emphasize the key larger points: that scientific understanding comes through expert assessments and the consensus of collective independent studies, and that the balance of scientific evidence clearly points to the reality of human-caused climate change.

The hearing, as one might have expected, emphasized theater over substance and myth over fact. I spent much of my testimony refuting the various claims Soon and Legates made. I explained, for example, to Senator Allard why so-called CO_2 fertilization (the hypothesis that plants will take up greater amounts CO_2 in a warmer world, therefore mitigating its buildup in the atmosphere) was not the panacea that Legates and Soon had argued it was,[35] noting that any increases in carbon uptake tend to saturate quickly with further warming. When Legates and Soon asserted that satellite data contradict surface observations of warming, Jeffords knew I was anxious to rebut this standard of climate change denial, and he gave me the opportunity to point out that the claim had "pretty much been debunked in the peer-reviewed literature."

Inhofe claimed at one point that my testimony had somehow impugned the integrity of the two other witnesses. I took issue with the assertion and noted that I had simply questioned "their qualifications to state the conclusions that they have stated." There were a number of other testy exchanges. In refuting a Legates claim that models are unable to reproduce modern observations, I pointed out that they, in fact, "track the actual instrumental warming and the slight cooling in the northern hemisphere [from the 1950s to the 1970s]." Inhofe cut me off, however, directing me to comment instead on his projections of the cost of mitigating climate change (should it exist), based on estimates by one particular group.[36] (His projections were very high, and of course with no consideration of the costs of not reducing CO_2 emissions.) "I am not a specialist in public policy and I do not believe it would be useful for me to testify on that," I responded. While the hearing was ostensibly about climate science, Inhofe and his allies often appeared to want to talk about anything but that.

Partway through the hearing, Senator Hilary Clinton arrived. Her opening statement reflected a nuanced view of the issues. She expressed concern for the balance decision makers must make between quality of life in the here-and-now and the sustainability of our future. But she left no uncertainty about where she felt the burden of proof lay on climate change: "I just want to believe that I am making a contribution to ensuring that the quality of life for future generations is not demonstrably diminished. I would feel terrible if I participated, either as a willing actor or a bystander, in this potential undermining of our Earth's sustainability."

Clinton then turned to me, "So, Dr. Mann, let me ask you, what was the Earth's climate like the last time that there was atmospheric concentration of carbon dioxide at today's levels of 370 parts per million?" I took advantage of the question to provide what I felt was much-needed context and an appreciation for the profound influence human activity was now having on our climate. "We have to go back to the time of dinosaurs probably to find CO_2 levels that we know were significantly higher than CO_2 levels today," I said and went on to articulate what I felt to be the key point: that it is not the absolute levels of greenhouse gases, but the rate at which we are increasing them, that is the worry. Past changes had taken place over tens of millions of years, while the changes taking place today were occurring "on timescales of decades." It was these rare, brief opportunities potentially to increase understanding among the audience of policy makers in front of me that made me feel as if there was at least some silver lining in having participated in the hearing.

Midway through the hearing, Jeffords dropped a bombshell. He announced that his staff had received a note from Hans Von Storch announcing his resignation as chief editor of the journal *Climate Research*, in protest over the publication of the Soon and Baliunas paper. Von Storch was no scientific ally of mine. Indeed, as may be recalled, he and I had had disputes in the past regarding the relative merits of statistical climate reconstruction methods. But ally or not, Von Storch was outraged that such a transparently flawed paper had been published in his journal. His note, which Jeffords read aloud, was to the point: "My view . . . is that the review of the Soon et al. paper failed to detect significant methodological flaws in the paper. . . . The paper should not have been published in this forum, not because of the eventual conclusion, but because of the insufficient evidence to draw this conclusion."

Von Storch's resignation had been precipitated by the refusal of the journal's publisher, Otto Kinne, to allow him to publish an editorial expressing his view that the peer review process had clearly failed with the Soon and Baliunas paper.[37] Several other editors quit as well (ultimately six editors—half the editorial board—would quit in protest over the incident).[38] Despite the devastating development, Legates and Soon continued to defend the study at the hearing. Legates suggested that the editor who handled the paper, Chris de Freitas, was being unfairly criticized for what Legates argued was entirely normal editorial practice. When asked to respond by Jeffords, I pointed out not only that de Freitas had a highly controversial track record in his editorial decision making, but also that "my understanding is that Chris de Freitas, the individual in question, frequently publishes op-ed pieces in newspapers in New

Zealand attacking IPCC and attacking Kyoto and attacking the work of mainstream climatologists. . . . So this is a fairly unusual editor that we are talking about."

Despite the "tag-team" approach I was up against, I felt good about how the hearing went. Science seemed to have prevailed that day. When I stepped out of the hearing room, a swarm of journalists, staffers, and others surrounded me for comment on the affair. Among other things, I pointed out that the intellectual bankruptcy of the climate change denial campaign had been in full display in that hearing.

There would be further aftermath to the hearing, including my first taste of the sort of campaign that would become part and parcel of the attacks against my coauthors and me. De Freitas and the climatologist John Christy wrote letters of complaint to President John Casteen of the University of Virginia (where I was then teaching), claiming that I had unfairly maligned them in my Senate testimony.[39]

In my letter of response to Casteen, I refuted Christy's claim that I referred to his work as having been "debunked." It was quite clear by context that I had been referring not to Dr. Christy's work, but to the myth perpetuated by "climate change contrarians" that the satellite evidence he had analyzed in some way contradicted surface evidence of warming. This myth had indeed been dispelled or debunked in a National Academy of Science report published in 2000.[40]

With regard to de Freitas, who had claimed that I damaged his reputation, I pointed out that that there was nothing in my testimony that was either untrue or unfair. De Freitas's advocacy against policies to restrict carbon emissions and his frequent attacks against the IPCC made him a highly unusual editor for an academic journal. It was hardly my fault that his actions had led the editor in chief and half the editorial board to resign from the journal in protest. Nor was it my doing that various leading scientists had criticized de Freitas for the way he had conducted himself as editor at *Climate Research*, as reported in the mainstream press.[41]

Perhaps the single most troubling issue to arise from the Soon and Baliunas affair was that of apparent editorial malpractice. At the two journals that published versions of the paper, the peer review process appears to have been compromised to produce a study in the scientific literature that could be seized upon by those with a contrarian policy agenda. The importance of the integrity of the editorial and peer review process cannot be overstated. While it is not infallible and bad papers inevitably get published occasionally, the process is an essential component of the self-correcting machinery of science. Peer review is a necessary—if not always sufficient—condition for taking the conclusions of a study seriously.[42]

It is particularly pernicious when that process is compromised or co-opted for political ends. The repercussions of such lapses may be more dramatic in other areas, such as tobacco, medicine, or pharmaceuticals, where there is a more immediate connection with human health and the toll of allowing industry-funded ghostwriters to publish distortions in the medical literature can literally be measured in human lives. However, it is no more acceptable in other fields. In the case of climate change, the costs to humanity may not be as immediately evident or observable, but the long-term threat is every bit as real.

A Coordinated Attack

The crescendo in contrarian attacks aimed at our work (and that of other climate scientists) through the summer of 2003 was no accident. The attacks had a clear purpose: to block an imminent effort in fall 2003 to pass meaningful legislation to combat climate change. That legislation would take the form of the Climate Stewardship Act, jointly sponsored in the U.S. Senate by Joseph Lieberman (D-CT) and John McCain (R-AZ). The bill would require the EPA administrator to regulate greenhouse gases as well as to set up a market-based cap and trade system designed to curb greenhouse gas emissions of polluters. In addition to the events recounted earlier in this chapter, yet another attack, this time focused even more directly on the hockey stick, would be released on the eve of a crucial October 30 Senate vote.

The attack was launched in a paper released on October 27, published by the very same controversial journal, *Energy and Environment*, that had published one of the versions of the Soon and Baliunas study.[43] The right-wing economist Ross McKitrick was one of the two authors; the other was Stephen McIntyre of the climateaudit blog, a man who had no publication record in science or any apparent formal training in areas of science directly relevant to climate. McIntyre billed himself as a "semiretired minerals consultant"; as disclosed by investigative journalist Paul Thacker, he had close ties to the energy industry, having served as president of Northwest Exploration Co Ltd. before it became CGX Energy, an oil and gas exploration company that subsequently listed McIntyre as a "strategic adviser."[44]

The central claim of the McIntyre and McKitrick paper, that the hockey stick was an artifact of bad data, was readily refuted.[45] The paper's dramatically different result from ours—purporting an extended warm period during the fifteenth century that rivaled late-twentieth-century warmth—was instead an artifact of the authors' having inexplicably removed from our network two-thirds of the proxy data we had used for the critical fifteenth–sixteenth-century period.[46] The fact that their critique rested on what could most charitably be described as a misunderstanding of the data used in our study hardly mattered, though. For the time being, climate change deniers had everything they needed to do immediate damage. They had a published study purporting to call into question the basis of the scientific evidence for human-caused climate change, which they could promote with much fanfare just before a key U.S. Senate vote on climate change policy. In terms of the impending vote, what would it matter if the study were later shown to be based on a bogus analysis? Conservative media outlets had enough time to publicize the claims before the October 30 vote, but there was not enough time for an organized rebuttal on our part.

A press release entitled "Important Global Warming Study Audited—Numerous Errors Found; New Research Reveals the UN IPCC Hockey Stick Theory of Climate Change Is Flawed" was distributed to journalists on October 27 by Laura Braden Dlugacz[47] on behalf of TechCentralStation, an industry-funded Web site that was later singled out by Senators Olympia Snowe (R-ME) and Jay Rockefeller (D-WV) in demanding that ExxonMobil "end any further financial assistance" to groups "whose public advocacy has contributed to the small but unfortunately effective climate change denial myth."[48] On the morning of the October 30 Senate vote, an op-ed parroting the McIntyre and McKitrick claims appeared in *USA Today*—authored by Nick Schultz, editor of TechCentralStation.[49]

The previous evening, James Inhofe had gone on the attack, promoting these latest allegations in a speech delivered on the Senate floor. One of the more unintentionally humorous moments in Inhofe's diatribe came when he characterized me as being in charge of the IPCC.

Honored as I was by this instant promotion from a minor player (one of ten lead authors of one chapter of one of the three IPCC working group reports) to IPCC chair, I was sure the news would come as a surprise to Sir John Houghton of the United Kingdom, the actual IPCC chair at the time. When it came to Inhofe's attempts to push the McIntyre and McKitrick critique, however, he was frankly ineffective. He misstated what McIntyre and McKitrick had actually claimed, asserting that their reconstruction—with its spurious fifteenth-century warmth—showed how warm the medieval warm period really was (the fifteenth century falls in the heart of the Little Ice Age). The criticism of our work fell so flat that other senators appear to have simply ignored it. Later in the floor debate, Olympia Snowe presented a chart of the hockey stick as a key piece of evidence in support of the reality of human-caused climate change. It was difficult not to be amused to find our work used by both sides of the debate that night.

Was this display nothing more than the usual political theater? Or did the attacks against our work have a material impact on the final vote on the McCain-Lieberman bill? The final vote was 55–43 against the bill, far closer than many had expected, and dramatically closer than the unanimous 95–0 defeat of Senate support for the Kyoto Protocol in 1997.[50] A swing of just 7 nay votes would have led to passage of the McCain-Lieberman bill. Were there seven or more positively predisposed U.S. senators who were negatively swayed by the hoopla over the McIntyre and McKitrick claims? I'll never know whether those claims made any difference one way or the other.

Aside from peddling the specious allegations of the McIntyre and McKitrick paper, the *USA Today* op-ed claimed that "Mann never made his data available online, nor did many of the earlier researchers whose data Mann relied upon for his research. That by itself raises questions about the U.N. climate-change panel's scientific process." We were able to prove to the *USA Today* editors that the claim was in fact entirely false, pointing them to the location on our public Web site where the data had clearly been available for years. The claim was retracted by *USA Today*, albeit in an obscure correction notice published two weeks later.[51] *USA Today*'s senior science journalist, Dan Vergano, would further correct the record with a far more objective account of this latest manufactured controversy, but this, too, was printed well after the Senate vote.[52] Meanwhile, deniers were playing the controversy up for all it was worth. An e-mail a colleague sent on to me indicated that Fred Singer had sought to pressure *Nature* into "withdrawing" the MBH98 article based on the McIntyre and McKitrick paper.

As would often prove to be the case with future such attacks, my spirits were lifted by supportive letters and notes from colleagues. A couple stood out in this instance. One was an exchange with the climate scientist Stephen Schneider, long a personal hero of mine, who exhorted me to stand my ground. Another was a short note from Walter Munk of the Scripps Institution of Oceanography/University of California-San Diego. I'd studied Munk's seminal work on ocean circulation in introductory graduate courses on physical oceanography. Munk had also developed the concept of acoustic tomography (using the behavior of sound waves as they travel through the ocean to infer the thermal structure of seawater) and was the first scientist to rigorously explain a simple, but rather enigmatic, observation: why the same side of the Moon always faces Earth. There was no living scientist in my field that I respected more. In early November 2003 in the aftermath of the latest developments, Munk sent a kind note praising me for the way that I had dealt with and responded to the attacks, and offered the sage observation that "the problems of climate reconstructions are sufficiently challenging when all work together in good faith. Little is served by introducing unnecessary elements of controversy."

Chapter 9

When You Get Your Picture on the Cover of . . .

A half-truth is a whole lie.

—Yiddish Proverb

A hit song of 1972 by the rock group Dr. Hook, "The Cover of the Rolling Stone," satirized musicians knowing they had "made it" when they appeared on the cover of *Rolling Stone* magazine.[1] In the global warming debate, it was the climate change deniers who would win that honor—in January 2010.[2]

David Versus Goliath

I instead got a dubious honor of a different sort some five years earlier: On February 14, 2005, I received a Valentine's Day present—a rather unwelcome one—from the *Wall Street Journal* (*WSJ*).[3] My likeness was featured as one of the paper's stylized, hand-rendered portraits on the front page accompanying the article "In Climate Debate, the 'Hockey Stick' Leads to a Face-Off" by environment reporter Antonio Regalado. The article reported the latest attacks on the hockey stick by Stephen McIntyre, whose likeness was also featured on the front page. McIntyre missed making the cover of the *Rolling Stone* five years later, however, as he was not chosen for its list of the planet's seventeen "worst enemies"—a list that included other now-familiar protagonists such as James Inhofe, Marc Morano, Fred Singer, and the Koch brothers (McIntyre *did* make a similar list of "climate villains" compiled by another publication around the same time[4]). Though separated in time by five years, the *Rolling Stone* and *WSJ* cover stories were emblematic of the dueling narratives of the two sides in the climate change debate.

While it was easy to dismiss the numerous denialist columns that had run over the years on the *WSJ*'s typically right-leaning opinion pages, contrarians could now point to an ostensibly more credible *WSJ news piece* in their efforts to discredit the hockey stick and, more important, what it represented. Regalado invoked an archetypical David versus Goliath narrative, describing how a simple "semi-retired minerals consultant" supposedly toppled "one of the pillars of the case for man-made global warming."

There were a few complications with that narrative that Regalado glossed over, however. The David in the scenario (McIntyre) had the backing of the most powerful industry on Earth,[5] while the Goliath (me) was a lowly assistant professor at a small public university. And the "pillar" in question—the hockey stick—was hardly the central piece of evidence for human influence on climate that deniers liked to pretend it was, though its iconic status had made it a frequent whipping boy.

The major deficiency in Regalado's article was that it conveyed the highly misleading impression that McIntyre had invalidated our work. He hadn't. The problems with the article were detailed by investigative journalist Paul Thacker, writing for the American Chemical Society's *Environmental Science and Technology*.[6] In the course of his investigation, Thacker interviewed a number of the scientists Regalado had contacted and interviewed, several of whom were not mentioned or quoted in the article. Among them was Dr. Jerry Mahlman, the former director of a NOAA climate modeling center.[7] Recounting his interview with Regalado, Mahlman told Thacker: "He had this cute little lead, 'Oh, I heard you're the guy that coined the term hockey stick.' I said, 'Guilty as charged.'" According to Thacker, "what began as an interview, quickly evolved into a spirited debate. Whenever [Mahlman] pointed out the importance of Mann's work Regalado would try to shift the discussion back to McIntyre and McKitrick."[8] Speaking of McIntyre and McKitrick, Mahlman recounted: "I told him that as far as I know they're quacks. That kinda riled him." Thomas Crowley, then at Duke University, was also interviewed but not mentioned in Regalado's article. Crowley told Thacker: "I did go into a long explanation for why McIntyre's work isn't great shakes, as some people would like to believe. That didn't come out in the article." Thacker went on: "The resulting bias in the article,

Figure 9.1: Getting Their Picture on the Cover . . .
A characterization of climate change deniers on the January 6, 2010 cover of *Rolling Stone*. [Cover courtesy of Rolling Stone issue dated, January 6, 2010. © Rolling Stone LLC 2010. All Rights Reserved. Reprinted by Permission.]

[Crowley] says, confirmed his suspicions that the *WSJ* slants their news on climate change."

Others who were cited were quoted selectively, at best. Francis Zwiers, chief of the Canadian Centre for Climate Modeling and Analysis, indicated that he had not—despite Regalado's assertion to the contrary—endorsed McIntyre's claim that the hockey stick was an artifact of the statistical conventions used.[9] He had gone "to great pains in the interview" to underscore that the technical issues under discussion did "not mean that the general form of the reconstruction (illustrating the unusual nature of the 20th century) is wrong," and he pointed to a number of independent reconstructions that provided a similar picture to that of the hockey stick.

I was the third person Regalado interviewed who emphasized that several independent reconstructions confirmed the hockey stick conclusions. Yet that crucial point was not even hinted at in the article. I also explained to Regalado that my group had published work showing that the hockey stick was not sensitive to statistical conventions in the manner implied by McIntyre. This fact was barely recognizable, however, in Regalado's innuendo-laden paraphrase (emphasis added): "Dr. Mann *says he can create the same shape* from the climate data using completely different math techniques."

Regalado also gave readers the false impression that my coauthors and I had something to hide. He quoted me as saying: "Giving them [McIntyre and McKitrick] the *algorithm* would be giving in to the intimidation tactics that these people are engaged in." I doubt I said that to Regalado since the algorithm was published with our original (MBH98) article, and there would thus be no need to give it to someone.[10] I might have said something like that regarding the *source code*, that is, the specific computer program I had written, and for good reasons: (1) Our source code wasn't necessary to reproduce and verify our findings. Scientists such as Eduardo Zorita who were engaged in research to assess our methods, for example, had independently implemented our algorithm without access to our source code. (2) Our source code was our intellectual property. The National Science Foundation (NSF), which had funded our work, had already established that we had more than met the standards of disclosure of data and methods expected of NSF-funded scientists and that the specific source code we had written was our intellectual property.[11] (3) While I was happy to provide source code—and had—to colleagues (including competitors) who were engaged in good faith attempts to assess our methods, important precedents were at issue here. Did we really want to head down the slippery slope of releasing proprietary materials indiscriminately? What other vexatious demands might be made of us and others? It was source code today, but where would it end? Short scripts, research notes, perhaps even private e-mail correspondence? As it would turn out, my worst fears along these lines would be justified.

There were other problems with Regalado's article. Thacker notes, for example, that Regalado inappropriately conflated the more nuanced scientific disagreements that our group had with Hans Von Storch (documented in chapter 7) with McIntyre's erroneous claim that the hockey stick itself was some sort of statistical artifact.[12] Regalado's most severe critic was, ironically enough, former *WSJ* front-page editor Frank Allen. Allen characterized Regalado's article as a "public disservice" littered with "snide comments" and "unsupported assumptions." According to Thacker, Allen could "not understand how the story got past the editors." Stated Allen, "it had this bizarre undertone of being investigative but it didn't investigate . . . it purported to be authoritative, and it's just full of holes."

Of Tribes and Trees

One of the most serious allegations in the *WSJ* article was the claim by McIntyre that the hockey stick was an artifact of the statistical conventions we had used. McIntyre and McKitrick had quietly dropped their erroneous original assertion (in their 2003 paper discussed in chapter 8) that the hockey stick was an artifact of bad data. Their new, albeit equally erroneous, assertion was that the hockey stick was an artifact of the conventions used in applying principal component analysis (PCA) to certain tree ring networks, which, they argued, "manufactured Hockey Sticks" even from pure noise. They had initially submitted this argument to *Nature* as a comment on the original MBH98 article. *Nature* rejected their submission for lacking merit,[13] but the journal *Geophysical Research Letters* (*GRL*) published the comment as an article two days before Regalado's February 12, 2005, *WSJ* piece ran.[14]

To investigate this newest claim and the reasons for its falsehood, we need to revisit the statistical concept of PCA introduced in chapter 4. As deniers have fastened onto McIntyre's claim in attacks against the hockey stick that persist to this day, it is important to get to the bottom of it. Interestingly, it wouldn't be the first time that it was necessary to delve into this seemingly arcane construct (or, at least, its close relative) to understand the origins of a societally relevant scientific controversy.

In *The Mismeasure of Man,* Stephen J. Gould explained how early-twentieth-century psychology researchers had misused statistics to argue for the existence of a unique and unitary measure of human intelligence—what they believed was a truly robust measure of IQ that could be applied across cultures to rank the relative intelligence of members of various races, cultures, and even individual tribes.[15] The researchers in question were Charles Spearman, who introduced the tool known as factor analysis to the field,[16] and his disciple, Cyril Burt. With factor analysis, as with principal component analysis (PCA), researchers can take a large two-dimensional dataset and break it up into a small number of leading patterns found in the data (typically, just a handful of the most important patterns can be used to characterize the lion's share of variation in the data). The distinction between PCA and factor analysis is minor, and for our purposes we can consider them the same thing.

Spearman (and Burt subsequent to him) had applied the tool to a large set of different potential measures of purported intelligence across various cultures. They were interested in seeing whether a single dominant pattern emerged in the data. If such were the case, the first pattern derived from PCA (the leading principal component, or PC#1) would explain the vast majority of variation in the data and thus, they thought, signal the existence of a single underlying quality that could be defined as "intelligence." What they didn't realize, however, was that the converse is not necessarily true; just because PC#1 explains a large amount of variation in the dataset does not mean that it has captured all of the significant variation in the data. This, in essence, was the fatal error Spearman (and then Burt) committed. Gould refers to the fallacy as "reification," which he defines as "our tendency to convert abstract concepts like 'intelligence'" into entities.[17] For our purposes, it amounts to the assumption that a physical (or biological) meaning can be attributed entirely to a single, individual pattern (e.g., PC#1) that falls out of a statistical analysis procedure.

Spearman indeed found that the first PC in his analysis explained a substantial percentage—about 50 to 60 percent—of the variation in the data, but that meant that nearly half of the variation in the data was still unexplained. Spearman nonetheless ascribed a deep psychometric significance to the PC#1 of his data and even gave it a name, the g factor. Spearman's g soon became widely celebrated as the Holy Grail of psychometric analysis, a unique, mathematically definable measure of human intelligence. The problem is that g was not in fact unique. If one were to adopt different statistical conventions, one could get a quite different answer using the same dataset.

To explore just how this might work, let us revisit the synthetic example we considered in chapter 4, but with a tweak, the details of which we'll return to later. We will first need to draw an analogy between the seemingly quite disparate problems of characterizing variations in global temperature (as with the synthetic example from chapter 4) and measures of human intelligence (as with the Spearman analysis discussed above). As different as these two problems might seem, they potentially lend themselves to strikingly similar data analysis procedures. Imagine that instead of different potential measures of what we might think of as intelligence (e.g., reading comprehension and facility with arithmetic), we have different spatial regions (e.g., western and eastern hemispheres) of temperature variations, and instead of looking for trends across cultures in measures of intelligence, we are interested in trends across time in temperature.

Let us idealize Spearman's example and imagine that all of the variation in the putative intelligence data are described by just two principal components. PC#1—Spearman's g—would then reflect a pattern that is similar across all measures of intelligence (from reading comprehension to arithmetic) and show a clear trend across races and cultures (e.g., high performance in the Western world and poor performance in those cultures that were thought of as "primitive"). PC#2, by contrast, would show a more complex pattern of variation across races and cultures, and varying performance among competing measures of intelligence. Drawing an analogy with the temperature problem, then, PC #1 would be like the uniform global warming with its clear trend over time for all regions, while PC#2 would reflect a more complex pattern in space and time that constitutes the regional and temporal departures from that trend. In both cases, Burt's putative measures of intelligence and our synthetic temperature example, each of the patterns in the data are important, and one would be in error to throw out either one and focus exclusively on the single remaining pattern as if it summarizes all of the important information in the data.

The above situation is portrayed in the right half (panels g–l) of Figure 9.2. PC#1—resolving 55 percent of the variation in the data— shows a uniform warming trend (panels i and j), while PC#2—resolving the remaining 45 percent of the variation in the data—is a roof-shaped temperature variation that plays out oppositely in the West (panel k) and East (panel l). Interestingly, this is precisely the same data shown and analyzed back in chapter 4, yet the order of the two PC components has reversed, as the percentages of variation they explain has changed. In our original analysis (left side of figure, panels a–f) PC#1—resolved 60 percent of the variation in the data—and was associated with the roof-shaped graph and its inverse (panels c and d), while PC#2— resolving the remaining 40 percent of the variation—was associated with the global warming trend (panels e and f).

What has allowed for this reversal of fortune for the two patterns? In fact, just a mild change in statistical convention. Rather than centering the temperature data about their long-term (full hundred year) average as in the original analysis in chapter 4 (we can call this long-term centering of the y-axis), we have now centered the data about their average during only the latter half of the hundred-year time interval (the most recent fifty years—we can call this modern centering). In other words, we've shifted the baseline for the data. This leads PC#1, associated with the trend pattern in the modern-centering convention (panels i and j), to now take on negative values over most of the hundred-year interval be-cause the temperatures are measured relative to the relatively warm baseline of the latter half-century, instead of the long-term average.

If we had chosen to retain only the leading pattern from the data (PC#1), we would have kept the global warming pattern—or, returning to the Spearman example, our g. However, by eliminating the second pattern, PC#2, we would have removed all of the complication and nu-ance in the data, all of the information that tells us that we really can-not represent all of the important variation in the temperature dataset by a single, globally uniform ramp in temperature, or, returning again to the Spearman example, by a single unitary measure of intelligence that is consistent across cultures and races.

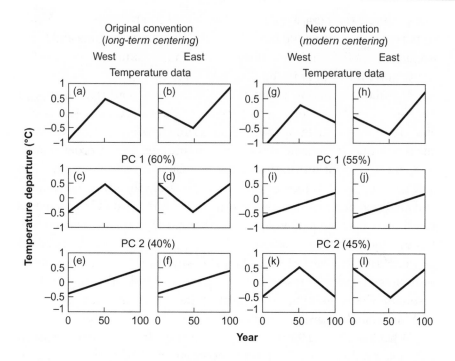

Figure 9.2: PCA Example: Spatial and Temporal Variations in Temperature Data (Revisited)
PCA example shown in chapter 4 (left—see page 46) with baseline defined as average temperature over the full one hundred years (long-term centered), compared with alternative PCA (right) where departures in temperature are measured relative to a baseline defined as the temperature average over only the second half of the time period, the final fifty years (modern-centered).

Indeed, it should now be apparent that, had we used a different convention (the original long-term centering of the data), retaining only the first pattern (panels c and d) would have given us an entirely different result! In that case, retaining only PC#1 would yield no evidence for the global warming pattern (in the temperature example) or, alternatively, for g (in the Spearman example). One modest change in our centering convention, then, and that pattern seemingly evaporates from the data.

Is there any way to resolve this conundrum? Indeed there is: We must accept that there is no g! That is to say, there is no single pattern that characterizes overall intelligence across cultures (or, in the temperature example, there is no single pattern of variation that characterizes the temperature history at all locations). There are two patterns of near equal importance in the dataset, and their relative prominence will depend on the precise convention that we use. The only fail-safe approach is to recognize that both patterns are necessary to describe the data. We must not overstate the importance of PC#1 alone. We must, in short, avoid the pitfall of reification.

Principal component analysis simply provides a convenient way of efficiently summarizing information in a large dataset. Any appropriate application of PCA must retain enough PCs to describe all significant patterns in that data. Retaining only a single leading pattern will not in general achieve that goal. As Gould explains in *The Mismeasure of Man*, by retaining only PC#1 and unjustifiably throwing out the rest of the variation in the data, Spearman, and Burt after him, discarded significant information that conflicted with their hypothesis of a single, culturally universal, unique measure of intelligence. Such data, Gould notes, include culturally specific factors (e.g., the relative demands in a particular culture on individuals to develop, say, verbal as compared to arithmetic skills) that belie the concept of a simple, objective, universal metric of innate intelligence. Indeed, had they adopted just a slightly different convention, keeping only PC#1 would have led them to an entirely different conclusion.

With Spearman and Burt, the arcane tool of PCA had been misapplied to putative metrics of human intelligence to support theories of a racial basis for intelligence. With McIntyre (and colleague McKitrick), it was—as we shall now see—misapplied to sets of tree ring records to support a critique of climate change research. If there is a lesson in this curious confluence, it is that scientific findings that rest on such technical complexities are prone to abuse by those with a potential ax to grind. Inappropriate decisions made in the statistical analysis can have profound consequences for the results. Given the complexities, it's easy enough to make mistakes. For those with an agenda, it is even easier to overlook them or, worse, exploit them intentionally.

Hiding the Hockey Stick

While the specifics were of course different, McIntyre and McKitrick in their critique of our work had in essence committed the same statistical error as had Spearman and Burt. The MBH98 set of various proxy data, as noted in chapter 4, was heavily weighted toward tree ring data. Had we not taken appropriate precautions to deal with that issue, our reconstruction would have been largely determined by the tree ring data alone—no doubt, something that our critics would have jumped on us for. So we used PCA to represent the dense networks of tree ring data in terms of smaller numbers of representative patterns of variation in each region (North America, Eurasia, etc.).

We employed a standard, objective criterion for determining how many PCs should be kept for each region. This criterion is known as a "selection rule," and it is derived using the very same sorts of Monte Carlo techniques I described in chapter 1. One creates various surrogate datasets that in key respects have the same attributes as the actual data (same size, same overall amplitude of variation, etc.), but that are randomized in a way that destroys any significant structure in the data. By comparing how much variation is resolved in the actual PCs of the data relative to the PCs of the randomized surrogate data,[18] one obtains an objective answer to what we have already seen to be the crucial question: How many significant PCs are there in the actual data? Such a criterion might indicate that one needs to keep the PCs that resolve the leading 50 percent of variation in the data, but one could alternatively find that as much as 90 percent or as little as 10 percent of the variation in the data must be retained. The precise answer will depend on the characteristics of the data at hand.

For this procedure to be valid, the random datasets must be treated with precisely the same statistical conventions as the original data. This issue is a nontrivial one because, as we have seen, there are different possible conventions for how the data might be centered and, as we have also seen, this choice can play a crucial role in determining the relative ordering of the various patterns in the data. We chose to use the same twentieth-century base period we had used for the instrumental data for centering the proxy data when performing the PCA step—a modern-centering convention.[19] The same convention was, therefore, as it needs to be, also used for all of our random surrogates in determining how many PCs to keep.

The North American tree ring data, as previously described, played a particularly important role in our analysis. Applying our selection rule to these data, using a modern centering convention indicated that the leading two PC series should be retained. PC#1 emphasized the tree ring data from high-elevation sites in the western United States, which, as discussed in chapter 4, contained a key long-term temperature signal, the hockey stick signature of a cold Little Ice Age interval followed by pronounced twentieth-century warming. PC#2 emphasized lower-elevation tree ring series, which showed less of a twentieth-century trend.

McIntyre and McKitrick used a different PCA convention in their 2005 paper. They centered the tree ring data over the long term (1400–1980). That's fine—in fact, long-term centering is actually the traditional convention, and given the fodder our less traditional modern centering convention has provided for climate change deniers, I wish we'd used the long-term centering from the start. It doesn't make any difference which convention you use; you get the same final answer in the procedure as long as you do the analysis correctly.

McIntyre and McKitrick got a dramatically different answer by *not* doing the analysis correctly. Their error is easy to understand using our synthetic PCA example. As we saw above, using a modern-centering convention (centering over the final fifty of the one hundred years in the example), the global warming pattern was carried by PC#1, describing 55 percent of the variation in the data, while PC#2—the oscillation pattern—described the remaining 45 percent. Let's imagine that our selection rules told us that we should retain any PC explaining at least 45 percent of the total variation in the data. We would end up keeping both patterns, PC#1 and PC#2, resolving all of the important information in the dataset (the long-term trend and the oscillation).

Suppose instead that we used that same retention criterion (PCs resolving at least 45 percent of the variation in the data) that had been derived for the modern centering, and misapplied it to PCA results where the alternative, long-term centering (centering the data over the entire hundred years) had been used. Since the global warming pattern in that case showed up as PC#2, explaining only 40 percent of the variation in the data, it ends up on the cutting room floor, just missing our threshold for retention. By misapplying a selection rule derived for one convention (modern centering) to PCA results based on a different convention (long-term centering), we would end up erroneously throwing out the proverbial baby with the bathwater.

That's precisely what McIntyre and McKitrick did with the North American tree ring data. They applied retention criteria that we had obtained using our convention (modern centering) to PCs calculated from the tree ring data based on a different convention (long-term centering). Through this error, they eliminated the key hockey stick pattern of long-term variation.[20] In effect, McIntyre and McKitrick had "buried" or "hidden" the hockey stick. They had chosen to throw out a critical pattern in the data as if it were noise, when an objective analysis unambiguously identified it as a significant pattern.[21] It was essentially the same error they had committed in their 2003 paper, wherein the key proxy data were simply thrown out[22]—it's just that here, they were thrown out in a way that was not as obvious.

The Rebuttals

While we had refuted a number of the McIntyre and McKitrick claims ourselves in the peer reviewed literature,[23] analyses by a number of other groups over time would independently confirm that the various claims McIntyre and McKitrick had made were false or misleading. The first such analysis was conducted by National Center for Atmospheric Research (NCAR) scientists Eugene Wahl and Caspar Ammann who, in May 2005, announced findings that "dispute McIntyre and McKitrick's alleged identification of a fundamental flaw that would significantly bias the MBH climate reconstruction toward a hockey stick shape." They concluded that "the highly publicized criticisms of the MBH graph are unfounded."[24]

Wahl and Ammann demonstrated that the hockey stick was not an artifact of PCA conventions and that the basic result is robust as long as key proxy records are not thrown out (either explicitly, as in the original 2003 McIntyre and McKitrick paper, or implicitly through the use of erroneous selection rules, as in their 2005 paper). Wahl and Ammann closely reproduced the original hockey stick reconstruction with the data available in the public domain, even after accepting—for the sake of argument—all potentially valid points made by McIntyre and McKitrick (e.g., adopting their long-term centering convention, but using correct PC selection rules). They showed that, had McIntyre and McKitrick subjected their alternative reconstruction to the statistical valida-

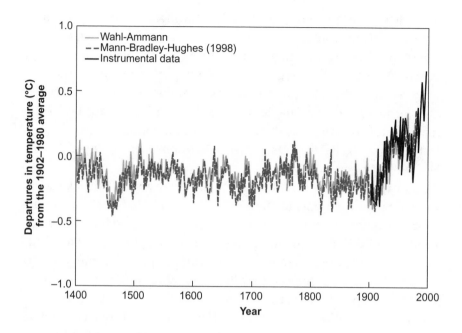

Figure 9.3: Which Is Which?
Wahl and Ammann superimpose their analysis on the original MBH hockey stick reconstruction.

tion tests stressed in MBH98 and MBH99 (and nearly all related studies), it would have failed these critical tests. McIntyre and McKitrick, in short, had not only failed to reproduce the hockey stick by eliminating key data, but their own results, unlike the MBH98 reconstruction, failed standard statistical tests of validity. (Wahl and Ammann provided all data and code used online so interested individuals could check any of these findings for themselves.)

As for McIntyre and McKitrick's claim that our centering convention "manufactures Hockey Sticks" out of noise, Wahl and Ammann showed that this too was false.[25] McIntyre and McKitrick had inflated the amplitude of apparent long-term natural variations by using an inappropriate statistical model for random noise. That made it deceptively easy for them to generate long-term trends (including hockey stick-like structures) from what they claimed was standard "red noise" (the type of noise that is characteristic of natural climate variations, which has more long-term variation than the highly erratic "white noise" typically encountered in other branches of science; i.e., the wiggles look smoother). In fact, what they had used was not simple red noise at all, but something that had far greater long-term persistence (even smoother wiggles) than red noise, and that lacked any sound scientific basis.[26] This inappropriate noise model allowed them both to greatly underestimate the reliability of actual proxy reconstructions and to grossly overstate the potential for our statistical conventions to "manufacture Hockey Sticks" from noise.

Other groups found fault with the McIntyre and McKitrick claims. In two separate comments on the McIntyre and McKitrick paper published in *Geophysical Research Letters*, Hans Von Storch and Eduardo Zorita of GKSS in Germany (using climate model simulations)[27] and Peter Huybers of MIT/Woods Hole Oceanographic Institute (analyzing proxy records)[28] showed that PCA conventions had nothing to do with the shape of the actual hockey stick reconstruction. Huybers provided an additional demonstration that McIntyre's attempt to dismiss the reliability of our reconstruction was flawed.[29] Other colleagues using different methods and data continued to produce independent reconstructions that were broadly similar to that of the hockey stick.[30] All of these confirmations of our findings took years to work their way through the system of scientific publication. In the meantime, the McIntyre and McKitrick claims became a staple of denialists and contrarians.

One example was retired astrophysicist Richard Muller.[31] Muller used his column in MIT's prominent journal *Technology Review* to defend the deeply flawed Soon and Baliunas medieval climate study in late 2003.[32] A year later,[33] he echoed McIntyre and McKitrick's claim that the hockey stick was an artifact of our PCA procedure—a claim that hadn't yet been published, let alone subject to the independent scrutiny under which it would later wither. With the imprimatur of an MIT journal behind it, Muller's piece helped the "hockey stick is a statistical artifact" view penetrate into respected academic circles.

The most prodigious of the anti-hockey stick attacks, however, was a thirty-plus page polemic that appeared in the Dutch magazine *Natuurwetenschap & Techniek* in January 2005, authored by an individual named Marcel Crok. The title of the article was rather peculiar, in a telling sort of way: "Proof That Mankind Causes Climate Change Is Refuted: Kyoto Protocol Based on Flawed Statistics." Evidently, I was such a powerful force that I now had control over both space and time: Somehow, the hockey stick (first published on April 22, 1998) and I must have traveled four months back in time and halfway around the world to influence the Kyoto Protocol, which had been adopted in Kyoto on December 11, 1997 (Superman meets H. G. Wells?). A plot of the hockey stick, compared against the bogus "corrected" version of McIntyre and McKitrick, was featured near the beginning of the article, and bore the caption "Mann versus McIntyre"—the David versus Goliath narrative rising up again. Crok, like Muller before him, simply repeated the discredited McIntyre and McKitrick claims about bad data handling, PCA conventions, and so forth. But he added a new touch of innuendo, characterizing the directory that was titled "censored" in our public data archive as if it were indicative of a nefarious plot rather than simply an appropriately descriptive name for a folder containing analyses wherein individual series were censored—a statistical term for "left out"—to determine which records were playing the most critical role in yielding a skillful reconstruction (see chapter 4).

Battle of the Blogs

As we entered the first decade of the third millennium, the new media of the Internet age began to outpace traditional print media as the primary means through which Americans obtained their information. Climate change deniers fully exploited this trend, increasingly making use of the internet as a means for purveying climate change disinformation. As individuals searched the Web for information on climate change, increasingly what turned up were contrarian litanies from "news" sites such as junkscience, TechCentralStation, and newsbusters to climate change contrarian Web sites like John Daly's Waiting for Greenhouse, Patrick Michaels's World Climate Report, and Fred Singer's SEPP. Performing a Google search on "climate change" or "global warming" invariably turned up denialist sites as the leading hits, with legitimate sources of scientific information buried further down the list, a casualty of the virtual flood of climate change disinformation that now saturated the Internet.

An increasingly large cross-section of the public was thus being exposed to biased, sometimes dishonest reporting of what was known about climate change. There was no reliable online outlet available for climate scientists to correct the record in real time. That situation worried many of us, including my friend and colleague Gavin Schmidt of the NASA Goddard Institute for Space Studies and other colleagues such as Stefan Rahmstorf of the University of Potsdam, Germany, and Eric Steig of the University of Washington. We'd discussed amongst ourselves possible ways of better communicating accurate science to the public and had even touched on the possibility of doing some sort of blog. Nothing had yet materialized from these discussions—that is, until a friend working with a Washington, D.C., nonprofit organization called Environmental Media Services contacted me in August 2004, mentioning that she had acquired a small budget to help climate scientists communicate to the public. Things were starting to come together.

My three colleagues and I agreed to take the offer of free Web space and support—and assistance with publicity for the launch—from the organization. No compensation was offered for the effort, nor would we have accepted it if it had been. This was, after all, a labor of love on our part. We would exercise all choices over content and maintain editorial control of the site. What we envisioned was a central resource where interested members of the public, journalists, and policy makers could see what actual working climate scientists have to say about climate-related issues of the day. There was already a very good model in another field for this sort of collective group science blog: Panda's Thumb (an homage to the late Stephen J. Gould's wonderful book by that name). Panda's Thumb was run by a group of more than a dozen evolutionary biologists enjoined in a common effort to communicate the science of evolution in the face of creationist antiscience attacks. Our mission would be similar. We would address analogous antiscientific attacks against our own science while providing a relatively nontechnical ongoing assessment of the scientific understanding of climate change.

The name—RealClimate—followed naturally. We launched the site on December 10, 2004, accompanied by a press release distributed to news outlets far and wide. In the end, a dozen or so of us participated in the effort. Each was an active, publishing climate researcher from the United States or Europe, and we were at varying stages of our careers. We also invited articles from a large number of guests. While our posts were not peer reviewed in the conventional sense—indeed a formal peer review process would be at odds with the necessarily fast reaction time of a blog—we tried to emulate the process internally. When one or more of us wrote a post, we would solicit feedback, criticisms, and suggested revisions from other members of the group.

We wanted readers to learn from the comment threads, rather than be misled, upset, or even repulsed by them. We thus chose to enforce a policy of moderation in our discourse. Polite debate and healthy challenges were permitted. Insults and regurgitated denialist talking points were not. The etiquette, as Gavin Schmidt liked to analogize, was similar to that of a dinner party. You are welcome to stay at the party as long as you don't shout obscenities or insult the other guests and hosts. The blog provides a rare opportunity for those genuinely interested in the science to communicate directly with the experts in the field.

When we launched the site, we got considerably more attention than we had anticipated. Within the first month, news articles about Real-Climate appeared in the *Los Angeles Times*, the *Washington Monthly*, both *Le Monde* and *Libération* of France, the *Pittsburgh Post-Gazette*, *New York Newsday*, *MSNBC*, the *American Prospect*, and even the ultraconservative *Washington Times*. We were pleased, too, by the warm welcome from leading scientific journals. *Nature* devoted an entire editorial to our rollout,[34] *Science* provided us much appreciated publicity,[35] and the editors of *Scientific American* chose us as one of the twenty-five best science-oriented Web sites our first year, 2005.[36]

A primary purpose of RealClimate was to provide the context so often missing in mainstream media coverage of the science. It's the conundrum—discussed in chapter 6—that the methodical and incremental nature of scientific discovery does not lend itself well to the "news peg" and the twenty-four-hour news cycle. Many of our posts were aimed at providing the background needed to appreciate and better understand whatever climate change-related stories might be in the headlines on a particular day.

Another purpose, of course, was to help fight the climate change disinformation campaign. The climate contrarians had huge amounts of industry funding and a seemingly infinite network of advocacy groups and PR professionals to spread their message. We had—well—ourselves, assisted by well-wishers and interested citizens. Scientists are typically untrained in the art of public relations and ethically bound to be truthful, but often driven, with great passion, to try to inform the public about their science.

Our launch couldn't have been better timed. Our inaugural post by Gavin Schmidt—and to date our most popular and most often cited—addressed Michael Crichton's new science fiction (or, more accurately, antiscience fiction) work, *State of Fear.* The book had, as its central premise, Senator Inhofe's worldview that climate change is an elaborate hoax. To be fair, that premise is no less realistic than the prospect of making dinosaurs using frog DNA, and in that sense it was par for the course for Crichton when it came to scientific authenticity. The book attempted to mix a fictional narrative with ostensible reviews of climate science, though the "science" was distorted beyond all recognition. "Michael Crichton's State of Confusion"[37] quickly became the definitive debunking of Crichton's book and was widely linked or reprinted across the Internet.

RealClimate went live just in time to respond to Senator Inhofe's floor speech at the opening session of the U.S. Senate in 2005.[38] As we pointed out in our post,[39] Inhofe was still promoting myths about the paleoclimate record but now had as his newest "expert" on the topic Crichton himself, who had volunteered this morsel of fiction (among others) in his book: "We are also in the midst of a natural warming trend that began about 1850, as we emerged from a 400 year cold spell known as the Little Ice Age." Actual scientific studies, we pointed out in our piece, had come to the opposite conclusion—that the cooling of the Little Ice Age was natural, while the recent warming trend was not.[40]

It was probably inevitable that Inhofe and Crichton would at some point directly join forces. The coalescence took place at another Senate hearing, called by Inhofe in late September 2005.[41] In our RealClimate post "Inhofe and Crichton: Together at Last!" we reviewed the various false or misleading claims made in the hearing, such as the supposed "1970s global cooling" scare. During the course of the hearing, Crichton even took a couple of swings at the hockey stick and me, for which he was admonished by Senator Barbara Boxer (D-CA).[42] Boxer then proceeded to place into the Senate record my rebuttal of Crichton's erroneous claims, followed by Gavin's "Michael Crichton's State of Confusion." It was Inhofe, though, who summarized his own position best (if unintentionally) when he concluded the hearing by promising to "sit back and look at [the issue of climate change] in a non-scientific way."

Our RealClimate efforts appear to have been appreciated by our peers. In February 2005, just a couple months after we launched the site, I briefly mentioned our new effort while paying homage to Stephen Schneider, the premier climate science communicator whom I and roughly two hundred other scientists had traveled to Stanford University to honor on the occasion of his sixtieth birthday. My passing reference to RealClimate, much to my surprise—and delight—elicited a spontaneous, enthusiastic round of applause from the august crowd assembled for this event. Since RealClimate began, a number of other climate change blogs have emerged as well. They appear, in various ways, to have been inspired by our initiative. Some maintain a parallel mission,[43] while others complement RealClimate by dealing with economics or ethical considerations of climate change.[44]

Writing and editing for the site was a welcome distraction for me, an opportunity to expound on the larger issues surrounding climate science and to engage in communication of the science to the public. It was an opportunity to step back, look at the bigger picture, and gain some needed perspective. It had the further benefit of providing a vehicle for debunking some of the misinformation propagating across the Internet regarding the hockey stick[45] at a time when a new line of attack was seemingly always around the corner. And the hockey stick ordeal would indeed soon take yet another unexpected turn.

Chapter 10

Say It Ain't So, (Smokey) Joe!

Continuous research by our best scientists . . . may be
made impossible by the creation of an atmosphere in
which no man feels safe against the public airing of
unfounded rumors, gossip, and vilification.

—President Harry S. Truman (September 13, 1948)

In *The Demon-Haunted World*, Carl Sagan recounts an episode from
the McCarthy era involving the distinguished American physicist Edward U. Condon.[1] Condon had been a participant in the Manhattan
Project to develop an atomic bomb; the director of the National Bureau
of Standards; and president of the American Physical Society. In 1948,
he came under attack by the House Committee on Un-American
Activities, chaired by Representative J. Parnell Thomas (R-NJ). During
the course of the hearings, Thomas variously referred to Condon as
"Dr. *Condom*," the "weakest link" (i.e., for American security],[2] and even
the "missing link." Most remarkable, though, was this accusation by an
inquisitor at a later hearing: "Dr. Condon, it says here that you have been
at the forefront of a revolutionary movement in physics called [read
slowly] 'quantum mechanics.' It strikes this hearing that if you could be
at the forefront of one revolutionary movement, you could be at the
forefront of another."[3] Harry Truman delivered the remarks quoted
above standing at Condon's side in an opening address to the 1948 annual meeting of the American Association for the Advancement of Science. Condon, it might be added, was never found guilty of any wrongdoing; J. Parnell Thomas later served a term in Danbury Federal Prison
on an unrelated conviction of fraud.

A Dangerous Mix

The scientific process—left to operate freely—is inherently self-correcting, even if the gears may at times turn more slowly than we would like. Science is unlike almost any other human endeavor in that regard. Scientists are inherently skeptical—of a claim, a hypothesis, a finding, a new piece of data. They want to see the evidence and form their own opinions about it. This skepticism leads scientists to vigorously challenge each other—in papers, at scientific meetings, and in their private exchanges—when they believe they have good reason or data to question or refine another's conclusions or methods. The arguments that hold up in this process eventually prevail; those that don't, fall to the wayside. If people with ulterior motives—be they political, religious, social, or otherwise—attempt to rig the rules governing scientific inquiry in their favor, they threaten that objective, self-correcting process. Most pernicious of all have been the efforts of those who feel threatened by scientific findings to stifle scientific investigation, to censor scientific knowledge, to intimidate and attack scientists for their work, and to interfere with their ability to do their work or communicate their findings.

History is replete with cautionary tales of the dangers of political interference with the process of scientific investigation, from the trial of Galileo by the Roman Catholic Inquisition for his advocacy of the Copernican or heliocentric system of astronomy to recent creationist attempts to undermine the teaching of evolution in schools. The phenomenon of Lysenkoism in the Soviet Union during the early to mid-twentieth century involved a mistaken, ambitious scientist supported by powerful political forces suppressing the work of other scientists. Trofim Lysenko was a Russian agronomist who claimed to have developed revolutionary new techniques for increasing agricultural yields. He rejected the mainstream scientific theory of Mendelian genetics and instead argued that characteristics acquired during an organism's lifetime could be passed on genetically to subsequent generations. While Lysenko never claimed that his views of plants generalized to human beings, such a notion was convenient from the standpoint of Soviet ideology. With Stalin's political support, Lysenkoism became accepted Soviet dogma, despite the consensus of world scientific opinion against it. This promotion of bad science for political reasons was to the detriment of a whole generation of Soviet scientists and the field of modern genetics.

Academic freedom is a fundamental principle of higher education, meant to ensure open and vigorous intellectual discourse. Given the crucial role that academia has historically played in advancing basic scientific research, the role of academic freedom in supporting the progress of science cannot be overstated. The principle of academic freedom holds that scholars—both students and faculty—must be free to express their viewpoints, however unpopular they may be, without fear of retribution by their institutions. This principle, importantly, extends to the most controversial areas of intellectual inquiry, whether religion, politics, or policy-relevant science. In short, academics—including scientists—must be allowed to follow the path along which their intellectual inquiries take them, even if their findings and views might appear inconvenient or threatening to outside special interests.

Academic freedom is, of course, not without boundaries. Academics, like other citizens, must avoid the proverbial "yelling 'Fire!' in a crowded theater." They must obey principles of ethics in their research. They must make clear that their opinions on controversial matters reflect their own personal views, not those of their institution. Finally, there is no allowance for behavior constituting academic misconduct, such as fraud, falsification, plagiarism, and defamation.

The Soviet phenomenon of Lysenkoism is often held up as the poster child for the dangers posed when open scientific inquiry is not protected. Lysenko quickly rose to a position of great influence in Soviet science policy; then, as director of the Soviet Academy of Sciences, he outlawed challenges to his theory of acquired genetic characteristics. Scientists who continued to challenge his views were purged of their academic positions, in many cases imprisoned, and in some cases killed. In the absence of any internal scientific challenges, Lysenko's ideas were implemented as agricultural policy on collective farms throughout the Soviet Union, and later in Communist China, where they may have contributed to widespread famine and mortality.[4]

The dangers that arise in the absence of academic freedom are clear. What measures can be put in place to ensure it, at least within academic institutions themselves? One answer—in a word—is tenure. Tenure grants job security to those academics fortunate enough to have a tenure-track position who can withstand the trials of a probationary period. During the probationary period, typically six years long, academics must establish a record of demonstrated scholarship, teaching excellence, and commitment to service. At the end of that period, the university or college evaluates whether their record merits a permanent position.[5] This process is intended to protect academic freedom in much the same way that a lifetime appointment is intended to preserve the independence of judges in the federal judiciary system. Ideally, tenure gives the academic an opportunity to pursue potentially risky, unpopular, or controversial topics, and to challenge entrenched interests without the fear of loss of livelihood.

Early summer 2005: My own tenure case at the University of Virginia was up that year, and no decision had yet been announced. Meanwhile, an important energy bill with climate change policy implications (it would establish incentive structures for various modes of energy production) was expected to come up for a vote in Congress in the months ahead. What better time than now for the foes of a proactive climate policy to seize upon anything that might be used to suggest a "scandal" regarding the perceived scientific underpinnings of global warming? If they could entangle me in the affair, it would be icing on the cake.

Smokey Joe Barton

That year, Joe Barton (R-TX) was the powerful head of the House Energy and Commerce Committee. In April, he had been honored by the Annapolis Center, a group supported by fossil fuel interests, for his work "supporting rational, science-based thinking and policy-making."[6] His hometown paper, the *Dallas Morning News,* had given him the nickname "Smokey Joe" for his efforts to exempt cement plants in his district from stricter antismog rules.[7] One of the largest recipients of fossil fuel money in the U.S. House of Representatives, he would be named one of *Rolling Stone's* seventeen most notorious climate change "polluters and deniers" five years later.[8] He would also be accused of unethical conflicts of interest regarding his personal stake in natural gas wells.[9] But it was his infamous apology to British Petroleum in June 2010 (over the Obama administration holding it accountable for its role in the Deepwater Horizon oil spill in the Gulf of Mexico) that finally made Joe Barton a household name.[10]

Figure 10.1: Barton's Apology
Washington Post satire of Joe Barton's apology to BP for the Obama administration's plan to hold the corporation responsible for its actions. [Washington Post, June 22, 2010. TOLES © 2010 The Washington Post. Reprinted with permission of UNIVERSAL UCLICK. All rights reserved.]

On June 24, 2005, five years before that infamous apology, I received an intimidating letter from Joe Barton (also signed by Representative Ed Whitfield of Kentucky, Republican chair of the subcommittee on House investigations). Similar letters were sent to my colleagues Ray Bradley and Malcolm Hughes, to the IPCC chairman Rajendra Pachuari (a citizen of India), and to National Science Foundation director Arden Bement. It was the Serengeti strategy in action several times over. Seizing upon the criticism of our research in the *Wall Street Journal* as justification for an open-ended investigation of us and our work,

Barton's letter to me opened, "Questions have been raised, according to a February 14, 2005 article in *The Wall Street Journal*, about the significance of methodological flaws and data errors in your studies of the historical record of temperatures and climate change." The letter repeated discredited criticisms of our work and the straw-man argument that our hockey stick papers constituted a central pillar of the IPCC and the scientific case for human-caused climate change.

Barton's wording was carefully crafted to make it appear to us (and perhaps others) that he had the power to subpoena extensive materials: "In light of the Committee's jurisdiction over energy policy and certain environmental issues, the Committee must have full and accurate information when considering matters relating to climate change policy. . . . To assist us as we begin this review, and pursuant to Rules X and XI of the U.S. House of Representatives, please provide the following information requested below on or before July 11, 2005."

In fact, Barton had no such subpoena power—that could only be granted by a full vote of the committee. Such realities did not, however, stop him from issuing burdensome and intrusive demands of me particularly, and also of my coauthors. Barton's letters appeared intended to send shivers down our spines, and certainly did succeed to an extent. Moreover, he sidelined us with vexatious demands, the response to which both proved a major time sink and required us to solicit legal advice and representation. Needless to say, I would have rather been spending my time meeting my teaching, advising, and professional obligations and advancing various scientific research projects. Responding to such intimidation tactics was most certainly not what I had bargained for when I chose to go into science.

Barton demanded extensive materials dating back through my entire scientific career. Among them were to "List all financial support you have received related to your research, including, but not limited to, all private, state, and federal assistance, grants, contracts (including subgrants or subcontracts), or other financial awards or honoraria." He further demanded : (1) "the location of all data archives relating to each published study for which you were an author or co-author," (2) "such supporting documentation as computer source code, validation information, and other ancillary information," (3) "when this information was available to researchers," (4) "where and when you first identified the location of this information," (5) "what modifications, if any, you have made to this information since publication of the respective study," and (6) "narrative description of the steps . . . to replicate your study results or assess the quality of the proxy data." He insisted that I "Explain in detail your work for and on behalf of the Intergovernmental Panel on Climate . . . including the process for review of studies and other information, including the dates of key meetings . . . the steps taken by you, reviewers, and lead authors to ensure the data underlying the studies forming the basis for key findings of the report were sound and accurate . . . the requests you received for revisions to your written contribution . . . and . . . the identity of the people who wrote and reviewed . . . portions of the report."[11]

There was a swift and dramatic response to Barton's letters, but probably not the one he was expecting or hoping for. Prominent members of the scientific community, along with leading national and international scientific organizations, denounced his actions in the strongest terms. The European scientific community was the first to speak out. On July 7, the European Geophysical Union (EGU), the largest professional geosciences society in Europe, issued a position statement: "We urge the Committee of Energy and Commerce and its Chairman to withdraw the highly inappropriate letters of June 23 and instead schedule a hearing of . . . experts . . . discussing the available scientific evidence on . . . climate change." The EGU statement stressed some basic facts that continued to be conveniently ignored by our critics: "we would like to point out that the . . . IPCC . . . finding that the increase in 20th century northern hemisphere temperatures is 'likely to have been the largest of any century during the past 1,000 years' [is] based on multiple lines of evidence, not just . . . Mann et al. . . . and . . . the results of . . . Mann et al. . . . have been confirmed by an independent team of scientists with freely available computer code and data." The same day, the journal *Nature* published an editorial, "Climate of Distrust," expressing concern that "by requesting information on research that does not fit his world view, Barton seems determined to use his political influence to put pressure on the scientific process."[12] *Nature*'s words evoked the scientific community's deep-seated concerns over precisely the sort of political tampering with science that history so clearly warns us against.

Over the next week, a number of leading U.S. scientists, science groups, and organizations, including the president of the National Academy of Sciences and the executive publisher of the journal *Science*, added their voices, expressing similar concerns.[13] A group of twenty leading U.S. climate scientists (including a Nobel Prize winner, a National Medal of Science winner, eight members of the National Academy of Sciences, and the president-elect of the AAAS) also wrote to Barton and Whitfield to express their concern.[14] They pointed out, among other things, that though the hockey stick was *not* the primary line of evidence for human-caused climate change anyway, "the essential points of the Mann et al. study that the late twentieth century likely included the warmest decades in the last millennium are supported by numerous other studies." They also stressed the threat to science itself Barton's actions represented: "We . . . note that much of the information that you have requested . . . is unrelated to the stated purpose of your investigation. Requests to provide all working materials . . . stretching back decades can be seen as intimidation . . . and thereby risks compromising the independence of scientific opinion that is vital to the preeminence of American science as well as to the flow of objective advice to the government."

Over the next month, more than thirty other scientific organizations would write similar letters. Among them were the two largest American scientific societies in my field, the American Geophysical Union (AGU) and the American Meteorological Society (AMS), followed by the American Astronomical Society, the American Chemical Society, the American Mathematical Society, and many others. In late August, *Science*'s editor in chief, Donald Kennedy, wrote a particularly strongly worded editorial in *Science* in which he bluntly declared, "I'm outraged at this episode, in which science becomes politics by other means."[15] Barton's activities also got a cool reception in the mainstream media. The *Houston Chronicle* (July 20, 2005) ran an editorial "Truly Chilling: Rep. Barton's Harassment of Scientists, Disdain for Fellow Lawmakers a Disservice." The *Denver Post* (July 22, 2005) followed with "An Attack on Sound Science," the *Washington Post* (July 23, 2005) with an editorial entitled "Hunting Witches,"[16] and the *New York Times* (July 23, 2005) with an editorial characterizing Barton's attack as an "inquisition." Dozens of similarly themed editorials, op-eds, and news articles continued to bring Barton bad press in the ensuing weeks.[17]

It appeared that Barton did indeed—forgive the pun—feel the heat. The most telling sign was the contrast between the threatening rhetoric of the letters he had issued in June and the tone of his chief of staff's letter to the editor of the *Toledo Blade* responding to some of the news coverage in late July.[18] That letter offered an amusingly revisionist history of the entire affair: "A few weeks ago, the chairman of the House Energy and Commerce Committee, Rep. Joe Barton, wrote a letter to three honest men and asked for some facts. In this case, the men were scientists and the facts are contained in the data that support their global warming theories. . . . Fortunately, scientists are almost uniformly proud of their work and downright eager to explain it to the people who pay for it—the taxpayers—and it looks like the three climatologists are no different." From the sound of it, you'd think Barton had simply invited my colleagues and me to chat with him about climate change over cocktails.

Battle of the Committee Chairs

Barton was also getting battered in the political arena—and, most surprisingly, from both sides of the aisle. The initial challenge came on July 1 from congressman Henry Waxman (D-CA), who wrote: [O]n June 23, 2005 you wrote to three of the world's most respected experts on global warming. . . . These letters do not appear to be a serious attempt to understand the science of global warming. Some might interpret them as a transparent effort to bully and harass climate change experts who have reached conclusions with which you disagree."[19] A few weeks later, two Democratic members of Barton's committee, Jay Inslee (D-WA) and Jan Schakowsky (D-IL), wrote a letter expressing their "concern with the letters you sent to some of the world's most renowned climate scientists."[20] They indicated, with some degree of sarcasm, their appreciation of Barton's newfound interest in the topic of global climate change after not having "held a single hearing [on the subject] in the past 11 years." They were nonetheless "very concerned that the tone of [Barton's] letters indicate that these were not requests for hearing background materials, but rather an attempt to intimidate Dr. Michael Mann and discredit peer reviewed scientific research."

The critical development, however, came in mid-July, when a clash erupted, pitting two Republican House committee chairs against each other: Barton, chair of the Energy and Commerce Committee, and Representative Sherwood Boehlert (R-NY), chair of the Science Committee. Boehlert was an old-school moderate who represented a rural district in upstate New York. He was widely recognized as a friend to both science and the environment, and he clearly viewed Barton's actions as a threat to both. Firing a shot across Barton's bow, Boehlert wrote a letter criticizing Barton in remarkably strong terms.[21] He questioned Barton's jurisdiction over matters involving science and pointed out that the only allegation that could have justified congressional notice—that my coauthors and I had purportedly refused to share the data used in our studies—had already been "soundly rejected" by the National Science Foundation, adding that Barton was "well aware of that."

Boehlert was far more blunt than even Waxman had been: "I am writing to express my strenuous objections to what I see as a misguided and illegitimate investigation you have launched concerning Dr. Michael Mann, his co-authors, and sponsors. . . . My primary concern about your investigation is that its purpose seems to be to intimidate scientists rather than to learn from them, and to substitute Congressional political review for scientific peer review," something he believed "would be pernicious." Boehlert's remarks were eerily evocative of Harry Truman's admonition over "the creation of an atmosphere in which no man feels safe" that began this chapter, warning that Barton's attack "raises the specter of politicians opening investigations against any scientist who reaches a conclusion that makes the political elite uncomfortable." He added, "The only conceivable explanation for the investigation is to attempt to intimidate a prominent scientist. . . . The precedent your investigation sets is truly chilling. Are scientists now supposed to look over their shoulders to determine if their conclusions might prompt a Congressional inquiry no matter how legitimate their work? . . . If Congress wants public policy to be informed by scientific research, then it has to allow that research to operate outside the policy realm." Barton, in Boehlert's view, was seeking to "erase that line between science and politics."

The political theater of two Republican House chairs going head to head proved irresistible to journalists, and it drove much of the ensuing media coverage that summer, most of it unsympathetic to Barton's view. An article entitled "Barton Blasted by Peer," for example, appeared in Barton's hometown paper, the *Dallas Morning News*, on July 18. The only support Barton could find was in right-wing venues like the *Boston Herald*, the *Washington Times*, and the Richard Mellon Scaife-owned

Pittsburgh *Tribune-Review*. The *Dallas Morning News* did afford Barton an opportunity to justify his actions in the form of an op-ed, "How Bad Is Global Warming," on July 31, but Boehlert corrected the record with a letter to the editor on August 10, noting that the sole basis Barton gave for his attacks was the "work of two Canadians, one an assistant professor of economics, another a retired mining executive, neither of whom are climate scientists and whose claims have been widely contradicted by respected experts." Had "the Committee asked those Canadians questions about their work?" Boehlert asked, and then answered: "No. . . . What this is all about is a threat to the open pursuit of science."

Boehlert's passionate objections to Barton's actions were apparently contagious. By the end of August, a second prominent Republican, Senator John McCain (R-AZ), had entered the fray, denouncing Barton's actions in similarly blunt terms in a letter to the *Chronicle of Higher Education*.[22] "Scientists must be allowed to conduct their work unfettered by political or commercial pressures," McCain said, drawing an only thinly veiled comparison between Barton's actions and Lysenkoism, "We have only to look at the failures of biological science in the former Soviet Union to understand the scientific and political costs of interference." McCain went on: "[T]he message sent by the Congressional committee to the three scientists was not subtle: Publish politically unpalatable scientific results and brace yourself for political retribution, which might include denial of the opportunity to compete for federal funds."[23]

We Reply

My coauthors Ray Bradley and Malcolm Hughes and I provided our own formal responses to Barton's letter on July 14, 2005.[24] Each of us made some of the same basic points. We noted that the findings of our own specific published studies (MBH98 and MBH99) had been independently replicated by other scientists using the available data and algorithm description, and that numerous studies since (nearly a dozen, in fact!) using different types of proxy data and alternative statistical methods had confirmed our key conclusion: that it is likely that the late twentieth century is the warmest period of at least the past one thousand years. We also noted that all of those studies are just one small piece of evidence in a solid scientific case that humans are now altering the climate, a case that has been established entirely independently of any paleoclimate evidence.

The tone of Ray Bradley's letter was dismissive, but witty. He began: "It is good to know that your committee is keenly interested in understanding the basis for President George Bush's recent statement [that] '. . . the surface of the Earth is warmer and [that] an increase in greenhouse gases caused by humans is contributing to the problem.'" After describing to Barton and Whitfield how science is actually done and how scientific knowledge actually advances, Ray explained that "it does not move forward through editorials or articles in the *Wall Street Journal* . . . it does not advance through ad hominem attacks on individual scientists in the Congress of the United States."

Malcolm Hughes took a somewhat more deferential tack in his letter, gently pointing out the erroneous nature of Barton's various assertions, and ending with a cordial invitation: "Our university is a major center of research on past environments and, should you ever have the opportunity to visit Tucson, we would be pleased to give you an introduction to the wide and fascinating range of work being done here."

As I was the primary object of attack, I took the matter somewhat more seriously.[25] In the process of crafting my response, I consulted with authorities kind enough to provide me legal advice, on a *pro bono* basis, as to what rights I had in the face of a hostile attack by a congressman. David Vladeck of Georgetown University Law School, a leading legal scholar who had argued several cases before the U.S. Supreme Court and who had considerable experience with such matters, was particularly helpful.[26] In my remarks, I stated for the record that I was choosing to respond to Barton voluntarily, and that I recognized no authority on his part to demand any materials from me; this was important, as Boehlert had already challenged his jurisdiction over the matter. I then proceeded to respond to each of Barton's claims, assertions, and questions. I began by affirming the overwhelming body of evidence—of which our work was only a small part—that human-caused climate change is actually occurring. I stressed that our own finding that recent decades were unusually warm had been independently reaffirmed by many other researchers. The IPCC Third Assessment Report's conclusions, I noted, were based not on my work alone, but on multiple studies coming to these same conclusions. Further, the criticisms of our work he had cited had been rejected by other scientists.

I then addressed the details of his claims and allegations. I pointed out, for one, that the data he was demanding that I make available were already available in the public domain, and had been for years. Furthermore, the NSF general counsel had affirmed—in response to Stephen McIntyre's allegations—that my coauthors and I had in fact disclosed all of the data and materials expected of any researcher and that the computer code—the actual computer program that we had written to implement our algorithm—was our intellectual property.[27] Although I had no desire to give in to tactics of intimidation, I also wanted to make clear that we had absolutely nothing to hide, so I added the source code to our public archive and pointed out to Barton in our letter that it was available there.

Despite all of the push-back by the scientific community, the castigation by prominent politicians in his own party, and the exceedingly bad press he was getting in the mainstream media, Barton forged ahead. In November 2005, Boehlert, seeking perhaps to preempt Barton's misguided further efforts, formally commissioned the U.S. National Academy of Sciences to review the science behind paleoclimate reconstructions, including our own work. He even extended an olive branch to Barton, asking him to join with him in commissioning the study. Barton, however, would not be deterred. He rejected Boehlert's offer, and instead announced in February 2006 that he had commissioned his own investigation by a statistician from George Mason University named Edward Wegman, much of whose research was defense-related and classified, involving the use of so-called data mining for intelligence gathering purposes.[28]

The Dog Days of Summer

As it happens, back in March 2005, months before the Barton affair had begun and before my tenure case at the University of Virginia (U. Va) had proceeded to completion, I had already privately tendered my resignation from U. Va and accepted a position at Penn State University—with tenure.[29] I would start there in August. None of this was public in June, when Barton sent out his letters. As for the energy bill, the Energy Policy Act of 2005 would pass Congress and be signed into law by President Bush in early August. Among the provisions in the original bill favored by environmentalists were higher vehicle fuel efficiency ("CAFÉ") standards and greater incentives for non-greenhouse-gas-generating energy sources. Both provisions were stricken from the bill before passage, though I doubt the Barton controversy had anything to do with it. At the time, I was in China to attend a scientific conference and to do some sightseeing with my wife. Our first child was due in December. I remember wondering while walking a section of the Great Wall of China what sort of world our new child and her generation would inherit.

Chapter 11

A Tale of Two Reports

Far better an approximate answer to the right question,
which is often vague, than an exact answer to the wrong
question, which can always be made precise.

—John Tukey (1962)

The appearance of two dueling reports within weeks of each other in early summer 2006 constituted the next major development in the Hockey Stick battle. The first of the reports, *Surface Temperature Reconstructions for the Last 2,000 Years*, commissioned by Representative Sherwood Boehlert (R-NY), was published on June 22 and carried the imprimatur of the National Academy of Sciences (NAS).[1] The NAS committee was chaired by a leading expert in statistical climatology, Gerald North of Texas A&M University, and consisted of a blue ribbon panel of a dozen experts with diverse relevant expertise, including leading climate scientists, paleoclimatologists, and statisticians, and several members of the National Academy itself. Among the panel members was John Christy, widely regarded as a skeptic in the climate change debate. The panel in turn solicited input broadly from an array of relevant experts, including me and the most prominent critics of our work. The report was subjected to the Academy's extremely rigorous peer review process (there were thirteen peer reviewers and two review editors).

The second of the two reports, the so-called Wegman Report commissioned by Congressman Joe Barton (R-TX), was released on July 14.[2] It was written by statistician Edward Wegman of George Mason University and two handpicked associates: David W. Scott, a statistician from Rice University, and Yasmin H. Said, a Wegman graduate student.[3] Neither Wegman nor his two coauthors possessed any training in physical science, let alone in climate science. The report had no backing from any recognized scientific organization, there was no evidence of attempts to solicit input from leading scientific experts, and only lip service was paid to the idea of formal peer review.[4]

The Academy Speaks

The basic message of the NAS report was clear: Our original conclusions were broadly supported by the evidence: "The basic conclusion of Mann et al. (1998, 1999) . . . that the late 20th century warmth in the Northern Hemisphere was unprecedented during at least the last 1,000 years . . . has subsequently been supported by an array of evidence that includes the additional large-scale surface temperature reconstructions and documentation of the spatial coherence of recent warming . . . and also the pronounced changes in a variety of local proxy indicators." The report concluded that "based on the analyses presented in the original papers by Mann et al. and this newer supporting evidence, the committee finds it plausible that the Northern Hemisphere was warmer during the last few decades of the 20th century than during any comparable period over the preceding millennium." Contrarians in the climate change debate tried to exploit the committee's use of the word *plausible*, which might appear weaker than the word *likely* that the IPCC used in its report and we used in our work. However, subsequent clarifications by the panel members indicated that indeed they had reached a conclusion similar to ours and that of the IPCC.[5]

In a press release, the NAS committee asserted that there was "high confidence that [the] planet is warmest in 400 years," "less confidence in temperature reconstructions prior to 1600," and "little confidence" prior to A.D. 900. While deniers seized on every single qualifier to suggest that the NAS was somehow retreating from our findings or those of the IPCC, the panel made it clear that their conclusions were consistent with those of MBH99. They noted that our work was "the first to include explicit statistical error bars" and reminded readers of the original MBH99 findings that "the error bars were relatively small back to about A.D. 1600, but much larger for A.D. 1000–1600," explaining that "the lower precision during earlier times is caused primarily by the limited availability of annually resolved paleoclimate data."[6]

There were some minor points of contention. For instance, the panel expressed some skepticism about conclusions regarding the warmth of individual years and decades. The MBH99 conclusion that 1998 was likely the warmest year and the 1990s the warmest decade of the past millennium rested largely, however, on the proposition—which the panel endorsed—that the past few decades were likely the warmest in a millennium. The warmest year and decade of an unusually warm interval (the past few decades) were, our reasoning went, likely unusually warm themselves. The panel also seemed to neglect the most recent peer reviewed literature, such as the work of Wahl and Ammann and others who demonstrated in greater detail than in previous work that the entire issue of centering in PCA that McIntyre and McKitrick had raised was a red herring; it didn't make any difference to the results obtained in the end.[7]

Some of the more important of the panel's findings came to light during its press conference on the date the report was released.[8] The report authors made clear that they backed the key conclusions of our original work. Panel chair Gerald North stated that "We roughly agree with the substance of their findings." *New York Times* reporter Andrew Revkin, noting that we had indeed emphasized the importance of uncertainties and caveats in our original millennial hockey stick analysis (MBH99), asked the panel who, if anyone, may have been responsible for any overstating of our conclusions. North opined that "the community probably took the results to be more definitive than Mann and colleagues intended."

The panel emphasized the seminal nature of our now nearly decade-old studies. Panel member Curt Cuffey characterized that work as "the first analysis of its type . . . it was a really remarkable contribution" that ended up "teaching us a lot about how climate has changed." Like any groundbreaking scientific work, he said, "it's not surprising that they could have probably done some detailed aspects of it better." Panel member Peter Bloomfield of North Carolina State University, a fellow of the Royal Statistical Society, added that ours was "a first of its kind study." We "had to make choices" that included "how the data were processed," the "initial preparation, the inversion of the calibration equations," and "the selection of variables that were to be used in those equations." Our "methods were all quite reasonable choices," he said, and to the extent that some of the choices might not have been the optimal ones, he added that they "didn't have a material effect on the final conclusion." Bloomfield volunteered he "would not have been embarrassed by that work" if he'd been an author of it himself. North expressed similar thoughts.[9]

The NAS report was widely reported to be an affirmation of our work. Even contrarian pundit Roger A. Pielke Jr. characterized the report as a "near-complete vindication for the work of Mann et al."[10] The Associated Press (AP),[11] first off the block on June 22, the day the report was released, announced that the National Academy of Sciences had determined that "recent warmth is unprecedented for at least the last 400 years and potentially the last several millennia."[12] The *New York Times* headlined the report "Science Panel Backs Study on Warming Climate,"[13] while the *Washington Post* announced "Study Confirms Past Few Decades Warmest on Record."[14] Meanwhile, in the United Kingdom the BBC characterized the findings as "Backing for 'Hockey Stick' Graph."[15] *Nature*, which had published our original hockey stick article, declared, "Academy affirms hockey-stick graph."[16] In the days and weeks ahead, dozens of similarly themed articles would appear in publications around the world.[17]

CNN's Lou Dobbs, once considered a contrarian in the climate change debate, characterized the NAS findings[18] as having "supported the conclusions of the scientists."[19] I had been invited to appear on the show along with my two coauthors Ray Bradley and Malcolm Hughes. I was unable to make it in time, however, having been detained by the extreme weather—the irony of which Dobbs commented on in his show. The following month, I appeared along with fellow climate scientists Gavin Schmidt and Alan Robock. Shortly before the three of us were to go on the air, Dobbs entered the green room and explained that he didn't even want to broach the scientific debate. We'd moved past that—the scientific debate in his view "was over." Instead, he wanted to discuss *solutions*.

Of course, some individuals sought to downplay, cherry-pick, or simply misrepresent the NAS findings. The *Wall Street Journal* could only muster up the anemic "Panel Study Fails to Settle Debate on Past Climates."[20] Written by Antonio Regalado, the same reporter who had publicized McIntyre's attacks against the hockey stick on the paper's front page two years earlier, the coverage of our formal vindication by the NAS was buried on page B2. The article invoked the classic straw man: "An expert panel . . . said that the key conclusion of a widely cited study of past temperatures is 'plausible' *but not proved*" (my emphasis), as if any scientific proposition can ever be "proved." Giving new meaning to the word *spin*, Congressman James Inhofe announced that "Today's NAS report reaffirms what I have been saying all along, that Mann's 'hockey stick' is broken," a claim that prompted Al Gore to remark that deniers like Inhofe "will seize on anything to say up is down and black is white."[21] This didn't stop the perpetuation of Inhofe's false assertion. In December 2009, for example, James Sensenbrenner (R-WI)—a leading climate change denier in the U.S. House of Representatives—claimed in a hearing that the hockey stick had been discredited by the 2006 NAS study.[22] Testifying at the hearing, presidential science adviser John Holdren politely pointed out that the NAS study had in fact confirmed our conclusions.[23]

The NAS panel members unambiguously rejected the accusations of scientific misconduct that had been leveled by some of our detractors. The *New York Times* reported that "several members of the panel reviewing the study said they saw no sign that its authors had intentionally chosen data sets or methods to get a desired result."[24] North announced: "I can tell you that is my own opinion and I think it's probably true across the board here, I certainly did not see anything inappropriate." Bloomfield stated he "saw nothing that spoke to me of any manipulation," adding that the hockey stick study was "an honest attempt to construct a data analysis procedure." One might think that this would have put an end to the accusations once and for all. One would be wrong.

Barton Bites Back

The Wegman Report, commissioned by Joe Barton and published several weeks after the NAS report, seemed a transparent effort to further spread the attacks against our work. It uncritically repeated the old and tired McIntyre and McKitrick claim that the hockey stick was an artifact of the conventions used in a statistical (PCA) analysis, while ignoring various already published peer reviewed papers that refuted that claim. The more extensive and authoritative NAS review, for example, had specifically dismissed the notion that PCA conventions had any substantial impact on our findings. As Bloomfield had put it at the NAS press conference, "the committee, while finding that the issues are real, [found] they had a minimal effect on the final reconstruction."

The only thing the Wegman Report offered that was actually new was a dubious "network analysis" of my publishing history with other scientists. Wegman used that analysis to attempt to demonstrate that there was an ostensible conspiracy in climate science, a tight-knit cabal of coauthors and peer reviewers, with me at the center. It was like "Six Degrees of Kevin Bacon" applied to climate scientists. There is a related concept in mathematics, known as the "Erdos number," which has a become an informal measure of one's academic status. Paul Erdos was an eccentric, extremely productive mathematician who coauthored many papers with large numbers of other mathematicians. The Erdos number represents an author's degree of separation from Erdos in his or her published work. If you published with Erdos, you have an Erdos number of one. If you published with someone who published with Erdos, it's two. And so on. (My Erdos number appears to be four.[25]) As a result of the Wegman claims, there was now some discussion in the blogosphere of a new measure—the "Mann number."[26]

It might have seemed amusing, were it not for the thinly veiled accusation at the center of Wegman's network analysis and the thorough lack of understanding of how modern science is conducted embodied in the employment of that analysis. Climate science, like many multi-disciplinary fields, requires broad collaboration with researchers across many areas. Any well-published climate scientist would show a wide-ranging pattern of connection with other researchers in the field. While I was flattered that Wegman and company seemed to think that I was at the nexus of climate research, the same type of analysis would have shown a very similar pattern for any of a large number of climate scientists. Wegman's analysis erroneously conflated our coauthors (who were known) with our peer reviewers (who, given that they are anonymous, were not). In fact, the two groups likely overlap little if at all: Most editors avoid soliciting peer reviews from an author's collaborators. Wegman's reasoning—that the network analysis revealed me to have been at the center of the climate research agenda for the past decade, thus somehow undermining any independent review of our findings— also defied the very principle of causality. The fact that I had worked with such a broad group of other scientists was largely a product of my earlier work (MBH98/MBH99), which yielded, among other things, useful benchmarks for later comparisons with the results of theoretical climate model simulations. It was ridiculous for Wegman to argue that these subsequent collaborations could somehow have influenced the assessment of earlier work that laid the groundwork for them.

Wegman used the social network analysis to support the bizarre claim that there is "too much reliance on peer-review" in our field. This claim, of course, goes against every principle of scientific practice; peer review is a key ingredient in the self-correcting processes of science. Perhaps this dismissal of the importance of peer review was, at least in part, an attempt to innoculate the Wegman Report against the criticism that, unlike the NAS report, it was not peer reviewed in any formal sense.

The Heat Is On

One week after the release of the Wegman Report, on July 19, as temperatures soared near or above the century mark over much of the country—including Washington, D.C.—during what would later come to be known as the 2006 North American heat wave, Barton called a hearing of the subcommittee on Oversight and Investigations. Entitled "Questions Surrounding the 'Hockey Stick' Temperature Studies: Implications for Climate Change Assessments," the hearing was set up for a day the committee had already learned I would be unavailable.[27] I did my best, nonetheless, to follow the developments from afar.[28]

Wegman and McIntyre were witnesses, along with a number of climate scientists. I had recommended the committee invite paleoclimatologist Tom Crowley in my absence. His climate modeling experiments had not only reproduced the hockey stick pattern, but had also attributed the anomalous recent warmth to human causes. Other scientists who testified at the hearing were the NAS committee chair Gerry North, NOAA scientist Tom Karl, and Hans Von Storch.

With assistance from Wegman and McIntyre, Barton used the hearing to advance one very specific, trivially true, but in reality totally inconsequential claim: that centering conventions of principal component analysis can influence the character of the first principal component obtained in a PCA of a dataset.[29] On that narrow point, North, as well as everyone else involved in this debate (including us!) were in agreement. Wegman and McIntyre were conspicuously silent, however, on the only issue that actually mattered: whether these conventions had any material impact on the result of the MBH analysis, correctly performed. Even if one accepted Wegman and McIntyre's claims, would it make a difference? North and the NAS had already provided a definitive answered to that question: No. But Barton—a firm subscriber to the "up is down, black is white" alternative reality of climate change denial—was intent on finding some credible witness who might be willing to argue otherwise.

Perhaps Barton was hoping that he had found such an individual in Hans Von Storch. After all, Barton had specifically cited Von Storch's criticism of the hockey stick in his original, July 2005 letter. Von Storch, as it turns out, was the one witness who spoke clearly and unambiguously to this point at the hearing: Asked about the effect of PCA centering conventions on the MBH98/MBH99 reconstruction, Von Storch replied "the effect is very minor . . . it really doesn't matter here."[30] The impact of a respected climate scientist viewed by some as a critic of our work dismissing with such dispatch the essence of the Barton, Wegman, and McIntyre criticisms was profound.

What about Wegman's other chief claim; that that my collaboration with a wide group of other researchers—forty-three coauthors in total—somehow compromised the underlying validity of our work? North dismissed the notion out of hand, explaining that my broad collaborations were something he would "probably look very favorably on if I were considering him for tenure." North used the example of physicists responsible for the great discoveries of the early twentieth century—Einstein, Bohr, and Heisenberg—to underscore the absurdity of Wegman's "social network analysis," noting that Wegman would have come to precisely the same conclusion had he done a similar analysis of any one of these great scientists. Yet in reality, North explained, these individuals were fiercely competitive and disagreed intensely with each other on key scientific matters.[31] As North put it later: "They [Wegman and Barton] make it sound like a love nest—well, it isn't. If you know anything about science, it's more like a contact sport."[32]

Wegman didn't help his own credibility with his testimony. He became flustered when asked repeatedly by ranking member Bart Stupak (D-MI) why his analysis focused only on a set of North American tree ring data, while the actual MBH analyses involved a global proxy dataset. He committed gaffes that called into question his basic understanding of the central topic at hand. As reported by Richard Harris of

NPR, "[T]he limits of Wegman's expertise became painfully clear when he tried to answer a question from Illinois Democrat Jan Schakowsky about the well known mechanism by which carbon dioxide traps infrared radiation—heat—in our atmosphere."[33] Harris quoted Wegman's revealing response, "Carbon dioxide is heavier than air. Where it sits in the atmospheric profile, I don't know. I'm not an atmospheric scientist to know that. But presumably, if the atmospheric—if the carbon dioxide is close to the surface of the earth, it's not reflecting a lot of infrared back." The greenhouse effect, however, has nothing to do with "reflecting" radiation. And it is well known to atmospheric scientists that CO_2 is well mixed in the atmosphere by the same turbulent motions that keep the major atmospheric constituents (oxygen, nitrogen) at constant ratios. The atmospheric CO_2 concentration today is 390 ppm not only at the surface, but also at five and fifteen kilometers above the ground.

Wegman wasn't the only one generating gaffes at the hearing. Whitfield, in his opening remarks, repeated the self-evidently preposterous claim that the 1997 Kyoto Accord was based on the 1998 hockey stick paper.[34] In the end, there had been much political theater, but very little scientific substance at the hearing that day. As Richard Harris put it, "if anyone showed up at this hearing room to hear a true scientific debate on global warming they ended up instead with just a political debate often far afield from the facts." It was a poor enough showing by Barton and his supporters that the Democrats on the committee smelled blood. They would readily agree to a follow-up hearing with me testifying, as Barton had originally requested, in the hope that they could turn the tide on the Republicans, making the topic of human-caused climate change, which the committee had for so long avoided under Barton's chairmanship, front and center in the discussion.

That hearing was held the following week on July 27.[35] Wegman and McIntyre were again present. Three other scientists were also invited to testify: Ralph Cicerone, the newly minted president of the National Academy of Sciences; Jay Gulledge of the Pew Center on Global Change; and John Christy. Ranking member Stupak used his opening statement to point out that Barton, having failed to discredit our work the previous week, appeared now to be trying to change the topic: "it appears that these critics have lost interest in simply attacking Dr. Mann's work. . . . Now the purpose of today's hearing is to cast doubt on all scientific evidence of global warming."[36] Stupak then went on offense: "if we are going to discuss the larger issue of global warming, which many of us on this side would be happy to do, we need to put more time and effort into putting together a series of well thought out hearings with adequate time for witnesses and staff to prepare." Henry Waxman (D-CA) then pounced, accusing Barton of trying to use very same Serengeti strategy deployed by the fossil fuel industry against Ben Santer in the mid-1990s against me: "The Chairman seems to think that if he can discredit one climate scientist, Dr. Mann, he can cast doubt on all the climate change research." With respect to the larger issue of climate change, he accused Barton of employing the same tactics the tobacco industry used to cast doubt on the linkage between smoking cigarettes and cancer[37]—something Waxman would know about, having presided over a series of hearings on the topic in the 1990s.

This time there was a sharper focus on whether the Wegman Report in any way invalidated the hockey stick.[38] Stupak, noting that "other climatologists have recreated Dr. Mann's work and have come to the same conclusions using both similar and different data sets and methodologies," suggested the motive behind Wegman's social network analysis was somehow to invalidate the wealth of independent confirmatory studies: "Dr. Wegman . . . will try to discredit all of these studies with an un-supported hypothesis questioning the independence of a large group of scientists' work."

Jay Gulledge pressed the point in his opening statement. Gulledge noted that "the Wegman report failed to accomplish its primary objective . . . [it] did not at all assess the merits of the criticisms directed toward the MBH reconstructions." He stressed that the Wegman/McIntyre critique would only have merit if correcting for any putative errors "yields results significantly different from the original result that can no longer support the claim of unusual late 20th century warmth." Gulledge then took the issue to its logical conclusion, noting that while "the Wegman Report takes no steps to make such a determination . . . fortunately, a different group, one well qualified both statistically and climatologically to tackle this question of merit, had already performed the task several months before the Wegman Report was released. The study by Wahl and Ammann." He went on to explain that that study had both demonstrated the robustness of the MBH hockey stick and the flimsiness of the criticisms advanced by McIntyre and now the Wegman Report.

Having failed to make any objective scientific case against the hockey stick, Barton and his colleagues turned instead to insinuation. Congresswoman Marsha Blackburn (R-TN) first quipped that she "remembered" when she was growing up in the 1960s that climate scientists were worried about another Ice Age—evidently she hadn't studied her talking points very closely; the claim is supposed to apply to the *1970s*. She then asserted that supposed flaws in the hockey stick weren't caught early on because the peer review "was not an independent and separate review outside of" the "social network" of our fellow scientists.[39] Asked by Blackburn to confirm her assertion, Wegman replied, "I believe that is the case." Blackburn could have been naive enough to believe what she was saying, but I was surprised that Wegman—an academic—would make such a claim. To support it, he would have to have known who the reviewers of our article were, which was impossible (assuming that the confidentiality of the review process had not been compromised) because they were anonymous. And he would have to conclude that *Nature,* one of the world's premier science journals, was somehow unqualified to select independent and unbiased referees. The whole argument was preposterous; and more than anything else it spoke to the intellectually bankrupt nature of the attack against us.

The final line of the attack involved once again criticizing us for a supposed lack of openness with regard to the way we conducted the MBH98 and MBH99 studies. I pointed out that the allegation that we had not released our data was simply false, while the criticisms against us for not initially publishing our computer code were at best disingenuous, since the NSF had already established that such materials were considered proprietary, and independent researchers had implemented our algorithm and confirmed our results without it.[40]

This line of attack—criticizing my coauthors and me for a supposed lack of scientific openness—ended up backfiring on our critics. John Christy's scientific claim to fame, as noted earlier, was a satellite-based record of atmospheric temperature trends that appeared to contradict surface evidence of warming. That Christy's record, which had been held up as a central pillar in the case for climate change denial, had been shown to be an artifact of faulty computations[41] generated curiously little outrage on Joe Barton's part, however. In his testimony at the hearing, Christy lectured the audience about scientific openness, presenting himself as a paragon of virtue when it came to sharing source

Figure 11.1: Barton's Congressional Hearing
My testimony before the House Energy and Commerce Committee in July 2005.
Mostly obscured is John Christy. National Academy of Sciences President Ralph
Cicerone is seen looking on in the background.

code: "When asked by others [scientists at Remote Sensing Systems (RSS) in California], we provided sections of our code and relevant data files. By sharing this information, we opened ourselves up to exposure or a possible problem which we had somehow missed, and frankly this was not personally easy. On the other hand, if there was a mistake we wanted it fixed."

Christy seemed to have no idea what he had just walked into: Henry Waxman (D-CA) apparently had been waiting for precisely this moment:

WAXMAN: Thank you. I want to ask Dr. Christy about this because you stated that you provided your computer code to other researchers when it has been requested, and you specifically mentioned providing your code to Remote Sensing Systems or RSS. Is that accurate?

CHRISTY: We provide[d] the part of the code that was in question.

WAXMAN: Well, I contacted RSS about your testimony and Mr. Frank Wentz sent me a letter last night, and he wrote to say, "Dr. Christy has never been willing to share his computer code in a substantial way," and he provides the text of a 2002 e-mail exchange between RSS and yourself. And according to this letter when asked for your code, you replied, "I don't see how sharing code would be helpful because there are at least seven programs that are executed (several thousands lines of code) and we would be forced to spend a considerable amount of time trying to explain coding issues of the spaghetti we wrote." In light of this letter, Dr. Christy, I would be interested if you care to clarify your testimony because Mr. Wentz wrote further, "I think the complexity issue was a red herring. My interpretation of Dr. Christy's response is he simply didn't want us looking over his shoulder, possibly discovering errors in his work. So we had to take a more tedious trial-and-error approach to uncovering the errors in his methods . . ." What do you say about that? That sounds inconsistent with what you have told us.[42]

If Christy had been invited to testify by Barton's staff, he ended up being the unintended star witness for the "defense" rather than the "prosecution." This exchange didn't settle the underlying arguments over source code release, which requires some balance between intellectual property rights and scientific access and is still being debated today.[43] It did, however, demonstrate the hypocrisy of climate change deniers pretending to occupy the moral high ground on this issue.

Waxman lobbed a series of softball questions that served to discredit the case advanced by Barton and his sympathizers. He asked me to correct a number of serious misconceptions betrayed by Wegman's previous testimony. These involved Wegman's apparent lack of awareness of basic climate science principles such as the greenhouse effect and the nature of greenhouse gases,[44] his misunderstanding of the distinction between natural and human-caused climate change,[45] and his perplexing claim that future human impact on the climate will be both negligible and harmless.[46] Wegman's plea for greater involvement of statisticians in climate studies, as I pointed out in follow-up testimony, was also disingenuous, given the history of more than a decade of close collaboration between climate scientists and statisticians.[47]

Waxman also gave me an opportunity to correct Wegman's mischaracterizations of my own work, in particular his claim that my colleagues and I had "circled the wagons" with respect to our work in this field. I pointed out that we had contributed to fundamental advances in the field of paleoclimate reconstruction in the decade since we had begun our original work, developing and applying new techniques such as the more sophisticated "regularized expectation-maximization method" and testing them rigorously using climate model simulations. Finally, given the evidence that emerged that Wegman had collaborated closely with Stephen McIntyre in preparing his report, Waxman wondered if Wegman had ever made even a single effort to contact me. The answer was no.

There were some lighter moments at the hearing. Jay Inslee (D-WA) asked Ralph Cicerone to ponder a sort of *It's a Wonderful Life* scenario with me as the George Bailey character, posing the rhetorical question of whether the evidence regarding the reality of human-caused climate change would be affected if I had never been conceived by my parents. My parents report having been amused.[48] And I'm fairly certain that this was the first congressional hearing at which a representative uttered the phrase "regularized expectation-maximization." You could just hear statistics junkies everywhere doing their best *Beavis and Butthead* impressions: "Did he just say *regularized expectation-maximization? Heh, heh, heh.*"

Shortly after the first hearing, Stanford physics professor David Ritson, who had previously published on proxy-based climate recon- structions,[49] identified some serious problems with the Wegman Re- port. In an e-mail sent days after the first hearing to Wegman and co- authors, and copied to NAS panel chair Gerald North and me, Ritson expressed concern that Wegman et al. appeared curiously to have re- peated precisely the same error that underlay McIntyre and McKitrick's original claim[50] that the hockey stick was a statistical artifact. Ritson forwarded the message twice over the next two weeks, without any re- sponse from Wegman or his two coauthors.[51] A member of Waxman's staff contacted Wegman to ask why he had not responded to Ritson's inquiries. Wegman claimed[52] that he had never received the e-mail[53] and that he was under no obligation to reveal details about his calcula- tions or methods anyway.[54] He was simply too busy now to respond to Ritson, he said, but would eventually set up a Web site where the materi- als would be available. He also asserted that his code fell under the cat- egory of "classified research" and thus might be protected by the U.S. Navy. The irony of Wegman's stonewalling was not lost on Waxman.[55] More than a year later, Wegman had refused to respond to any of Rit- son's subsequent inquiries and had failed to deliver on any of what he had promised.[56] In Ritson's words, "This is a sad commentary on people who have so strongly and publicly attacked others for supposed failures to provide such information, and their report must accordingly be judged in this context."[57]

In the end, the conventional wisdom was that Barton and gang fared poorly in the two days of hearings that summer. Nonetheless, these were stressful times for me. In the past year, I had spent much of my time responding to baseless allegations, preparing for hostile ques- tioning, and fending off one attack after the next, all over alleged mis- deeds I had not committed. I was, however, buoyed by the outpouring of support I received. The previous summer, when I'd just received Barton's threatening letter and was still in transition between jobs, I received a very welcoming note from the president of Penn State University, Gra- ham Spanier, reassuring me that he and Penn State were firmly behind me. There was support from members of the scientific community, col- leagues, administrators, friends, family, and indeed citizens I had never met. I received scores of letters and e-mails that year from friends

I hadn't heard from in years, and from total strangers who just wanted to thank me for standing my ground.

In March 2006 I was invited to give the Margolin Lecture at Middlebury College to a standing-room-only crowd. The timing of the lecture couldn't have been better. The Middlebury College hockey team had just won its division in the NCAA tournament. As my host introduced me, he gave me a special memento: a Middlebury hockey stick signed by each member of the tournament-winning team. The Middlebury hockey stick is still proudly displayed in my office. I sometimes joke to visitors that I keep it there for defense.

I also had the opportunity to meet Sherwood Boehlert in person just after the July 27 congressional hearing. Having announced his retirement from the House earlier that year, he was being honored for his long-standing support for the Baseball Hall of Fame, located within his district in upstate New York. He graciously took several minutes out of this scheduled event to talk with me. We had a pleasant, if brief conversation, talking among other things about our mutual fondness for Big Moose Lake in the Adirondacks. I relished the opportunity to personally thank him for all he had done. As I shook his hand, though, I couldn't help but feel a certain wistfulness, as if I were saying farewell to the last member of a vanishing tribe—the pro-environmental Republican.

The latest salvo in the climate wars had now come to a close. Things would soon enough heat up again, however.

Chapter 12

Heads of the Hydra

A lie can travel halfway around the world while the truth
is putting on its shoes.

—Mark Twain

In 2007, climate science was on somewhat of a winning streak. Al
Gore's summer 2006 documentary *An Inconvenient Truth*, love it or
hate it, had introduced a far greater number of Americans to the science of climate change than ever before. Many media outlets were now
treating the scientific evidence more seriously, not inserting a contrarian in every piece for "balance." Even Frank Luntz of the infamous
"Luntz memo," which had coached climate change deniers on their
messaging back in 2002, had now come to accept the reality of human-
caused climate change.[1] This same summer, the Intergovernmental Panel
on Climate Change released its Fourth Assessment Report, and with the
widespread and positive media coverage of the report's findings, there
was finally a feeling within the climate science community that our science had become generally accepted. For the time being, anyway, climate change denial had been relegated to the fringe.

On October 12, 2007, there was a further development: The IPCC
was co-awarded the Nobel Peace Prize, along with former Vice President Al Gore. The several hundred of us who had served as lead authors
were each sent a plaque acknowledging our sharing of the honor. At an
IPCC meeting celebrating this award,[2] I was honored to be one of two
individuals (the other being Ben Santer) singled out by IPCC scientific
working group chair Susan Solomon for special commendation for the
personal sacrifices made in the name of the IPCC. It was a good time to
be a climate scientist. It was an extraordinary experience to feel part of
an international effort collectively lauded in this way.

AR4—What Say You?

The 2007 IPCC Fourth Assessment Report (AR4) was considerably more definitive in its conclusions than its predecessors. "Most of the observed increase in globally averaged temperatures since the mid-20th century," the report said, "is very likely due to the observed increase in anthropogenic greenhouse gas concentrations."[3] The increased confidence was well justified. The climate models used in this latest assessment were considerably more refined than earlier generation models. They could be run at much finer spatial scales, better resolving surface topography, regional wind patterns, and current systems. The modeling of physical processes such as cloud formation had grown more sophisticated. Many of the models were now producing realistic looking El Niño events, generating a natural pattern of alternation between warming and cooling in the eastern tropical Pacific every few years that resembled the real-world phenomenon—a strong indication of physical realism in the models' treatment of climate dynamics. We now had six more years of observations since the IPCC Third Assessment Report (TAR) of 2001, and, with temperatures trending upward and other climate variables following suit, every additional year of data added to the overall significance of the observed climate changes. This was particularly relevant for certain measurements, such as the extent of summer sea ice in the Arctic, for which direct satellite observations only went back a few decades, and for which every additional year of measurement highlighted the striking rate of decline.

Though encouraged by IPCC colleagues to take on the role of lead author for two different chapters, I had chosen to sit out the AR4. Since my own scientific findings had come under attack by climate change deniers, I felt it was important that the IPCC be able to provide a thoroughly independent assessment of the underlying scientific issues. Moreover, I certainly didn't want to serve as a lightning rod. I was satisfied to have served as an expert reviewer, providing constructive criticism of the report. Unlike the previous report, which had combined all observations (whether instrumental or paleoclimate) into a single observations chapter, AR4 devoted, for the first time, an entire chapter to paleoclimate, including a substantial section devoted to the "The Last 2000 Years." It was a wise choice, as it allowed controversial matters to be aired more fully.

With an assessment of the existing literature that was both more thorough and more up-to-date than the 2006 NAS assessment, the IPCC was in a position to come to more definitive conclusions regarding the evidence from paleoclimate reconstructions. There were now more than a dozen published reconstructions of Northern Hemisphere average temperature. They all came to a common conclusion: The most recent decades were warmer than any comparable period in the past as far back in time as the reconstructions went. Several of the reconstructions now extended further back than the original hockey stick, as far back as two thousand years in some cases.[4] Moreover, a number of reconstructions were based on fundamentally different sources of proxy information than those used in the hockey stick, such as boreholes and, even more recently, information from mountain glaciers around the world, yet still came to essentially the same conclusion. Such additional evidence led the authors of the AR4 report to reach even stronger conclusions regarding the unprecedented nature of recent warmth than either their earlier 2001 assessment or the 2006 NAS report:

> The TAR pointed to the "exceptional warmth of the late 20th century, relative to the past 1,000 years." Subsequent evidence has strengthened this conclusion. It is very likely that average Northern Hemisphere temperatures during the second half of the 20th century were higher than for any other 50-year period in the last 500 years. It is also likely that this 50-year period was the warmest Northern Hemisphere period in the last 1.3 kyr [1300 years], and that this warmth was more widespread than during any other 50-year period in the last 1.3 kyr.[5]

Despite this strengthening of our original conclusions in the new IPCC report, in the distorted up-is-down, black-is-white world of climate change denial, the IPCC had supposedly "dropped the hockey stick"![6] The claim was entirely false. The hockey stick graph simply was not featured in the Summary for Policy Makers as it had been in the TAR. Why would it be? The purpose of the policy maker's summary was to highlight key new findings, not old ones. The original MBH99 hockey stick was indeed one of the dozen reconstructions shown in the IPCC paleoclimate chapter (see figure 12.2), which collectively formed

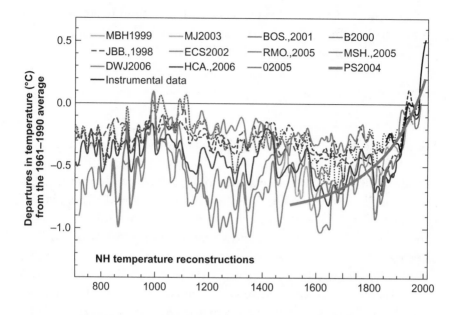

Figure 12.1: Expanding Spaghetti Plot
Figure from the IPCC AR4 assessment (working group 1 report, chapter 6) showing a dozen different reconstructions of Northern Hemisphere average temperature in past centuries along with the instrumental record through 2006.

the basis for the strengthened conclusions of AR4 regarding the anomalous nature of recent warming.

Falling Pillars

By the time the AR4 report was published in summer 2007, the case for legitimate climate change skepticism was on thin ice, so to speak. The four main pillars of denial that contrarians continued to cling to were:

1. Climate models are crude, untested, and unvalidated.
2. The instrumental record of global temperature is unreliable
3. Other data contradict the claim that Earth is warming.
4. Even if Earth is warming, it could be due to natural factors.

Various strands of purported evidence had been used to support these broader pillars. By 2007, most of those strands had weakened, and the pillars mostly crumbled, or toppled entirely.

PILLAR 1

Let us consider, as things stood then in 2007, the first pillar: the claim that climate models are unreliable. As the statistician George Box famously once put it, "all models are wrong, but some are useful."[7] Climate models are no different. They don't capture individual clouds, and even when they hypothetically could sometime in the future, they won't capture individual molecules, and so on *ad infinitum*. Critics will always be able to point to something that climate models can't resolve. But that's a red herring, the common fallacy that "because we don't know everything, we know nothing." As noted previously, climate models had already passed critical tests by the late 1980s and early 1990s, and since then they had become increasingly realistic in their ability to reproduce key features of the climate system. Many of the climate models used in the AR4 assessment produced, for example, very realistic-looking El Niño events. Enough credibility had been established to take the models' basic predictions seriously: warming surface temperatures, more widespread continental drought, vanishing Arctic sea ice, and rising sea levels given ever-increasing greenhouse gas concentrations. *Judgment: pillar crumbling if not yet toppled.*

PILLAR 2

Consider now the second pillar of climate change denial that still remained in 2007: that the global surface temperature record is ostensibly untrustworthy.[8] Traditionally, claims of contamination in the surface temperature record have been based on arguments regarding the so-called *urban heat island bias*—the observation that urban locations have typically warmed more than outlying rural areas owing to different thermal characteristics of urban and rural environments (for example, asphalt absorbs and retains more solar heating than a grassy field). To the extent that such locations are used more frequently in constructing the temperature record, they could in principle lead to a warming bias. Other types of changes in instrumentation and siting of meteorological stations—for example, the placement of urban weather stations in artificially cool locations such as parks—could instead lead to a cooling bias. Just how the potential warm and cold biases play out is not *a priori* obvious. By 2007, though, a number of recent studies had assessed the extent to which any urban heat island bias might still remain in the global temperature record.

A study in 2003 by NOAA scientist Tom Peterson and collaborators indicated that the cool park effect largely mitigates any urban heat bias in the U.S. measurements.[9] A subsequent 2004 study by the United Kingdom Meteorological Office's David Parker[10] tested for the presence of an urban heat island effect by comparing temperatures on windy and calm nights across the globe, which should show a difference if there is urban heat island effect,[11] and he found none. There were even more basic reasons for rejecting the claim that the surface temperature record was compromised by urban heat island effects. The global warming trend is seen not only in land measurements but also in ocean surface temperatures, where obviously no urbanization is occurring.[12] The ocean warming isn't as large as the observed land warming, but this is expected from basic physics and predicted by all climate models, the primary reason being that heat efficiently diffuses down into the interior of the oceans, leading to less warming at the surface. *Judgment: pillar toppled.*

PILLAR 3

What about the related third pillar: that other independent evidence contradicts surface evidence of warming? The source of this claim was a set of analyses by one pair of scientists, John Christy and Roy Spencer, of one particular dataset.[13] Those data come from instruments, known as microwave sounding units (MSUs), installed on polar orbiting satellites. The MSUs measure, from space, the intensity of microwave radiation emitted by Earth's atmosphere, which, in turn, can be related to temperatures in the atmosphere. Different channels of the MSU measure different frequencies of radiation, and an appropriate combination of information from the different channels (called 2LT) provides an estimate of temperatures in the lower atmosphere. For more than a decade, the Christy and Spencer MSU record was held up as the decisive refutation of global warming—a role that Christy, in particular, seemed to relish. NPR, for example, highlighted Christy's role in the climate change debate thusly in 2004: "His major contribution has been to analyze millions of measurements from weather satellites, looking for a global temperature trend. He's found almost no sign of global warming in the satellite data, and is confident that forecasts of warming up to 10 degrees in the next century are wrong."[14]

In the late 1990s, a number of problems had already been identified in the Christy and Spencer analysis. One issue—analogous to the segment length curse encountered in dendroclimatology—was the way Christy and Spencer merged what were distinct overlapping satellite records to form a single, ostensibly continuous record. The net effect of their method was to underestimate any long-term warming trend.[15]

Still other researchers found that Christy and Spencer had failed to account for the slow decay of the orbit of the satellites, which yet again worked in the same direction, tending to impose spurious apparent cooling.[16] Christy and colleagues, however, claimed that fixing another offsetting error, related to a required correction for the diurnal heating by the Sun, left the record basically unchanged.[17] Other authors later demonstrated an additional problem: that Christy and Spencer's claimed measurement of lower atmospheric temperatures was biased because they were actually averaging in some temperature information from the stratosphere to their estimates of lower atmospheric temperature.[18] Greenhouse warming is a kind of zero-sum game; the lower atmosphere warms at the expense of upper atmosphere (stratosphere) cooling. Erroneously averaging in stratospheric temperatures led Christy and Spencer to introduce artificial cooling into their estimates.

Then, in 2005, the independent team of Carl Mears and Frank Wentz of the Remote Sensing Systems (RSS) group in California determined through reverse engineering that there must be an error in the correction that Christy and Spencer had applied back in the late 1990s, as discussed earlier, to account for diurnal heating by the Sun.[19] As it turns out, their correction had the wrong sign. This was now the fourth consecutive error identified in Christy and Spencer's MSU analyses that had worked in the same direction, that is, that masked the true warming taking place in the atmosphere. Christy and Spencer admitted the error, albeit in a somewhat cryptic reply.[20] Using the same MSU data, Mears and Wentz produced their own estimates, which were free of these errors and which showed a warming of the lower atmosphere that was in fact entirely consistent with observed surface warming. This one is perhaps the easiest call of all. *Judgment: pillar toppled.*

PILLAR 4

Now, we come to the fourth and final pillar: that warming, even if it is taking place, might very well be a natural occurrence, independent of the effects of human activity. Support for the argument that climate change could be natural was said to be provided by a medieval warm period as warm as, or warmer than, today. Such a period of comparable warmth in the not-so-distant past, the logic goes, would imply that modern warming could be natural, too. The logic is flawed, however: The mere existence of a past warm period says nothing about the cause of the current warming. Even a modestly (say, 1 percent) brighter Sun during preindustrial times, for example, could have led to conditions warmer than today. Reconstructions of past temperatures such as the hockey stick (and the many other reconstructions coming to similar conclusions) indicated that peak medieval warmth did not rise to modern levels of warming in any case. That finding may have taken away a convenient contrarian talking point. But the finding was hardly necessary to render implausible the argument that natural variability could account for modern warming. To reach that conclusion it was sufficient to show, as many studies now did, that models could not reproduce the anomalous warming of the past century from natural factors alone. Only human impacts could explain that warming. *Judgment: pillar toppled.*

Heads of the Hydra

The complete or near collapse by 2007 of the pillars of defensible climate change skepticism represented a critical juncture in the debate over the science. Would climate change contrarians throw in the towel and at least concede the reality of human-caused climate change? Would they engage constructively in the discourse, focusing their efforts on the legitimate remaining uncertainties, such as the uncertain nature of climate change projections and the worthy debate to be had regarding what to do about the problem? Or would they retrench and continue to contest the ever-accumulating evidence supporting the reality of the climate change problem? The question is of course rhetorical; we already know the answer.

In Greek mythology, one of the tasks of Hercules was to destroy the nine-headed creature known as the Hydra. He found to his dismay that every time he cut off one of the heads, two would grow back in its place. So it was now with climate change denial. For every talking point that was refuted, two more would be offered. Moreover, the same arguments were eventually recycled, no matter how many times they were refuted in the peer reviewed literature. Whether or not a talking point is scientifically or even logically defensible is immaterial. If it has misinformed or confused an appreciable number of observers, it has served its purpose in manufacturing doubt and confusion.

Far-right media outlets, Internet disinformation sites, and contrarian blogs are still peddling factoids that were discredited in the actual scientific literature years ago, but that nonetheless endure in the contrarian mythology. Any bit of new information that might be twisted to reinforce the denialist canon is jumped on, while contradictory evidence is ignored or denounced. In the year or two following the release of the AR4 report, for example, the contrarian canard *du jour* was that the globe was in reality cooling—or at least not warming. The claim derived its credibility from contrarian folklore such as "climate scientists were predicting global cooling in the 1970s" and the now discredited "no warming" claim of Christy and Spencer.

The "globe is cooling" myth (and its sister, "global warming has stopped") soon gained new life through distortions of a 2008 article in *Nature*. Scientists from the Max Planck Institute for Meteorology had used climate model simulations to demonstrate that long-term greenhouse warming could be masked for timescales as long a decade or more by the impacts of natural variability.[21] While some scientists have disputed the magnitude of the effect, there is no debate within the mainstream scientific community over the validity of that basic premise. However, the finding was misrepresented by contrarians, and frequently misunderstood in popular accounts that presented the work as if it called into question the reality of human-caused climate change,[22] even though the authors were explicit that their study did nothing of the sort.[23]

Reenergized "global warming has stopped" claimants seized upon the sort of cherry-picking that would put even the very best fruit farmer to shame: An unusually strong El Niño event raised global temperatures in 1998, making it the warmest year on record in one of the three global temperature assessments, that of the University of East Anglia's Climatic Research Unit (CRU), while 2005 beat it out for the record in the other two assessments.[24] Contrarians noted that if they used the CRU record and chose the starting year as 1998, they could make the misleading argument that global warming had stopped because temperatures were cooler in subsequent years. Calculating the trend over the subsequent eight years, say, one could use the CRU record to seemingly argue against a net warming trend. This tells us nothing about global warming, however. It is meaningless to talk about trends over time intervals of a decade or less. Year-to-year fluctuations in global temperature are simply too large to establish a statistically significant warming (or cooling) trend over such a short time frame. One can find many instances in which the undeniable warming trend of the past century would not be evident in such short subintervals.[25]

Despite the thorough falsehood of the "global warming has stopped" claim, it would increasingly be hyped by climate change deniers during the run-up to the December 2009 Copenhagen climate summit. The claim was the centerpiece of a September 30, 2009, *Washington Post* op-ed, "For Alarmists, Ugly Truths on Global Warming," by conservative columnist George Will, for example, and shortly thereafter in an October 9, 2009, news story entitled "What Happened to Global Warming?" by contrarian BBC reporter Paul Hudson. The claim was even featured in the book *Superfreakonomics* released in October 2009. The "global warming has stopped" blitz of misinformation and confusion led AP reporter Seth Borenstein to take the initiative of digging a bit deeper. Borenstein engaged in some old-fashioned investigative journalism and contacted leading statisticians, asking them, in a blind experiment (they were given the global temperature series, but weren't told the source of the data or what it represented), if there was indeed any evidence to justify the claim that the upward trend in the series of measurements had stopped. The answer in every case was no.[26]

Heads of the Hydra of climate change disinformation are often quickly generated thanks to a single, deeply flawed paper that has slipped through the cracks of peer review. Take, for example, a December 2007 paper by David Douglass, John Christy, Ben Pearson, and S. Fred Singer[27] claiming to contradict an earlier 2005 *Science* article by Ben Santer and collaborators[28] establishing a human role in the warming of the tropical atmosphere. It took only a week for other scientists to demonstrate[29] that the Douglass et al. paper's principle claim arose from a simple misunderstanding of the concept of statistical uncertainty.[30] Within a year, Santer et al. had published a devastating critique,[31] showing that Douglass et al.'s approach gives clearly nonsensical results when applied to a climate model simulation where the cause of temperature trends are known. This was not before Douglass and his coauthors were able to milk a large amount of publicity with their claims, however. Shortly after the publication of their paper, for example, S. Fred Singer held a press conference at the National Press Club in Washington, D.C., where the authors asserted that their article established the "inconvenient truth" that "Nature rules the climate: Human-produced greenhouse gases are not responsible for global warming."[32] The paper was promoted on Fox News and featured in Singer's March 2008 NIPCC report that ABC News had characterized as "fabricated nonsense."[33]

Another head of the Hydra was a 2009 paper by John McLean, Chris de Freitas, and Bob Carter published in the *Journal of Geophysical Research*.[34] In the innocuously titled "Influence of the Southern Oscillation on Tropospheric Temperature," the authors claimed that El Niño drove essentially all variations in global temperature—a distinctly odd claim since almost nothing in climate science has been studied more closely than the relationship between El Niño and global climate. It was well known—and in fact had been demonstrated most recently in an article in *Nature*[35]—that, while El Niño, along with volcanic eruptions, did explain a fair amount of the short-term, year-to-year variability in global temperatures, it could not account for the warming trend. Had McLean et al. somehow discovered something that had eluded the entire climate research community for decades? The claim was indeed extraordinary. And the evidence? Not so much.

The study's principal findings were, yet again, the product of a surprisingly basic error. The authors hadn't, as it turns out, actually analyzed the statistical relationship between El Niño and global temperatures. They had instead analyzed the relationship between El Niño and the *rate of change* in global temperatures. That, combined with some additional unwarranted processing of the data,[36] ensured that in the end all McLean et al. had done was to confirm the well-known fact that El Niño explains a fair share of the year-to-year fluctuations in global mean temperature. Their analysis provided no basis for any conclusions regarding climate change. Most of these facts were pointed out by various climate bloggers within a few days of the publication of the paper. It took nearly a year, however, for a peer reviewed refutation to appear in the literature.[37]

In the meantime, the authors once again generated substantial publicity for their claims. Climate science "swiftboater" Marc Morano (see chapter 5) used his Climate Depot blog to hype the study. In a press release boldly entitled "Nature, Not Man, Is Responsible for Recent Global Warming," study coauthor Bob Carter claimed that the findings left "little room for any warming driven by human carbon dioxide emissions."[38]

The Hockey Fight Continues

The gaze of the Hydra remained largely focused, however, on the denialists' bête noir, the hockey stick. Despite the fact that the NAS, the IPCC, and more than a dozen independent peer reviewed scientific studies had now not only reaffirmed the key conclusions of our work, but in fact extended them further back in time, the denialosphere was still fond of claiming that the hockey stick had been "totally discredited" or "broken." Most of the attacks represented some version of the myth that the hockey stick was a statistical artifact, combined with a studied neglect of the numerous confirmatory independent studies. Some of the attacks were new, however.

In late 2007, the home journal of climate change denial, *Energy and Environment*, published a paper by Craig Loehle that purported to present a new two-thousand-year reconstruction of global temperature.[39] By contrast with the hockey stick studies—and every other peer reviewed scientific article on the subject—Loehle claimed that medieval warm period temperatures were warmer than "20th century values." Had the paper somehow identified key new sources of information or a more appropriate methodology to overrule the findings of all other recent studies?

The paper, in fact, suffered from serious problems that would presumably have been identified had it been submitted to a peer reviewed scientific journal and reviewed by individuals with the relevant paleo-climatological expertise.[40] Loehle was evidently unaware of the dating convention in paleoclimatology that in "BP" (nominally, "before present"), "present" actually refers to the standard reference year of A.D. 1950. By assuming that "BP" instead meant "relative to A.D. 2000," Loehle erroneously shifted many of his records forward by fifty years, in essence portraying the warmth of the records in the mid-twentieth century as if it pertained to the end of the twentieth century. This error thus had the effect of erasing all of late-twentieth-century warming. Most paleoclimate reconstructions, including the original hockey stick, show peak medieval warmth to be comparable to that of the early and mid-twentieth century. It is only the late twentieth century that stands out as anomalous.

Among other problems, many of the sediment records Loehle used in his analysis had chronologies that were determined by just a few radiocarbon dates distributed over the past two thousand years. The dating in these records is consequently uncertain by as much as four hundred years or so, precluding their use in reconstructing centennial timescale temperature variations.[41] There were several records that Loehle wrongly assumed to reflect temperature but instead reflected some other quantity,[42] and he inappropriately averaged records that had different temporal resolutions. Loehle did issue a correction that appears to have dealt with some of the most glaring problems,[43] but the fundamental problem remained that his estimates were insufficient to allow for a meaningful comparison of past and present global temperatures.[44] Yet even so, had he performed the critical step of statistical validation emphasized in all serious paleoclimate reconstruction studies, and had he published the work in an actual scientific peer reviewed journal, the paleoclimate community might not have so readily dismissed it.

Loehle's approach was laudable by comparison with that of many of the contrarians. He did attempt to make a positive contribution, putting his own reconstruction out there to be scrutinized and criticized. While the reconstruction didn't stand up to the scrutiny (and the venue for its publication was dubious), he made an attempt to contribute to the scientific discourse in a meaningful and constructive manner. This can be contrasted with many others who are more than happy to take potshots at peer reviewed studies from their blogs but are unwilling to produce a reconstruction themselves, or even to provide evidence that genuinely contradicts the current scientific consensus that recent warmth is anomalous.

When the Loehle paper came out in late 2007, we were in the process of finalizing our own revised set of surface temperature reconstructions. I had presented the preliminary results of that work more than a year earlier at the December 2006 AGU meeting in San Francisco. A decade after we'd begun the original hockey stick work, we had now amassed a much expanded set of more than a thousand proxy records, thanks to the laborious work of numerous paleoclimatologists who had developed many new and different kinds of proxy records in the intervening years. If there were numerous important new tree ring records, there were also many new coral and ice core records, and new proxy records such as "speleothems" (stalagmite/stalactite layers containing calcite, whose oxygen isotopes can be analyzed for climate information) as well as very-high-resolution lake and ocean sediment records. We had compiled more than twelve hundred proxy records worldwide, more than fifty of which extended back to A.D. 1000, and more than thirty back to A.D. 500.[45] Meanwhile, we had been conducting extensive tests of statistical reconstruction methods using synthetic (pseudoproxy) networks derived from climate model simulations built to have properties similar to actual proxy data networks.[46] These tests demonstrated that, given the available proxy data, our methods were likely to capture past temperature variations faithfully.

We were now in a position to address explicitly the main recommendations offered in the 2006 NAS report for furthering progress in paleoclimate reconstructions of the last two thousand years. We had made use of a greatly expanded dataset, adequate for reconstructing past temperatures even without the much-maligned tree ring data, and employed complementary methods that had been thoroughly tested and validated with model simulation experiments. We used two different statistical approaches—the regularized-expectation maximization method we had been working with in recent years,[47] and the simpler composite approach discussed in chapter four—neither of which employed the PCA centering at the heart of McIntyre's attack on the original hockey stick. We published our hemispheric temperature reconstructions in the *Proceedings of the National Academy of Sciences* (*PNAS*) in early fall 2008.[48] A year later, we published a follow-up article in *Science* describing the spatial patterns of past temperature changes.[49]

With far more ice core and sediment records now available, we were able to obtain a meaningful reconstruction of the Northern Hemisphere average temperature for the past thirteen hundred years without using tree ring data at all. If tree ring data were used, the reconstruction could be extended, with some reservations, back over the past seventeen hundred years.[50] The amplitude of century-to-century variations in the new reconstruction were somewhat larger than in other previously published reconstructions (including the original hockey stick), with somewhat greater peak medieval warmth.[51] That warmth, however, still did not approach the warmth of the most recent decades. Recent warmth, it now at least tentatively appeared, was unprecedented for nearly the past two millennia, and perhaps longer.

Stephen McIntyre wasted little time in launching a series of attacks on the *PNAS* paper, employing—it would seem—the strategy of throwing as much mud against the wall as possible and hoping that some would stick. Teaming up with his former coauthor Ross McKitrick, he submitted a short letter to the editor of *PNAS* claiming that our reconstruction used "upside down proxy data."[52] That was nonsensical; as we pointed out in our response,[53] one of our methods didn't assume any orientation, while the other used an objective procedure for determining it.[54]

McIntyre also appealed to the conclusions of the 2006 NAS report to claim that our continued use of the very long bristlecone pine tree ring records was inappropriate. Yet this was a misrepresentation of what the NAS had concluded. The NAS panel expressed some concerns about so-called strip bark tree ring records, which include many of the long-lived bristlecone pines. These trees grow at very high CO_2-limited elevations, and there is the possibility that increases in growth over the past two centuries may not be driven entirely by climate, but also by the phenomenon of CO_2 fertilization—something that had been called attention to and dealt with in MBH99 (see chapter 4). The NAS report simply recommended efforts to better understand any potential biases by "performing experimental studies on biophysical relationships between temperature and tree ring parameters." Such would be the focus of a paper published in *PNAS* by Mathew Salzer and coauthors the following year,[55] demonstrating that the much-maligned bristlecone pines were good temperature proxy records after all, and those records supported the conclusion of anomalous recent warmth.[56]

McIntyre settled then on a more specific avenue of attack: our use of a small group of sediment records from Lake Korttajarvi in central Finland. But this was quite inconsequential and, ironically, we were the ones who had raised concerns about these particular data in the first place, not McIntyre. We had included them for consideration only to be complete in our survey of proxy records in the public domain. In the online supplementary information accompanying publication of our *PNAS* article, we had both noted the potential problems with these records and showed that eliminating them made absolutely no difference to the resulting reconstruction.[57] McIntyre had thus attempted to fabricate yet another false controversy, in this case out of an issue that was noted first by us, and was explicitly shown by us not to make any difference. Paleoclimatologist Tom Crowley perhaps summarized it best: "McIntyre . . . never publishes an alternate reconstruction that he thinks is better . . . because that involves taking a risk of him being criticized. He just nitpicks others. I don't know of anyone else in science who . . . fails to do something constructive himself."[58]

Our own findings, of course, hardly existed in isolation. There was now an impressive array of reconstructions, a veritable "hockey team,"[59] and contrarians such as Stephen McIntyre had in recent years wasted no time in expanding their attacks to the ever-growing set of reconstructions. In addition to the dozen reconstructions shown in the IPCC AR4 report, there were recent studies led by Martin Juckes in the UK,[60] Darrell Kaufman in the United States,[61] and Fredrik Ljungqvist in Sweden,[62] each using different data and methods, collectively suggesting that recent Northern Hemisphere warmth is unprecedented not just for the past millennium, but likely for the past two.

And so both the climate wars and the hockey stick battle continued. It was increasingly unclear that any amount of evidence or additional work would satisfy the critics. After all, the attacks against the hockey stick fundamentally, as we have seen, were not really about the work itself. They were part and parcel of—forgive the pun—a proxy war against the science and its icons being fought by, or at least often funded by, powerful vested interests who found the scientific evidence for climate change inconvenient for political, financial, or philosophical reasons.

Chapter 13

The Battle of the Bulge

It is fatal to enter any war without the will to win it.

—General Douglas MacArthur (1952)

By early 2009, a troubling complacency had emerged among climate scientists. Perhaps it was the result of a confluence of seemingly game-changing events over the preceding few years: the rather conclusive and well-publicized Intergovernmental Panel on Climate Change (IPCC) Fourth Assessment, the increased public awareness resulting from the highly successful movie *An Inconvenient Truth* featuring Al Gore, the generally improved media coverage, and the images of devastation thrown up by several natural events—such as the 2005 hurricane season and hurricane Katrina in particular—that served as wake-up calls to the public of potential looming threats from climate change. On top of that, Barack Obama's recent election seemed to portend the prospect of a more science-friendly climate in Washington, D.C. To many of my colleagues, the climate wars had been won—in favor of the science. From here on out, there was still important climate science to be done, but in the public arena, it would all be about policy, solutions, and moving forward in confronting the twin challenges of climate change mitigation (to lessen the extent of further heating of the planet) and adaptation (to become more resilient in the face of impending climate changes already in motion).

However, it was increasingly clear to me that the climate change denial campaign was not simply going to fade away. There was too much at stake for the special interests behind the scenes. I wrote to colleagues in February 2009[1] about "complacency in the face of what is clearly a mounting, well organized, and well funded last attack, a Battle of the

Bulge if you will, against the science. . . . Most don't accept the serious-
ness of the battle we are going to face here."[2] Later, that September,
I wrote what turned out to be a sadly prophetic and ironic warning to
my colleagues at the University of East Anglia's Climatic Research Unit
(CRU) in the United Kingdom to "expect lots more attacks . . . over the
next several weeks as the U.S. senate debates cap & trade legislation."[3]

By early 2009, there were numerous indications that a major climate change denial offensive was gearing up. Not a day went by when
there weren't climate change disinformation pieces in the right-wing
media. The Heartland Institute was heavily promoting its climate change
denial conferences, and it had initiated a massive direct mail campaign
directed at scientists and engineers around the United States. Meanwhile, there was a growing drumbeat of character attacks against climate scientists.

Natural weather events, too, were playing to the advantage of climate change deniers, even though these events signaled little if anything
about climate change. There had been a series of relatively quiet Atlantic
hurricane seasons in the wake of the devastating 2005 season, and a
temporary lull in the yearly parade of successive record-breaking global
temperatures. Contrarians were always ready to exploit such vagaries of
natural climate variability and weather to their rhetorical advantage.
The mass media narrative of impending climate change catastrophe in
the preceding few years was an overreach, and it was growing a bit stale.
Maybe there was even some degree of buyer's remorse. Media outfits
were receptive to a new narrative, even if it was an old one retrofitted.
The one being served up by the deniers—that a cold spell in Peoria might
call into question the reality of global warming—fit the bill.

Myles O'Brien and the entire science and technology team at CNN,
who had done such a fine job of covering the science of climate change
in the past, were now gone. In their place was weatherman Chad Meyers.
I happened to be watching CNN in my hotel room on December 18'
2008, while attending a conference in San Francisco when Lou Dobbs,
in a segment entitled "This Is Global Warming?" had Chad Meyers on
to talk about the heavy snowstorms taking place in the United States.[4]
While it would hardly seem surprising that snowstorms might happen
in mid-December, Meyers apparently thought it was grounds for dismissing the existence of global warming.[5] Promoting a favorite climate
change denier talking point that would make pollster Frank Luntz

blush, Meyers proclaimed: "You know, to think that we could affect weather all that much is pretty arrogant."

Meyers was immediately followed by a supposed expert named Jay Lehr, who added: "If we go back really, in recorded human history, in the 13th Century, we were probably 7 degrees Fahrenheit warmer than we are now and it was a very prosperous time for mankind." Whether the thirteenth century was a "prosperous time for mankind" is a matter for historians to debate. But the claim that the thirteenth century was warmer than today, let alone seven degrees warmer, was nonsense, with no support whatsoever in the scientific literature. Lehr didn't stop there, though: "If we go back to the Revolutionary War 300 years ago, it was very, very cold. We've been warming out of that cold spell from the Revolutionary War period and now we're back into a cooling cycle." What were the qualifications of the person making these preposterous claims? He was the "science director" for the Heartland Institute. How far Lou Dobbs's views had devolved from "the debate about the science is over—let's talk solutions" position he had taken during my appearance on his show less than two years earlier! Not only was he now using his show as a vehicle for climate change denial propaganda, but in fact he was promoting it himself. He eventually went so far as to applaud James Inhofe for being "utterly vindicated" in his denialist views on climate change.[6] To the extent that Dobbs's transformation was a barometer of the impact of the resurgent climate change denial campaign, it did not bode well.

Initial Strikes

Meanwhile, the climate change denial machine was fighting new efforts by the Obama administration to regulate greenhouse gas emissions. On April 17, 2009, EPA administrator Lisa Jackson issued an initial "endangerment finding" that greenhouse gas emissions posed a threat to human health through "hotter, longer heat waves that threaten the health of the sick, poor or elderly; increases in ground-level ozone pollution linked to asthma and other respiratory illnesses; as well as other threats to the health and welfare of Americans."[7] She was acting on the authority of a 2007 Supreme Court decision that found that the George W. Bush administration had failed to observe key provisions of the

Clean Air Act requiring the regulation of greenhouse gases as pollutants.[8] Jackson was, in essence, picking up where Christine Todd Whitman had left off nearly a decade earlier, before Myron Ebell and the Competitive Enterprise Institute (CEI) blunted her efforts.

Following a period of open comment, review, and response, the EPA issued, on December 7, 2009, its final, official finding that human-caused increases in greenhouse gas concentrations "threaten the public health and welfare of current and future generations."[9] Particularly telling, though, was what had transpired during the intervening eight months. The EPA had held a sixty-day public comment period on the initial proposed endangerment finding, and it was flooded with nearly four hundred thousand comments. It was also challenged by ten different petitions from a group of organizations that reads like a who's who of organized climate change denial, including the U.S. Chamber of Commerce, CEI, Peabody Energy, and the attorney generals of Virginia and Texas.[10] The petitions challenged the validity of the underlying science of climate change. Prominent among the claims promoted in the various petitions were that the initial EPA finding was based on the hockey stick and that the hockey stick had been discredited. In response, the EPA pointed out that its findings had not relied upon any particular study, but instead were based on consensus lines of scientific evidence, among which paleoclimate reconstructions were only one—and not even the most important.[11]

Months earlier, in June 2009, the CEI—with help from James Inhofe—had been promoting a report by a supposed "whistleblower" within the EPA named Alan Carlin. The CEI claimed the report was "suppressed" by EPA higher-ups who feared it "cast doubt" on the reality of climate change, the primary reason for regulating carbon emissions. Inhofe called for a congressional investigation of the matter. The reality, as reported by the *New York Times* based on internal EPA documents it had acquired, was quite different:

> It is true that Dr. Carlin's supervisor refused to accept his comments. . . . But the newly obtained documents show that Dr. Carlin's highly skeptical views on global warming, which have been known for more than a decade within the small unit where he works, have been repeatedly challenged by scientists inside and outside the E.P.A.; that he holds a doctorate in economics, not in

atmospheric science or climatology; that he has never been assigned to work on climate change; and that his comments on the endangerment finding were a product of rushed and at times shoddy scholarship, as he acknowledged Thursday in an interview.[12]

Carlin, CEI, and Inhofe nonetheless milked the manufactured scandal for all they could. Carlin appeared on the Glenn Beck show to attack the IPCC and promote the worn-out "Earth has been cooling" myth.[13] The "suppressed EPA report" claim was picked up by Fox News, the *Wall Street Journal*, and mainstream venues such as CBS News. Even the progressive Huffington Post blog reported the story. Already evident at this early stage was a subtle but undeniable shift in the tone of the public discourse. Viewpoints that had previously been relegated to the denialist fringe of the Internet were now being expressed in mainstream media venues.

In early October 2009, CEI threatened to sue the EPA over the endangerment finding.[14] Again with Inhofe's help, it demanded that the EPA reopen the public comment period based on an accusation that the raw data upon which the evidence for climate change was based had been destroyed. The matter in question was the loss several decades earlier of some old hard-copy documents describing a modest number of thermometer records from one institution, the Climatic Research Unit (CRU) in the United Kingdom. All the raw data were in fact still available, and several independent scientific groups had produced an instrumental temperature record nearly identical to CRU's. In no way, furthermore, did the instrumental temperature record, let alone EPA's endangerment finding, hinge on the CRU data. As climate policy expert Rick Piltz put it, CEI, driven by an "antiregulatory ideology" was now simply "grasping at straws."[15]

Other efforts to undermine the case for the reality of climate change were afoot. That summer, a small number of climate change deniers within the physics community attempted to force the largest professional society in the field of physics—the American Physical Society (APS)—to reverse its position on climate change. The effort was spearheaded by none other than S. Fred Singer, joined by Will Happer, a prominent denier in the Princeton physics department and chairman of the board of the George C. Marshall Institute—the same group that Frederick Seitz used during the 1990s to wage his attack against the science of global warming.[16] Singer and Happer were joined by two of Happer's Princeton colleagues as well as by Roger Cohen, ExxonMobil's retired manager for strategic planning, and one other physicist.

The group of six brandished a petition signed by fifty-four physicists demanding that the APS reconsider its November 2007 policy statement affirming the reality of human-caused climate change. For some perspective, fifty-four is but one half of one percent of the APS, and smaller than the number of faculty members in the physics department at my own institution, Penn State University. The unimpressively small number didn't stop the *Wall Street Journal* from immediately seizing upon the development in a June 2009 editorial subtitled "The Number of Skeptics Is Swelling Everywhere."[17] APS President Cherry Murray appointed a high-level subcommittee of the APS to consider whether any changes in the policy statement were warranted, something that any responsible head of an organization would do under the circumstances. By early November, the subcommittee had rendered a decision; it rejected the petition. Nonetheless, the climate change deniers had already gotten quite a bit of mileage out of the repeated claim that a large group of physicists supposedly denied the reality of climate change.

The rhetoric that climate change deniers were using was at the same time growing increasingly inflammatory. In early October 2009, for example, commentator Marc Sheppard attacked climate scientists as "lying perpetrators of fraud" in "UN Climate Reports: They Lie," a piece he wrote for a Web site called, without irony, the American Thinker.[18] Just one month earlier, South Carolina Congressman Joe Wilson had shouted out "You lie!" during President Barack Obama's address to Congress. Wilson's outburst was widely condemned as an unprecedented breach of congressional etiquette and a sign of the disappearance of good faith in the discourse over public policy. The use of identical rhetoric by climate change deniers signaled a further disappearance of any residual good faith that might have remained in the climate change debate.

When in Doubt, Play Hockey

As the contrarian attacks against climate science were ramping up during fall 2009, attention returned once again to their most reliable target—the hockey stick. Only now, there wasn't just a lone hockey stick but, as we have seen, a "hockey team" of well over a dozen independent reconstructions, all pointing to the same conclusion—that the recent warming of the planet was indeed anomalous in a long-term context. This didn't slow the denialists in their baseless attacks on every single confirmatory study that was published.

When *Science* in early September 2009 published an article by Darrell Kaufman and his colleagues showing the most dramatic hockey stick yet—a two-thousand-year reconstruction of Arctic temperature changes[19]—Stephen McIntyre and his forces went on the attack on the Internet,[20] immediately trumpeting the false claim that the work was compromised by bad data, despite the fact that whether or not the authors used the data in question made no difference to the result they obtained.[21] A more vicious attack was reserved for later that month. The matter concerned a tree ring temperature reconstruction for Russia's Yamal region that Keith Briffa and colleagues had published some years earlier; it once again showed recent warmth to be anomalous in a two-thousand-year context. At a time when Briffa was known to be seriously ill and not in a position to respond to any allegations, McIntyre publicly accused him of having intentionally cherry-picked tree ring records to get a particular result.[22] Moreover, he demanded that Briffa turn over all of the individual underlying tree ring records in his possession. Yet correspondence later found between McIntyre and Briffa's Russian colleagues (who had supplied the tree ring data in the first place) revealed[23] that they, not Briffa, had chosen which tree ring records were appropriate for use in reconstructing temperatures and that McIntyre had the data all along![24]

To support his "cherry-picking" allegation, McIntyre had produced his own composite reconstruction—which happened to lack the prominent recent warming evident in Briffa's reconstruction. How did he accomplish this? By deleting tree ring records of Briffa's he didn't seem to like, and replacing them with other tree ring data he had found on the Internet, which were inappropriate for use in a long-term temperature reconstruction—an example of why there should be a healthy dose of skepticism with any such claims made by armchair "scientists."[25]

McIntyre and his supporters then used this manufactured scandal to generate a frenzy of climate change disinformation, leading to headlines in the *National Review* and *Daily Telegraph*, premised on the rather absurd proposition that the criticisms of the Briffa et al.'s Yamal tree ring analysis—even if they were valid (which they weren't)—could somehow undermine all of climate science. As my colleagues and I put it at the time, "Apparently everything we've done in our entire careers . . . all of radiative physics, climate history, the instrumental record, modeling and satellite observations turn out to be based on 12 trees in an obscure part of Siberia. Who knew?"[26] Most climate reconstructions either didn't use the Yamal series in question anyway[27] or were undetectably altered if the Yamal series was entirely eliminated from the pool of proxy data used.[28] A familiar pattern was emerging in the ongoing attacks by climate change deniers. The issues they raised never actually undermined any of the major conclusions of the work they were criticizing. Instead, the purpose seemed simply to cast doubt among those who don't know better or don't have the time to get to the bottom of the matter in question—to generate much heat, but little or no light.

This episode is a revealing case study in the anatomy of a climate change denial smear campaign.[29] First, bloggers manufacture unfounded criticisms and accusations. Then their close allies help spread them. McIntyre's colleague Ross McKitrick writes an op-ed piece in the right-wing National Post more or less accusing Briffa of fraud: "Whatever is going on here, it is not science."[30] Individuals such as Marc Morano,[31] Anthony Watts,[32] Thomas Fuller of the Examiner,[33] UK Telegraph blogger James Delingpole,[34] and CEI's Chris Horner[35] spread the allegations through the Internet echo chamber. That is all the justification that apparently is needed for commentators such as Andrew Bolt of Australia's Herald Sun[36] to eventually propel the unfounded accusations onto the pages of widely read daily newspapers. Some of the distortions were almost comical. James Delingpole, for example, announced the episode as proof that global warming is a "MASSIVE lie" [sic].

Horner and Delingpole both claimed that the Yamal revelations, by discrediting the hockey stick, even undermined the dramatic scene in An Inconvenient Truth where Gore used a stair lift to ascend to the tip of the spiking curve depicted. There were just two problems with that proposition: The Yamal series was not used in the hockey stick, and Gore—in the stair lift scene—was showing the dramatic recent spike in atmospheric CO_2 over the past million years, not temperature over the past thousand years. But in the climate change denial playbook, facts must never get in the way of a good smear opportunity.

Serengeti Resurgent

Ben Santer provided an apt description of the phenomenon I've termed the Serengeti strategy:[37] "There is a strategy to single out individuals, tarnish them and try to bring the whole of the science into disrepute." With no scientific leg to stand on, manufactured claims of incompetence and malfeasance, ladened with innuendo and vilification, have emerged as the denialist weapon of choice. The purpose is multifaceted: to subject climate scientists to intrusive demands for materials, making it difficult or impossible, and at the very least less enjoyable, for them to carry out their work; to intimidate climate scientists through public campaigns of ridicule and harassment; to thereby serve notice to other scientists of what will be store for them if they too speak out on the topic of human-caused climate change; and to keep alive the false notion that human-caused climate change is still a matter of scientific controversy.

This strategy has increasingly involved the abuse of vexatious Freedom of Information Act (FOIA) demands to harass scientists and impede their progress in research and ideally to find something, even if just a stray comment in personal correspondence, that can be twisted into a weapon against them. FOIA was signed into law by President Lyndon Johnson in 1966. Sometimes referred to as "open records" or "sunshine" laws, FOIA allows members of the media and the public access to governmental documents—that is papers, letters, private e-mail correspondence—within certain restrictions. FOIA in principle applies to the documents of federal scientists—though a number of exemptions exist that, among other things, exempt documents whose disclosure would constitute a breach of privacy. Similar FOIA and open documents laws apply at the state level, in the United Kingdom, and in some other countries such as Canada and Germany.

In November 2008 McIntyre filed a FOIA demand to NOAA requesting not only data used in a recent paper by Ben Santer and coauthors (all of which was already, in fact, available in the public domain), but all e-mail correspondence between Santer and his coauthors.[38] This disturbingly intrusive demand for personal e-mail foreshadowed the tactics of climate change deniers that would follow some months later.

The assault against climate scientists intensified in 2009, especially as the December 2009 Copenhagen climate change summit approached, catching many scientists both bemused and unprepared for the character attacks to which they would be subject. While contrarians were going after Keith Briffa for his Yamal tree ring work in the latter half of 2009, they were also badgering Phil Jones and his colleagues at CRU with an escalating barrage of FOIA demands—sixty of them in one weekend alone.[39] One colleague, Eric Steig—a climate scientist at the University of Washington specializing in polar climates (and fellow RealClimate blogger)—was bewildered by the blistering attacks he was subject to following publication in January 2009 of a paper in *Nature* on which he was lead author.[40]

The paper (of which I was a coauthor) used a combination of data analysis and model simulation results to conclude that Antarctica on the whole has warmed over the past half century. That conclusion contradicts a favorite contrarian talking point: "if the globe is warming, how come Antarctica is cooling?"[41] The short answer, according to Steig et al. (and several more recent supporting studies), is that it isn't cooling; on the whole it is warming, particularly in the West Antarctic region so critical to the stability of the Antarctic Ice Sheet.[42]

The paper was immediately attacked by climate change deniers. Christopher Booker dismissed the work in a commentary in the *Telegraph* whose very title, "Despite the Hot Air, the Antarctic Is Not Warming Up," was mistaken.[43] Booker imagined that one of his favorite targets—Michael Mann—must be to blame: "The fact that Dr. Mann is again behind the new study on Antarctica is, alas, all part of an ongoing pattern. But this will not prevent the paper being cited *ad nauseam* by everyone from the BBC to Al Gore." Somehow I—a favorite new boogeyman of Booker and denialist fellow travelers—had been elevated to the principal scientist "behind" the new study, even though as fourth author of six I played a minor role.

Of the accusations of malfeasance that were made against Steig on contrarian Internet sites, which included blog postings with titles such as "Eric Steig Wears No Clothes"[44] and "Steig Professes Ignorance,"[45] Steig commented: "[I]t is actually rather bizarre that so much effort has been spent in trying to find fault with our Antarctic temperature paper. It appears this is a result of the persistent belief that by embarrassing specific scientists, the entire edifice of 'global warming' will fall."[46]

After Darrell Kaufman and his colleagues published their *Science* article in early September 2009 independently reaffirming the hockey stick pattern of temperature changes over the past two thousand years (in this case, for the Arctic region), Kaufman, too, was taken aback by how vituperative were the attacks of climate change deniers, and he commented on it in an op-ed piece in his local newspaper, the *Arizona Republic:*

> When our article, "Recent warming reverses long-term Arctic cooling," was published, I was shocked by the vicious reactions, including genuine hate mail. I was immediately cast as one of a "team" of corrupt scientists that fabricates data to perpetuate the "global-warming hoax," with the goal of advancing its own research funding and the U.N.'s political agenda. I was accused of withholding and "cherry picking" data to validate predetermined conclusions. The study was "audited" by bloggers who found minor errors; they discredited me and pronounced the entire study "debunked."
>
> As a scientist, I welcome input that improves my work, and when needed, I issue proper revisions. . . . [However] no formal rebuttal by those who had supposedly "debunked" the study has been submitted to the journal editors, the standard procedure for engaging authors in a measured discussion of a scientific study. Instead, accusations remain plastered across blogs, apparently for political rather than scientific purposes.[47]

It's *Fraud*, I Tell You!

A prime example of the hit pieces that purveyors of climate change denial were publishing in fall 2009 was "None Dare Call It Fraud: The 'Science' Driving Global Warming Policy," which appeared on October 16 on contrarian sites such as the Post Chronicle and Townhall.com.[48] The author, Paul Driessen, has been variously employed[49] by the Center for a Constructive Tomorrow (CFACT), the Center for the Defense of Free Enterprise, the Frontiers of Freedom, and the Atlas Economic Research Foundation, among others—a virtual cornucopia of industry-funded disinformation outfits.[50]

If the overall aim of Driessen's piece was to malign as many individual climate scientists as possible in one relatively short commentary, it was successful. There was the shopworn myth that "Dr. Ben Santer and alarmist colleagues" by themselves chose the "discernable human influence" wording in the 1995 IPCC Third Assessment Report. Next came the spurious claims that Phil Jones had "withheld temperature data and methods" and that Keith Briffa had "selected just twelve" tree ring records to "prove a dramatic recent temperature spike." Last, Driessen re-promoted the familiar falsehoods that I had "refused" to "divulge [my] data and statistical algorithms" and that I had "cherry-picked tree-ring data."

Driessen even managed to malign the collective community of American environmental journalists as (my emphasis) *"supposed* professionals" who were purportedly just advocates for "Mr. Gore and his apocalyptic beliefs." He promoted Alan Carlin's "suppressed" EPA report, and decried the "inconvenient truth" behind "global warming hysteria," which after all was simply a plot to "enrich Al Gore" and "alarmist scientists." He went on to pose the unintentionally pertinent question, "Still not angry and disgusted?" followed by a litany of fossil-fuel industry-friendly Web site links. He implored his readers to "get on your telephone or computer, and tell your legislators and local media this nonsense has got to stop," and ended with the baseless accusation his piece began with: "It may be that none dare call it fraud—but it comes perilously close." All in a day's work!

The repeated reference to "fraud" by Driessen and other hired guns was not coincidental. The language choice was taken straight from the Frank Luntz playbook, designed to perpetuate the view that climate scientists were not only wrong, but were in fact evil. My colleagues and I, their narrative held, were engaged in the perpetration of a fraud—a diabolical conspiracy that would not only make us rich, but would reward a cabal of powerful pseudo-environmentalist fat cats whose bidding we were supposedly doing. According to Dreissen, we were all implicated in an elaborate scheme designed to reward "alarmist scientists who get the next $89 billion in US government research money" and "financial institutions that process trillion$$ in carbon trades." It was really scientists and environmentalists who were somehow getting rich. Borrowing from the great Oz himself, Dreissen would have us *ignore that multinational corporation behind the curtain!* ExxonMobil was setting new records for corporate profits with each passing quarter, but it was really climate scientists (and their banker friends) who ostensibly were raking in the dough. The depressing fact is that many of Dreissen's readers probably believed this.

Climate change deniers during this period bandied about groundless accusations of fraud and malfeasance with reckless abandon. Though the charges struck us as defamatory, there was little or nothing that scientists could do to protect themselves. The climate change denial machine is bankrolled by some the nation's wealthiest and most powerful private interests, such as the Scaife Foundations and Koch Industries. They fund their own public relations firms and even their own legal foundations (e.g., the primarily Scaife-funded Landmark Legal Foundation[51] and Southeastern Legal Foundation[52]). For a scientist to take on an industry-backed adversary in court would be an all-consuming task that would literally take over his or her life, and there would be little assurance of victory in a long, drawn-out legal battle against deep-pocketed opponents. Indeed, the prospects for successful legal recourse is made all the more challenging by the fact that scientists who have been in the public spotlight—whether by choice or not—are considered, for legal purposes, "limited public figures," meaning that the threshold for establishing libel or slander is considerably higher than for a typical citizen.

And so the reckless charges against climate scientists continued unabated. In the United Kingdom, Douglas Keenan accused several climate scientists of fraud, including Phil Jones of CRU[53] and me. In late 2009, he sent me an e-mail warning[54] that he'd attempted (to no avail) to get the FBI to investigate me—something that would be funny, did it not seem so mean-spirited. Even Stephen McIntyre, who had been copied on the message, demanded Keenan desist in involving him in such matters in the future.[55]

Perhaps the most repugnant of Keenan's attacks, however, was reserved for the climate scientist Wei-Chyung Wang of SUNY-Albany. In 2007 Keenan publicly alleged that Wang had committed fraud in work that he had published two decades previously in two articles with Phil Jones (one in *Nature*, the other in *GRL*) estimating the impact of urbanization on temperature records from China. In an e-mail to Wang, copied to Jones, Kennan demanded: "I ask you to retract your GRL paper, in full, and to retract the claims made in *Nature* about the Chinese data. If you do not do so, I intend to publicly submit an allegation of research misconduct to your university at Albany."[56] At issue were twenty-year-old temperature station records that he had used to assess urbanization heat island effects in the temperature record, and that formed the basis of the statement in his paper that "The stations were selected on the basis of station history: we chose those with few, if any, changes in instrumentation, location or observation times."[57]

Keenan appeared to be arguing that because Wang couldn't produce the two-decade-old hard-copy documentation of station metadata, he must be guilty of fraud. SUNY, as any academic institution is essentially forced to do when confronted with allegations of misconduct on the part of a faculty member (regardless of how dubious the charges may appear), conducted an investigation. They found absolutely no evidence to support any claims of wrongdoing on Wang's part. However, that process played out slowly. In the meantime, the controversial journal *Energy and Environment* and its editor, Sonja Boehmer-Christiansen, allowed Keenan to use the journal to level the fraud charge against Wang.[58]

This development—the use of a quasi-academic journal to level sometimes groundless accusations against scientists—was troubling. It happened again in 2009 when an economist and climate change contrarian named Hu McCulloch alleged that he could not reproduce a tropical ice core record produced by Lonnie Thompson and colleagues, implicitly claiming either ineptitude or, worse, malfeasance. Thompson is a member of the U.S. National Academy of Sciences and recipient of the Heineken Prize, the Blue Planet Prize, and the highest scientific honor in our country, the Presidential Medal of Science. His tropical ice core data provided independent support for the conclusion that modern warming was unprecedented for the past two thousand years and were featured in *An Inconvenient Truth*, making them a particularly tasty target for deniers. Just as with the attacks against our work by McIntyre and McKitrick that were rejected by *Nature*, McCulloch was unable to marshal a credible enough argument to publish a comment in the original journal of record (in this case, *Proceedings of the National Academy of Sciences* [*PNAS*]). The McCulloch paper apparently was rejected by several journals before the author found one willing to publish it—*Energy and Environment*.[59]

I knew McCulloch's claim that the tropical ice core composite was "irreproducible" was false, as I was able to reproduce Thompson et al.'s results easily from their raw data. I was also able to identify

McCulloch's error—an incorrect assumption that the amplitude of variation in a series of measurements must be constant in time—in about a half hour of work.[60] The paper was laden with innuendo, the conclusion section stating, for example (emphasis added), "The Thompson et al. tropical composite . . . bears no replicable linear relationship to . . . series on which *they claim* it is based." On the basis of erroneous analysis, and with some help from the journal's editor Sonja Boehmer-Christiansen, McCulloch had leveled false allegations against one of our country's most respected climate researchers. This was not an isolated incident for McCulloch. In fall 2009, he made a similarly baseless accusation of plagiarism against my colleague Eric Steig.[61]

Suffice it to say that in recent years the Serengeti strategy had become finely honed. Isolate particular climate scientists from the community, saddle them with trumped-up charges of incompetence or impropriety, or both, make an example of them to discourage others from coming forward, and use the controversy to distract the public and policy makers from the real matter at hand—what to do about the potential looming climate change threat. Yet in the ramping up of attacks through fall 2009, we had still seen only a set of opening skirmishes in what could be characterized as the climate war's "Battle of the Bulge." The main battle loomed in the near future, and nothing could prepare climate scientists for what was to come.

Chapter 14

Climategate: The Real Story

If you give me six lines written by the most honest man,
I will find something in them to hang him.

—Cardinal Richelieu

The most malicious of the assaults on climate science would be timed for maximum impact: the run-up to the Copenhagen climate change summit of December 2009, a historic, much anticipated opportunity for a meaningful global climate change agreement.[1] The episode began with a crime committed by highly skilled computer hackers, followed by a massive public relations campaign conducted by major players in the climate change denial movement.

The Hacking

On November 17, 2009, RealClimate was hacked by someone operating via an anonymous server located in Turkey. At roughly 6:20 A.M. EST, the hacker had uploaded a large file, FOIA.zip, to the site. The title was a thinly veiled allusion to the barrage of frivolous and vexatious FOIA demands issued against the University of East Anglia's Climatic Research Unit (CRU) in the months preceding. The hacker temporarily disabled our editorial access to RealClimate. Once we were able to restore access, we found that the hacker had created a draft post clearly intended to go live shortly. It read: "We feel that climate science is, in the current situation, too important to be kept under wraps. We hereby release a random selection of correspondence, code, and documents. Hopefully it will give some insight into the science and the people behind it. This is a limited time offer, download now."

The e-mails highlighted were hardly "random." While thousands of e-mails involving dozens of climate scientists were stolen, the thieves had obviously filtered them for a few names, those of prominent climate scientists such as Ben Santer, Kevin Trenberth, Tom Wigley, Keith Briffa, Phil Jones—and me. The twenty e-mails featured in the post had been carefully chosen to highlight single words or phrases that, while innocent in their use, might sound damning taken out of context. The hacker labeled each of the highlighted e-mails with a lurid title such as "IPCC scenarios not supposed to be realistic" and "Mann: dirty laundry." The accompanying FOIA.zip file contained a cache of over a thousand e-mails, themselves selected from many thousands of private correspondences of scientists from CRU dating back to 1996. The e-mails had been stolen, it turned out, from a vulnerable CRU backup server.

While the underlying crime—the theft of proprietary materials from a major UK university—remains unsolved and is still under investigation by UK authorities, a few things came out in the months immediately following. The first link to the purloined information temporarily stashed on RealClimate was posted on Stephen McIntyre's climateaudit Web site.[2] Digital forensics reported by the *Guardian* indicated that the hacker was based in eastern North America;[3] that, combined with what was known about another incident earlier that year, led the *Guardian* journalist to question whether McIntyre might somehow have been involved,[4] a possibility that McIntyre strenuously denied. Saudi Web servers were among the first used to post the stolen materials,[5] and Saudi Arabia was the first country to call for an investigation of climate scientists in what came to be known as climategate,[6] leading some to suggest that Saudi Arabia—the world's single largest exporter of oil—may have played a role.[7] In any case, the lead Saudi climate change negotiator, Mohammad Al-Sabban, was quick to use the manufactured scandal to help block progress at the Copenhagen summit. The e-mails would have a "huge impact" on the negotiations, he had predicted, and he went on to make the jaw-dropping claim that "it appears from the details of the scandal that there is no relationship whatsoever between human activities and climate change."[8]

Although we had intercepted the hacker before our site could be used to complete the mission, it was clear that the point of no return had been passed. The same content would be uploaded two days later to a Russian FTP server[9] and then onto many other Web sites, including the Wikileaks site. Soon enough, this trove of stolen e-mail messages would be released to the world. What ensued in the weeks ahead was one of the best-coordinated "swiftboat" campaigns in modern history.[10] While the hackers appear to have had access to the materials for months, the release was clearly timed for the run-up to Copenhagen.

The Distortions

As soon as the e-mails were leaked, climate change deniers claimed that they provided a "smoking gun," "the final nail in the coffin of anthropogenic global warming."[11] The e-mails, they claimed, revealed that climate scientists had been fabricating and hiding data, subverting the peer review system, and evading requests for their data. Is it possible they were right?

It would indeed have been surprising if those looking to make mischief could not find some e-mail excerpts that could be made to appear damning, given the opportunity of searching thousands of e-mails for phrases to take out of context. To show how easy this was to do, one commentator mined the personal correspondence of Sir Isaac Newton and found phrasing that, taken out of context, could be interpreted to suggest (in modern terms) "conspiring to avoid public scrutiny," (2) "insulting dissenting scientists," (3) "manipulation of evidence," (4) "suppression of evidence," (5) "abusing the peer review system," and (6) "insulting [his] critics."[12]

Imagine how unpleasant it might be to have your private e-mails, text messages, or phone conversations mined by your worst enemy for anything that, taken out of context, could be used to make you look bad. Then imagine what it would be like to be expected to defend each and every instance of sloppy word choice or ambiguous phrasing that could be found. This is the position in which climate scientists whose e-mails had been hacked found themselves.

One e-mail Phil Jones of CRU sent to my coauthors and me in early 1999 has received more attention than any other. In it, Jones both made reference to "Mike's *Nature* trick" and used the phrase "to hide the decline" in describing a figure he had been asked to prepare for the World Meteorological Organization (WMO) comparing different proxy temperature reconstructions. Here was the smoking gun, climate change deniers clamored. Climate scientists had finally been caught cooking the books: They were using "a trick to hide the decline in global temperatures,"[13] a nefarious plot to hide the fact the globe was in fact cooling, not warming! Conservative firebrand Sarah Palin was on it, declaring in a December 9 op-ed in the *Washington Post* that "The e-mails reveal that leading climate 'experts' . . . manipulated data to 'hide the decline' in global temperatures."[14]

Note here how the climategate claim was being tied to the climate change denial fascinations of the moment, in this case that "the globe is cooling"—a myth fueled by a winter (late 2009) that had thus far seen cold temperatures and unusual snowfalls in the eastern United States, though other regions, such as the Arctic and Southern Hemisphere, had been unusually warm.[15] In fact, the WMO had—just a day before Palin's op-ed appeared—reported that 2009 would end up as one of the ten warmest years on record globally, and that the first decade of the new millennium (2000–2009) would go down as the warmest decade on record.[16] Some decline to hide! Nonetheless, following one particularly heavy snowstorm that winter, James Inhofe built an igloo on the National Mall with signs reading "Al Gore's New Home" and "Honk if you ♥ Global Warming" to mock concern over climate change.[17] Funny how silent he and other deniers went the following summer when D.C., like many cities around the United States and the world, were experiencing record-setting heat.

A little more than a week after the *Washington Post* had run Palin's op-ed claiming that scientists had been hiding a decline in global temperature, the paper allowed me to respond with an op-ed of my own.[18] I pointed out that Jones had written the e-mail in question in early 1999. He therefore could not have been referring to global temperature trends during the most recent decade (2000–2009) as Palin and others seemed to imply. Moreover, the e-mail had been written on the heels of the warmest year (1998) ever recorded in the instrumental record. It therefore could not have been referring to a supposed decline in temperature during that decade either! So what was Jones actually talking about?

The full quotation from Jones's e-mail was (emphasis added), "I've just completed *Mike's Nature trick* of adding in the real temps to each series for the last 20 years (i.e. from 1981 onwards) *and from 1961 for Keith's to hide the decline.*" Only by omitting the twenty-three words in between "trick" and "hide the decline" were change deniers able to fabricate the claim of a supposed "trick to hide the decline." No such phrase was used in the e-mail nor in any of the stolen e-mails for that matter. Indeed, "Mike's *Nature* trick" and "hide the decline" had nothing to do with each other.

In reality, neither "trick" nor "hide the decline" was referring to recent warming, but rather the far more mundane issue of how to compare proxy and instrumental temperature records. Jones was using the word *trick* in the same sense—to mean a clever approach—that I did in describing how in high school I figured out how to teach a computer to play tic-tac-toe or in college how to solve a model for high temperature superconductivity. He was referring, specifically, to an entirely legitimate plotting device for comparing two datasets on a single graph, as in our 1998 *Nature* article (MBH98)—hence "Mike's *Nature* trick."

The MBH98 proxy reconstruction ended in 1980, since many of our key proxy records had been developed in the late 1970s and early 1980s, and did not extend to the present. As a result, the reconstruction did not cover the critical last two decades of the twentieth century during which substantial additional warming had taken place. At the suggestion of the reviewers (as mentioned in the Prologue), we supplemented our plot of reconstructed temperatures in MBH98 by additionally showing the instrumental temperatures, which extended through the 1990s. That allowed our reconstruction of past temperatures to be viewed in the context of the most recent warming. The separate curves for the proxy reconstruction and instrumental temperature data were clearly labeled, and the data for both curves were available in the public domain at the time of publication for anyone who wanted to download them.

That, in short, was the "trick" that Jones had chosen to use to bring the proxy temperature series in his comparison up to the present, even though the proxy data themselves ended several decades earlier. There was one thing Jones did in his WMO graph, however, that went beyond what we had done in our *Nature* article: He had seamlessly merged proxy and instrumental data into a single curve, without explaining which was which. That was potentially misleading, though not intentionally so; he was only seeking to simplify the picture for the largely nontechnical audience of the WMO report. After its own inquiry, *Nature* editorialized on the flap over "trick": "One e-mail talked of displaying the data using a 'trick'—slang for a clever (and legitimate) technique, but a word that denialists have used to accuse the researchers of fabricating their results. It is *Nature*'s policy to investigate such matters if there are substantive reasons for concern, but nothing we have seen so far in the e-mails qualifies."[19]

What about "hide the decline"? Or to be more precise, the phrase "for Keith's to hide the decline," referring to Keith Briffa and the Briffa et al. temperature reconstruction.[20] As discussed briefly in chapter 7, this particular reconstruction was susceptible to the so-called divergence problem, a problem that primarily afflicts tree ring density data from higher latitudes. These data show an enigmatic decline in their response to warming temperatures after roughly 1960, perhaps because of pollution[21]—that is the decline that Jones was referring to.

While "hide the decline" was poor—and unfortunate—wording on Jones's part, he was simply referring to something Briffa and coauthors had themselves cautioned in their original 1998 publication: that their tree ring density data should not be used to infer temperatures after 1960 because they were compromised by the divergence problem. Jones thus chose not to display the Briffa et al. series after 1960 in his plot, "hiding" data known to be faulty and misleading—again, entirely appropriate.

MBH98/MBH99 and many other reconstructions since, by contrast, use few such tree ring density records and thus do not suffer—at least obviously—from the decline that Jones was referring to in his e-mail. Individuals such as S. Fred Singer have nonetheless tried to tar my coauthors and me with "hide the decline" by conflating the divergence problem that plagued the Briffa et al. tree ring density reconstruction with entirely unrelated aspects of the hockey stick. Singer, for example, claimed[22] that we had ended our reconstruction in 1980 to hide a lack of warming, when the reason—as clearly stated in MBH98[23]—was that many of our proxy data terminated at or close to 1980; there could be no reconstruction later than 1980 in our case.

Climate change deniers made other allegations based on the leaked e-mails. They claimed, for example, that the e-mails showed that CRU had destroyed original instrumental surface temperature records, or had blocked access to public data. Both charges were false, their apparent basis being that some hard-copy printouts had been lost during a move decades earlier, though the data themselves were all preserved, and that CRU was not at liberty to release a restricted subset of temperature records that some other countries had provided because of contractual obligations not to release those data to third parties. These allegations were nonetheless used to undermine confidence in the instrumental surface temperature record. It was convenient for the attackers to ignore the fact that the raw data were available in the public domain and the fact that other groups—in particular NASA and NOAA—had produced their own independent global temperature curves from those data. Not only did these other two records show the same basic trends as the CRU record (called HadCRUT3), but among the three it was the CRU record that showed slightly less warmth over the past decade.[24] Ironically, this was the record climate change deniers were relying on in their efforts to argue for a lack of recent warming.

Some critics also claimed that the e-mails revealed a culture of "gatekeeping," that climate scientists, myself included, were unfairly preventing skeptics from publishing in the peer reviewed literature. So claimed Patrick Michaels[25] of the libertarian[26] Cato Institute roughly a month after the CRU hack in a December 17 *Wall Street Journal* op-ed.[27] Peer

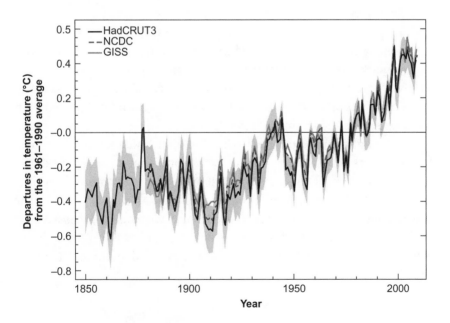

Figure 14.1: Competing Temperature Records
Three independent assessments of global temperature trends through 2009. The CRU ('HadCRUT3') record shows the least recent warming. [From the World Meteorological Organization.]

review, however, is by definition gatekeeping; it is intended to keep seriously deficient work from polluting the scientific literature.[28]

Deniers made numerous other false allegations based on misrepresentations of the content of the stolen e-mails. In one e-mail, I had quoted another colleague (Jonathan Overpeck, now at the University of Arizona) referring to the importance of "containing" the putative medieval warm period (MWP). It is obvious from reading the full e-mail that we were simply discussing the importance of extending reconstructions far enough back (more than one thousand years) so as to entirely "contain" the interval typically referred to as the MWP. It was hardly an effort, as some critics contended, to hide said interval. In another e-mail, Kevin Trenberth of the National Center for Atmospheric Research (NCAR) had stated: "The fact is that we can't account for the lack of warming at the moment and it is a travesty that we can't." Far from the secret admission of a global warming hoax that climate change deniers were claiming, Trenberth was pointing out the deficiencies in the ability of our current observing network to track how heat is being redistributed within the climate system (the atmosphere, oceans, and ice), a topic on which he had recently published.[29]

There were some ill-advised statements in some of the e-mails. Phil Jones asked colleagues to delete e-mails potentially subject to FOIA demands, for example.[30] Though the recent FOIA demands were vexatious, part of a campaign of harassment, his comment showed poor judgment. Let's keep in mind, however, that, first, nothing in one's training as a scientist prepares one for the kind of dishonest attacks to which he and others have been subjected, and second, that it can be incredibly exasperating to find yourself barraged by, for example, more than sixty frivolous FOIA demands in a single weekend, as was detailed in the preceding chapter. In any case, there is no evidence that Jones actually deleted any e-mails. Nor is there any evidence of any impropriety in his e-mails which, after all, are now all in the public domain as a result of the CRU e-mail theft.

Shaping the Narrative

Some pundits have likened the disinformation campaign built around the CRU hack to earlier public propaganda campaigns. *Climate Cover-*

Up author James Hoggan, for example, took note of its similarity to World War II propaganda in the way it employed "character assassination, innuendo, misdirection, and obfuscation" to create the appearance of "a sparkly 'scandal' easily digestible by a lazy, incompetent media."[31]

The climate change denial movement had absorbed the lessons contained in the 2002 Luntz memo on the importance of messaging and language and employed it in the East Anglia hacking episode to maximum effect. Critically, the deniers seized the initiative in framing the media narrative from the start, ensnaring journalists into emphasizing the out-of-context and—as we have now seen—distorted content of the stolen e-mails, rather than the crime through which they had been obtained. It was a sweet victory for climate change deniers that so much of the media adopted their framing, even to the point of accepting their moniker—climategate—to characterize the affair.

There were several layers of irony to this framing. First of all, many of the far-right groups working to manufacture or promote the climategate scandal advocated libertarian views of government that stress individual freedoms and warn against the intrusion of government into the affairs of individuals. Did they believe that it was somehow OK when individuals, rather than government, intruded into those affairs? Sarah Palin, among the first politicians to pounce on climate scientists after the CRU hack, expressed outrage when her own e-mails had been hacked during the 2008 presidential campaign and delight when the hacker was convicted of a felony offense: "Violating the law, or simply invading someone's privacy for political gain, has long been repugnant to Americans' sense of fair play."[32]

Indeed, the smear campaign surrounding the stolen CRU e-mails had more in common with the aforementioned swiftboat attacks against presidential candidate John Kerry than with the Watergate scandal that brought down the presidency of Richard Nixon. Award-winning filmmaker Richard Brenne perhaps summarized it best: "[Emphasizing] the content of the e-mails rather than the illegal hacking would be like someone emphasizing only what the Watergate burglars found without ever mentioning their crime or its cover-up."[33] Supporters of the climate scientists consequently argued for the use of a more appropriate name such as SwiftHack, Deniergate, KochScam, or Imhoax.[34]

As Morgan Goodwin of DeSmogBlog put it, "'Climategate,' or 'Swifthack' was a media story about a set of hacked emails that was pushed by a group of avid climate skeptics. . . . Collectively, they took a mountain of stolen material, condensed it into a well-packaged pitch, and sparked a scandalous story that reached virtually every major news outlet in the world."[35] A postmortem analysis of the affair by Goodwin reveals a well-choreographed media strategy whereby those promoting the story used coordinated messaging to seize control of the story and dictate the terms of its evolution.

Goodwin analyzed how the phrases denialists introduced in discussions on fringe Web sites were readily adopted over the following days by major media outlets.[36] Between November 20 and 26, for example, he followed the propagation of the phrase "smoking gun" from fossil fuel industry advocates such as Myron Ebell of the Competitive Enterprise Institute (CEI), Noel Sheppard of Newsbusters, and Steve Milloy, to the *New York Times*, Fox News, and CBS News. The phrase "final nail in the coffin [of global warming]" spread during the same period

Figure 14.2: Climategate
The essence of the so-called climategate episode is captured in a cartoon. [*Nick Anderson Editorial Cartoon* used with the permission of Nick Anderson, the Washington Post Writers Group and the Cartoonist Group. All rights reserved.]

from polemicist James Delingpole's *Telegraph* blog and Stephen Mc-Intyre's climateaudit, to conservative sites such as Free Republic and the Drudge Report, then onto Fox News and—in the form of an Inhofe op-ed—the *Wall Street Journal.* Calls for an investigation and preemptive cries of "whitewash" were propagated by Chris Horner and other bloggers, and eventually made it onto the pages of conservative and even mainstream newspapers.[37]

The repetitive use by deniers of phrases like "smoking gun" and "final nail in the coffin [of global warming]," Goodwin concluded, "would probably make GOP wordsmith Frank Luntz proud. By sticking to a small set of familiar, eye-catching, and dramatic-sounding memes, the denialosphere succeeded in getting many of the mainstream outlets covering the stolen emails scandal to follow suit by using the same language again and again. So, instead of investigating who was behind the theft, or exploring the context of some of the supposedly scandalous things said in private emails between a handful of climate scientists, many mainstream outlets followed the lead of the denier choir by repeating their version of events."[38]

The most disheartening aspect of "climategate" was the willingness of respected media outlets to uncritically parrot the accusations and innuendo being spun by the professional climate change denial machine. It was hardly surprisingly when Fox News headlined the affair as "global warming's Waterloo."[39] But within days of the hack, both the *New York Times* and the *Washington Post* ran multiple news articles adopting the denialist frame, focusing on the various out-of-context, cherry-picked words and phrases such as "hide the decline" and "trick." There was little or no discussion about the crime involved in hacking a university server and who might have been behind it, let alone who stood to benefit from it. Coverage by the UK press, including the progressive *Guardian*, was no better.

The networks waited longer to cover the story, but once they did, their coverage was for the most part similarly sensationalistic and slanted. Over the six-day period of December 3 to 9, 2009, I did interviews with each of the national television news organizations (ABC, NBC, CBS, and CNN twice—both *American Morning* and the *Campbell Brown Show*). The NBC *Today Show* and CNN's coverage were reasonable overall.[40] CNN did in fact go out of its way to correct some of the distortions being advanced by climate change deniers.[41] But my experiences with John Roberts, cohost of CNN's *American Morning*, were mixed. When Roberts first interviewed me in 1998 for *CBS Evening News* after our original *Nature* hockey stick article was published, he had pushed me toward overstating the implications of our study—something I resisted. It was somewhat ironic that Roberts's coverage of events surrounding the hacking of the East Anglia site was now suggesting that scientists had been overselling the evidence for climate change. While I found Roberts both fair and amiable in my personal dealings with him,[42] I was disappointed in how he had fallen for faux "balance," giving substantial air time to disingenuous claims of climate change deniers.[43]

The coverage by ABC and CBS Evening News was far worse. ABC correspondent David Wright repeated the outlandish claim that I had been "hiding the decline" in temperatures.[44] ABC News issued a correction on its Web site a couple days later, but few viewers would have seen it. CBS News, on February 4, 2010, ran a YouTube video ridiculing me and advancing the same "hide the decline" distortion, while briefly noting—as if it were simply incidental—that I had been exonerated of allegations of wrongdoing by Penn State University.[45] I had expected far better from the network of Edward R. Murrow and Walter Cronkite.

In early January 2010, a reporter named Fred Guterl interviewed me under the pretense of a piece to run in *Discover* magazine about the state of scientific understanding at the time of the December Copenhagen conference, including the impacts of climategate. I was taken aback when a fact-checker for *Newsweek* instead contacted me on Guterl's behalf, seeking to re-litigate now long-discredited claims of Stephen McIntyre regarding the availability of our proxy data. Guterl's article with the inflammatory title "Iceberg Ahead: Climate Scientists Who Play Fast and Loose with the Facts Are Imperiling Not Just Their Profession but the Planet" appeared in *Newsweek* two months later, on February 22, 2010.[46] The article resurrected the discredited hockey stick attacks, adding bogus new "climategate" allegations and low blows at James Hansen and IPCC chair Rajendra Pachauri for good measure. The article suffered from an increasingly familiar form of journalistic malpractice wherein the fact I'd been absolved of wrongdoing was acknowledged while simultaneously insinuating the very charges I'd been absolved of.

When I subsequently alerted Guterl that there were a number of serious and potentially defamatory errors in his *Newsweek* piece, to his credit, he made the corrections he was able to[47] and helped see to it that *Newsweek* published a short letter from me correcting the record.[48] I suspect that the problem at least partly lay with *Newsweek*'s editors, who had been taking a contrarian stance in their coverage of climate change.[49] Guterl's *Discover* piece, which appeared in March,[50] was little better, however; it was long on manufactured conflict and short on insight, with this lead: "Two eminent climatologists share much different views: Michael Mann—whose private emails were hacked—points a finger at skeptics. Judith Curry believes humans are warming the planet but criticizes her colleagues for taking shortcuts." In the interview, Curry made an interesting admission: "I live in Georgia, which is a hotbed of skeptics. The things I'm saying play well in Georgia."

A similar scenario of poor or biased coverage would replay itself repeatedly in the months ahead with a range of media, from the *Guardian* to the *Wall Street Journal.* Sometimes corrections and retractions were made, but they can never undo the damage done by the far more prominently placed original articles.

There were shining exceptions, of course. Among them were reporters for regional papers such as the *Philadelphia Inquirer* and *Allentown Morning Call* and local TV reporters, who for the most part strived for accuracy and context, rather than innuendo and false balance.[51] I participated in two hour-long NPR radio shows that brought a insightful and informative approach to the coverage of the "climategate" affair. On the *Diane Rehm Show* (November 30, 2009), I participated in a wide-ranging discussion about the understanding of climate change going into the Copenhagen conference, with a panel including three other participants: a professional climate change denier (Kenneth Green of AEI), a reporter (Stephen Porter of the *Wall Street Journal*), and the former chief of staff for President Bill Clinton, John Podesta. While Green did his best to play up the manufactured controversy of climategate, Rehm gave me more than a fair opportunity to refute his various falsehoods and half-truths. Porter and Podesta provided thoughtful, contrasting perspectives on the politics of the negotiations. In "On Point with Tom Ashbrook" (December 8, 2009), I participated along with *Washington Post* science journalist Juliet Eilperin in a discussion of the climategate affair and its aftermath. Both shows featured questions from callers, many of which were nuanced and thoughtful.

ABC News later provided more responsible coverage. In April I did an interview for *Nightline* with the same reporter, David Wright, who had done their awful December climategate piece. Airing on Earth Day in late April, the segment played up the conflict in viewpoints of contrarian-leaning on-air TV meteorologists and actual climate scientists.[52] However, it provided the critical context so often missing—that broadcast meteorologists are seldom trained as experts in climate science. The piece ended with my response to Wright's conflation of efforts to inform policy with policy advocacy. In response to his assertion that "it's your job to convince the public!"[53] I replied that "my job as a scientist is making sure that the public discourse is informed by an accurate understanding of the science."[54] The following month, *ABC Nightly News* ran another news piece that helped, at least in some small part, to shift the prevailing narrative away from the content of stolen e-mails toward the abuses being suffered by climate scientists.[55]

The Smear Campaign

Climategate was simply a vehicle for a more widespread and sustained smear campaign against climate science and climate scientists than was before possible. While the identity of those responsible for the initial break-in has not, as of this writing, been determined, it is clear that many of the usual suspects in the climate change denial movement played critical roles in fashioning how it would be framed and promoted. Indeed, the usual groups linked to fossil fuel interests appear to have been closely involved in the orchestration of the associated PR campaign.[56] Koch Industries and the Scaife Foundations played a particularly important role. One report showed that twenty or so organizations funded at least in part by Koch Industries had "repeatedly rebroadcast, referenced and appeared as media spokespeople" in stories about climategate.[57]

While these organizations may have been the principal architects of the climategate smear campaign, it is instructive to look at some of the individual spokespeople who led the charge. Marc Morano primarily acted as an aggregator, accumulating the various allegations and smears in a central location, his Climate Depot Web site and e-mail distribution list. The most often quoted attackers, however, were the usual contrarian protagonists, such as Patrick Michaels, Steve Milloy, Roy Spencer, Stephen McIntyre, Ross McKitrick, and Christopher Monckton.

The tenor of the public discourse had, in the months following the CRU hack, shifted in favor of the climate change deniers, and they seized full advantage. Suddenly, nothing was out of bounds. Otherwise responsible journalists were suddenly willing to act as little more than stenographers for the constant stream of bogus allegations being fed them. With each passing week, there was seemingly a new climate-related "gate" to be manufactured and promulgated by contrarians in the climate change debate.

Several of these "gates" related to minor errors in the IPCC AR4 report. There were two sentences on page 493 in the working group 2 impacts report regarding the projected rate of future decline of Himalayan glaciers that were in error. Rather than using the correct estimates as given in the report of the scientific working group (working group 1), the authors had relied upon an outside source. That source had made a transcription error, turning the projected time frame for disappearance of those glaciers from A.D. 2350 into A.D. 2035.[58] There was another error in a statement about sea level rise influences on the Netherlands in the working group 2 report, which actually came from the Dutch.[59] Neither error made it into the technical summary reports or the summaries for policy makers. Nonetheless, these minor errors in a three-thousand-page report were seized upon by deniers.

Suddenly, how quickly Himalayan glaciers would melt (ironically, nobody was even questioning that they *would* melt!) became the central issue of the IPCC report in the blogosphere and then the mainstream media. "GlacierGate" was born, breathed into life by denialists like Marc Morano, Christopher Booker, and James Delingpole of the *Telegraph*, and especially Jonathan Leake of the *Sunday Times*. Then there was "Amazongate" and "Africagate" based, respectively, on media distortions of what the IPCC actually said about Amazon drought, and African crop yields; Leake once again had a prominent role in promulgating these faux scandals.[60] And eventually there was "Pachaurigate," a baseless character attack against IPCC chair Rajendra Pachauri.[61] Such attacks on the IPCC played into domestic politics as well. Denial outfits in the United States were bent on overthrowing the EPA "endangerment finding" calling for regulation of greenhouse gas emissions by attacking its scientific basis, which, in substantial part, was the scientific findings of the IPCC. If the attackers could bring down the IPCC, they reasoned, the basis for an endangerment finding and all related calls for policy action would crumble. What more effective—if entirely cynical—way to bring down the IPCC than to bring down its figurehead, Rajendra Pachauri. Pachauri has been under a constant assault by climate change deniers ever since, with attacks becoming increasingly nasty and disingenuous.[62] Think of this as the lions of the Serengeti picking off the zebras at the edge of the herd.

Stefan Rahmstorf commented on the latest sequence of attacks by climate change deniers on RealClimate: "after the Himalayan glacier story broke, [they] have sifted through the IPCC volumes with a fine-toothed comb, hoping to find more embarrassing errors. They have actually found precious little, but the little they did find was promptly hyped into Seagate, Africagate, Amazongate and so on. This has some similarity to the CRU email theft."[63]

Again, unfortunately, it wasn't only the fringe media that took part in the attacks. Some of the most critical commentary came from a *Guardian* writer in the United Kingdom. The *Guardian* had in the past provided among the most accurate and responsible coverage of climate change in the UK media. But a series of articles by Fred Pearce, previously of the *New Scientist*, mangled the facts, distorting key details such as the chronology of events, to fit a preconceived narrative of conspiracy and corruption.[64]

One of the more egregious media distortions started with a BBC interview of Phil Jones in mid-February 2010. The BBC queried Jones, "Do you agree that from 1995 to the present there has been no statistically significant global warming?" This was a trap planted by climate change deniers.[65] Its origins probably lay in an e-mail from Richard Lindzen coaching climate change denier Anthony Watts on how to cherry-pick starting dates to get the desired trend.[66] From the temperature graphic shown earlier in this chapter, it is clear that the globe had warmed since 1995. But the precise amount depended on the particular temperature compilation, with Jones's (CRU) record showing the least net warming. As discussed in chapter 12, establishing significance of trends over short periods of time is difficult. Significance depends on how large the net trend over the time interval analyzed is by comparison to the amplitude of the year-to-year fluctuations (from things like

El Niño, volcanoes, etc.). For increasingly short time intervals, the former becomes increasingly small compared to the latter, and any trend, accordingly, becomes decreasingly significant. This was especially true if one just happened to choose 1995 as the starting year.

You might call this episode "the cherry-pick heard 'round the world." The goal was to get Jones on record agreeing that the statement was technically true—which he did—while ignoring the caveats and qualifications he raised[67] about how meaningless it is to talk about trends over short periods of time, the curiously selective use of a 1995 starting date, and the fact that the CRU record was the one that showed the least warming of the three major records. All too predictably, contrarian media outlets immediately trumpeted news of Jones's response with outlandish characterizations such as "World May Not Be Warming, Say Scientists," courtesy of the *Times* (the author, once again, being Jonathan Leake),[68] and "Climategate U-turn as Scientist at Centre of Row Admits: There Has Been No Global Warming Since 1995," courtesy of the right-leaning *Daily Mail*.[69] The story was then picked up in the United States by Fox News.[70]

There is some irony revealed by this incident. One might even say that the critics were engaged in an effort to, as one person put it, "hide the incline" in the global temperature series.[71] Some observers noted the asymmetric standards of honesty: "The denial movement will happily accuse scientists of playing tricks and engaged in fraud, yet at the same time engage in the very things they accuse the scientific community of."[72] Journalist Johann Hari made a related point in *The Nation*: "when it comes to coverage of global warming, we are trapped in the logic of a guerrilla insurgency. The climate scientists have to be right 100 percent of the time, or their 0.01 percent error becomes Glaciergate, and they are frauds. By contrast, the deniers only have to be right 0.01 percent of the time for their narrative."[73]

In some cases, the distortions were almost comical. For example, in February 2010, a paper on sea level rise that had previously been published in *Nature Geosciences*[74] was formally withdrawn[75] by the authors because of an error they had identified subsequently in their calculations. Fox News announced the development in this vein: "More Questions About Validity of Global Warming Theory."[76] In fact, the error in the calculations had led the authors to projections of future sea level rise that were too low![77]

After a while, many in the media seemed to realize they'd been had one too many times by the deniers, and they stopped taking the bait. In May 2010, for example, Marc Morano attempted to manufacture a new scandal, "MalariaGate," but he found few takers. On his Web site and through his broad e-mail distribution list, Morano misrepresented what the IPCC said about climate change influences on malaria and argued that a recent paper in *Nature*[78] had overthrown the IPCC thinking on the problem, which it hadn't. In the process, he took potshots at a Penn State collaborator of mine who had commented on the study, and me as well.[79] But media outlets ignored him. They had also ignored (with the predictable exception of Fox News) the latest Heartland Institute climate change denial conference that took place around the same time, despite the many efforts of the sponsors to manufacture publicity.

Political Intimidation

Especially disturbing, given the painful lessons of history, were the attempts of politicians to vilify and intimidate climate scientists in Congress and beyond. The attacks by congressional Republicans with close ties to the fossil fuel industry, though, were not unexpected.[80] Within days of the release of the hacked e-mails (November 23, 2009), the staff of Senator David Vitter (R-LA) released a letter claiming that global warming "could well be the greatest act of scientific fraud in history,"[81] essentially parroting James Inhofe's "greatest hoax" line. Not to be outdone, Inhofe called for congressional investigations of various climate scientists, including me, the very next day.[82] Inhofe blasted off threatening letters[83] to me, my coauthors Ray Bradley and Malcolm Hughes, and eight other prominent U.S. climate scientists, including IPCC working group I chair and Presidential Medal of Science winner Susan Solomon. He sent similar letters to representatives of a half dozen government agencies and a half dozen universities (including Penn State) and government research labs. These letters threatened impending investigations and vexatious demands for personal e-mail and other materials, though their real intent appeared simply to intimidate.

Similar attacks soon came from House Republicans. Within twenty-four hours, far-right Republican Congressman Darrel Issa of California, the ranking member of the House Investigations Committee, joined with Inhofe in calling for investigations of various scientists, including me.[84] Within a week, Jim Sensenbrenner (R-WI), the-now-familiar Joe Barton (R-TX), and the leader of attacks against Ben Santer in the 1990s, Dana Rohrabacher (R-CA), had joined in. A December 2 hearing of the House Select Committee on Energy Independence and Global Warming, at which Obama administration science adviser John Holdren and NOAA administrator Jane Lubchenco were called to testify, was indicative of the tone that would pervade many of these attacks. Sensenbrenner hijacked the proceedings with trumped-up climategate charges: "These e-mails show a pattern of suppression, manipulation and secrecy that was inspired by ideology, condescension and profit," and referred to "scientific fascism" and "an international scientific fraud." Holdren calmly and politely refuted Sensenbrenner's false allegations and mischaracterizations.[85]

In the movie *The Manchurian Candidate*, the McCarthyesque Senator John Yerkes Iselin announces on the floor of the U.S. Senate that "There are exactly 57 card-carrying members of the Communist Party in the Department of Defense at this time!" Inhofe couldn't find fifty-seven climate scientists to investigate, but he did manage to come up with a list of seventeen (including me, my collaborators, and leading American climate scientists such as Stephen Schneider, Susan Solomon, Tom Wigley, Kevin Trenberth, and Ben Santer) that he said should be criminally prosecuted, for various reasons he outlined in a lengthy report he released on February 25, 2010, on the Senate Environment and Public Works (EPW) minority Web page.[86] "Are you now or have you ever been a climate scientist?" is the way one commentator summarized it.[87]

A campaign of intimidation was meanwhile being waged against climate scientists by influential forces outside the halls of Congress who helped direct hate speech, threats, bullying, and taunting at individual climate scientists. Marc Morano's comment in March 2010 that climate scientists "deserve to be publicly flogged" seemed to set the tone.[88] That was just a few weeks after Fox's Glenn Beck had followed a litany of bogus allegations about the IPCC with the suggestion that climate scientists commit suicide.[89] Earlier, right-wing provocateur Andrew Breitbart had "tweeted": "Capital punishment for Dr James Hansen. Climategate is high treason."[90] The use of such vitriol and invective might be viewed as an effort to stoke the fires of irrationality, hate, and violence among sympathizers in the general population.

It certainly appears that this dynamic was in play. In the wake of climategate, several climate scientists reported receiving threatening messages.[91] Gavin Schmidt received "rude and crass e-mails" calling him "a fraud, a cheat, a scumbag and much worse," while Kevin Trenberth had a "9-page document of 'extremely foul, nasty, abusive' e-mails" that he'd received in just a few months. One colleague had been stalked by suspicious individuals and found a dead rat on his doorstep as the perpetrator drove off in a large yellow Hummer. A number of climate scientists (myself included) were accused of being part of a Jewish conspiracy to defraud honest Americans through a "climate change hoax" con job, with their names and photos posted on a neo-Nazi Web site.[92] Several scientists were at the receiving end of death threats so credible they were provided security detail.

Was this harassment orchestrated? A number of pundits believed so. "The purpose of this new form of cyber-bullying seems clear; it is to upset and intimidate the targets, making them reluctant to participate further in the climate change debate," in the view of Australian journalist Clive Hamilton.[93] He argued there was a stealth "Astroturf" campaign at work: "The floods of offensive and threatening emails aimed at intimidating climate scientists have all the signs of an orchestrated campaign by sceptics groups. The links are well-hidden because mobilizing people to send abuse and threats is well outside the accepted bounds of democratic participation; indeed, some of it is illegal. And an apparently spontaneous expression of citizen concern carries more weight than an organized operation by a zealous group."

Journalist Douglas Fischer noted in *Scientific American* that "Most of the e-mails appear to be the work of frustrated individuals, ranting into the ether. . . . But some appear to be the work of coordinated campaigns, and many, scientists say, appear to be taking their cue from influential anti-climate change advocates like Rush Limbaugh, Glenn Beck and ClimateDepot.com."[94] Fischer found a close parallel with intimidation campaigns against scientists in other areas where scientific work potentially threatened corporate special interests: "Researchers working on Atrazine, a widely used herbicide, bisphenol-a, a common plastic additive, and other environmental pollutants have received similarly intimidating e-mails and even threats." In the same article, *Climate Cover-Up* coauthor Richard Littlemore argued that while "determining whether any given e-mail is part of an organized campaign is difficult . . . it's not happenstance." "[T]he bullying doesn't start serendipitously or from scratch," he said. "It starts with a paid campaigner—Morano, . . . JunkScience.com/Fox News commentator Steve Milloy—and filters out from there. . . . They're the PR guys and they're in the game and taking money for what they do. They also wind up recruiting other folks."

It Gets Personal

A number of the hacked CRU e-mails that climate change deniers promoted most heavily involved me. Many of the buzzwords and phrases that were circulating (e.g., "trick," "hide the decline") were used to attack me as well, whether or not they even applied to me and my work. As we have seen in previous chapters, the predators of climate change denial had been stalking me for some time now, and climategate was an opportunity, they suspected, to go in for the kill.

One front of the attack involved the sending of voluminous e-mail and phone messages containing thinly veiled threats of harm against me and even my family. "You and your colleagues who have promoted this scandal ought to be shot, quartered and fed to the pigs along with your whole damn families," read one e-mail. Another read, "you should know the public will come after you," and "Six feet under, with the roots is were [sic] you should be doin [sic] your magic, how come know [sic] one has [edited] you yet, i was hopin [sic] i would see the news and you commited [sic] suicide."[95]

Most of these threats didn't seem credible, and while I did inform the authorities of them, by and large I brushed them off. One, however, could not be ignored. On August 18, 2010, I had to explain to colleagues in the Penn State University meteorology department, located in the "happy valley" of Central Pennsylvania, why there was police tape over the door to my office. The immediate answer was that the FBI had quarantined the room and sent a letter I'd received that afternoon off to their nearest testing facility to determine the nature of the white powder contained within it. At a more basic level, the answer was that this is simply what it means to be a prominent figure in the climate change debate in the United States today. The tests came back a few days later; luckily, the substance in the letter was cornstarch. The sender had nonetheless committed a felony crime.

There was also a second front of attack, one aimed at threatening my livelihood. In mid-January 2010, a group known as the National Center for Public Policy Research (NCPPR), which receives funding from the Scaife Foundations,[96] led a campaign to have my NSF grants revoked.[97] The perverse premise was that I was somehow pocketing millions of dollars of "Obama" stimulus money simply because I was a coinvestigator on several recently funded NSF grants.[98] These absurd distortions were—no surprise—promoted by Glenn Beck, Rush Limbaugh, and others of similar persuasion.

Perhaps a mere coincidence, but two weeks earlier, on New Year's Eve 2009, a former CIA agent named Kent Clizbe sent an e-mail message to dozens of my Penn State colleagues entitled "Climate Research Fraud—Whistleblower Rewards Program—Confidentiality Assured." In his message, Clizbe promised my colleagues monetary reward in return for accusing me of fraud in my government grant applications.[99] As far as I know, he got no takers.

The NCPPR, meanwhile, even managed to tap a reserve from the 1990s tobacco wars, Tom Borrelli, the former manager of Corporate Scientific Affairs at Philip Morris. Borrelli attacked me in an NCPPR-sponsored press release in mid-January 2010: "It's shocking that taxpayer money is being used to support a researcher who seemingly showed little regard to the basic tenets of science—a dispassionate search for the truth."[100] The irony in this line of attack did not go unnoticed by Mitchell Anderson of DeSmogBlog.[101] Anderson linked to a video,[102] released as part of the historic tobacco industry legal settlement,[103] showing Borelli denying the health threat of secondhand tobacco smoke. A few weeks later Deepak Lal, a senior fellow for the Cato Institute,[104] repeated false climategate allegations against me in a commentary that appeared in the *Business Standard*.[105] Lal, it turns out, also had a tobacco connection; he had been hired by British American Tobacco in the late 1990s to try to discredit a World Bank report on the health benefits of policies to control tobacco use.[106]

Leading congressional climate change deniers used their political clout to press this attack further. In early February 2010, Darrell Issa (R-CA) declared Penn State's initial inquiry, which on February 3 had reported[107] no evidence of scientific misconduct on my part, a "whitewash." He revived the campaign to have my NSF funding revoked, citing once again the climategate-related allegations of scientific misconduct. Fox News[108] and the *Washington Times*[109] publicized Issa's attacks. On February 16, Issa and Sensenbrenner joined forces to press the issue, complaining to the NSF: "conversations with NSF staff on February 4 revealed that no steps have been taken to freeze or withdraw the stimulus funds authorized for Dr. Mann's research . . . we encourage you to immediately initiate an independent investigation into this matter."[110]

Apparently unsatisfied by lack of progress on this front, and still digging for dirt, two Scaife-funded groups mentioned in the previous chapter, the Southeastern Legal Foundation and the Landmark Legal Foundation, had swung into action. The latter had already sued the University of Massachusetts and University of Arizona to obtain copies of my personal e-mails with my two hockey stick coauthors, while in May 2010 the former demanded extensive information from the NSF regarding grants that had been made to me as well as to several of my colleagues at Penn State, the University of Chicago, the University of Washington, the University of Arizona, and Columbia University.[111]

It began to strike me as curious that so many of the demands that I be investigated could be traced back to organizations with ties to the Scaife Foundations.[112] The Commonwealth Foundation, a Pennsylvania organization that is the recipient of considerable Scaife largesse,[113] for example, had been pressuring Penn State University to fire me since climategate broke in late November 2009. It managed[114] to get the sympathetic Republican chair of the Pennsylvania state senate education committee to threaten to hold Penn State's funding hostage until "appropriate action is taken by the university against associate [sic] professor Michael Mann."[115] Indeed, it was the Commonwealth Foundation attacks that essentially forced Penn State to launch its initial inquiry into the various allegations against me in December 2009 (similar inquiries and investigations of CRU scientists were initiated in the United Kingdom[116]). The Commonwealth Foundation kept the pressure on for months through a barrage of press conferences and press releases attacking me personally and criticizing Penn State for its supposed "whitewash" treatment of any number of supposed offenses.[117] It also ran daily attack ads against me in our university newspaper *The*

Collegian for an entire week in January and helped organize a protest rally against me on campus.[118] It is likely that these attacks forced Penn State's hand yet again, leading it, following the completion of the initial inquiry in February 2010, to move to a formal investigation, despite having found no evidence of misconduct in the initial inquiry phase.[119]

If the first two fronts of the assault against me involved intimidating me and my family and threatening my livelihood, the third was a related effort to discredit and humiliate me among my colleagues, peers, and community.[120] Marc Morano, through his large e-mail distribution list and Climate Depot Web site—financed by the Scaife-funded group CFACT—continued to spread malicious and false allegations about me and other climate scientists.[121] The allegations inevitably found their way onto Fox News,[122] the *Glenn Beck Show*, Rush Limbaugh's show, and numerous fringe news outlets. My friends, family, and colleagues, of course, could not help but hear or read the various lies that were being spread about my climate change colleagues and me. A new group run by fossil fuel industry advocate Steven Milloy even took out a sponsored Google ad that posed the question "Michael Mann: Defamed or Declined by 'Hide the Decline?'" which would appear in response to Google searches including either my name or climate change–related topics and key words.[123]

A YouTube video promoted by Koch and Scaife-funded groups was also created to ridicule me, and through me, climate change science.[124] The video featured a caricature of me repeating the line "Michael Mann thinks he's so smart totally inventing the hockey stick chart . . . hide the decline, hide the decline."[125] The video was purportedly created by a "grassroots" group calling itself Minnesotans for Global Warming, supposedly founded by a Elmer Beaureguard. A law firm giving me pro bono assistance determined, after some investigation, that "Elmer Beaureguard" was a fake identity and the Web site minnesotansforglobalwarming.com was the work of someone named Elroy Balgaar with, no surprise, apparent ties to conservative causes and the Republican Party.[126]

The episode evoked a similar event a few years earlier. In August 2006, a YouTube cartoon video mocking Al Gore went viral on the Internet, promoted in part by ads on Google. In the video, Gore is shown boring an army of penguins with his "An Inconvenient Truth" global warming lecture, and farcically blaming global warming for just about everything. The film was ostensibly created by an amateur twenty-nine-year-old filmmaker working out of his basement, but investigations by the *Wall Street Journal*[127] and ABC News[128] traced the video to the DCI Group, a Republican consulting and lobbying firm based in Washington, D.C., which derived its funding from a variety of corporate sponsors in the fossil fuel, pharmaceutical, fast food, and telecommunications industries, including oil giant ExxonMobil. Curiously enough, Google ads for the video abruptly disappeared once the DCI Group's role was exposed.

The law firm assisting me eventually sent a cease-and-desist letter to Balgaar informing him that he had misappropriated a copyrighted image of me and had used it in a defamatory manner,[129] while the owner of the copyright informed YouTube of the copyright violation, and the video was taken down. What happened subsequently was illuminating. No sooner had the video been taken down than a new group connected with Koch Industries and the Scaife Foundations,[130] the "No Cap and Trade Coalition," initiated a public relations offensive designed to portray poor Mr. Beaureguard as a defenseless victim. It even sponsored a press conference at the National Press Club featuring him, alongside Patrick Michaels and Myron Ebell.[131] Fox News, of course, covered it.[132]

Few if any mainstream media organizations took notice of the press conference, but the event did generate an unusually intense burst of vitriolic and derisive, generally anonymous e-mail and phone messages. They conveyed a remarkably consistent message, almost as if they had come from a common template. The message was this: If I sought legal action against the attacks of the industry-funded denial machine, I would—among other things—be challenged with endless, invasive "discovery" demands.[133] I should, in short, just put up with the attacks and "take it like a man."[134]

The assault would continue in the months ahead. Former British Prime Minister Tony Blair's chief scientific adviser Sir David King had once reported that individuals funded by the Competitive Enterprise Institute were heckling him at his public lectures on climate change.[135] I was now subject to the same thing. In early May 2010 when I gave a keynote lecture for the PennFuture global warming conference in Pittsburgh, for example, a small group of picketers—led by a man who identified himself with the Commonwealth Foundation—had assembled outside the conference center.[136]Affixed to a truck they had parked there was a billboard-sized display bearing an unflattering cartoon of me, with the phrases "hide the decline" and "Mann-made warming" underneath, and they were distributing T-shirts bearing the same image. (My host, PennFuture president Jan Jarrett, obtained one, which she awarded me before the audience as a war trophy of sorts.)

Meanwhile, Mr. Clizbe persisted in sending out periodic e-mails promising colleagues financial reward for claiming evidence of fraud in my research,[137] the Commonwealth Foundation continued to attack me in the local press, and the No Cap and Trade Coalition (under the guise of the Minnesotans for Global Warming, but with the ties between the two organizations now explicit)[138] continued to produce attack videos. For a February 2011 video entitled "I'm a Denier" and hosted on the Heartland Institute Web site, they even employed a doppelganger to play me dancing around with a hockey stick[139] in what Eric Alterman of *The Nation* described as "an awful, amateurish Monkees parody."[140] A group known as the Collegians for a Constructive Tomorrow, a sister organization to CFACT, engaged in similar efforts to ridicule me at the time.[141]

I could easily brush off many of the attacks. In fact, at some level, the attention was almost flattering. But to the extent that the industry-funded disinformation machine was taking direct aim not just at me—my livelihood, my reputation, my safety—but even my family, I was becoming incensed. And it wasn't just me. Many of my fellow climate scientists were outraged by the mean-spirited and dishonest attacks against us and our science. None had signed up for this sort of treatment. We'd seen the public polling data that suggested our credibility as a community had taken a hit. So too had efforts to confront the daunting challenges presented by climate change.[142] Could we allow this assault to continue? What could we do to defend ourselves?

Chapter 15

Fighting Back

This institution will be based on the illimitable freedom of
the human mind. For here we are not afraid to follow
truth wherever it may lead.

—Thomas Jefferson, on the University of Virginia (December 27, 1820)

Though hardly surprising to those who had taken a hard, honest look
at the matter, each of the investigations launched in response to the
climategate affair ultimately proved exculpatory. That applied not just
to me, but to the scientists of the University of East Anglia's Climate
Research Unit (CRU), and indeed all of the scientists who were em-
broiled in the manufactured scandal, as well as to the underlying science of climate change itself.

At the end of March 2010, the House of Commons Science and
Technology select committee of the UK Parliament issued its findings.
The report noted that CRU's "analyses have been repeated and the conclusions have been verified."[1] It exonerated Phil Jones of the various allegations of scientific misconduct against him, including those based
on out-of-context quotations such as "hide the decline" and "trick." Even
Nigel Lawson, the critic who had called for the investigation in the first
place, conceded that Jones's use of this terminology was entirely innocent.[2] While the report did give Jones a minor slap on the wrist over
issues involving making data available, it found that "the focus on CRU
and Professor Phil Jones, Director of CRU, in particular, has largely been
misplaced" and that his actions were "in line with common practice
in the climate science community." Next up was an external review of
CRU's scientific practices led by eminent geoscientist Lord Oxburgh.
Reporting in mid-April, the commission once again found no evidence
in the work of CRU scientists or their collaborators of any fudging or
destruction of data, or of scientific misconduct of any sort.[3] The quality

of their science was found to meet the highest standards. The only substantial criticism was that CRU scientists could have collaborated more broadly with outside statistics experts. The report denounced the attacks by climate change deniers in unusually strong terms.

Penn State, after four months of reviewing voluminous materials and interviewing numerous other scientists and experts, issued its final investigation report on July 1.[4] It found "no substance to the allegation against Dr. Michael E. Mann." More specifically, it found that "Dr. Michael E. Mann did not engage in, nor did he participate in, directly or indirectly, any actions that seriously deviated from accepted practices within the academic community for proposing, conducting, or reporting research, or other scholarly activities."[5] The National Science Foundation Office of the Inspector General, which has the final word in such matters, performed their own independent follow-up investigation. Their determination: "finding no research misconduct or other matter raised . . . this case is closed."[6]

Less than a week after Penn State reported its findings, the final of the CRU investigations was completed[7]—this one having been commissioned by the University of East Anglia to review "key allegations arising from the series of hacked emails from the Climatic Research Unit."[8] The commission, led by Sir Muir Russell, produced an unusually thorough, well-documented 160-page report. It too found no evidence of scientific wrongdoing on the part of any scientists who had been tarred with various climategate allegations. Refuting the claim that CRU had somehow rigged its global temperature estimates or deleted critical data, the committee went through the rather extraordinary measure of reproducing the global temperature record from scratch, using publically archived data and methods. CRU scientists were once again criticized for not having been as forthcoming with requests for data as they could have been, and Jones's 1999 World Meteorological Organization (WMO) report cover graph depicting past temperature trends was criticized as potentially "misleading" for merging proxy and instrumental data into a single curve—a conclusion nobody really disputed. These were minor issues. The panel's conclusions were overall overwhelmingly exculpatory.

Issued within the same week, these twin final acquittals received reasonably good media coverage, though nothing approaching in extent the widespread reporting of the original allegations. Some pundits, such as Howard Kurtz of CNN's *Reliable Sources* show, argued that many in the media had been irresponsible by not giving the acquittals the prominence in coverage that had been given to the original false allegations.[9] But CNN covered my exoneration on *The Situation Room*,[10] while CBS News—now partially atoning for its "hide the decline" piece from December—was among the first to report on my exculpation.[11] Positive coverage was provided by numerous major newspapers and magazines.[12] Even Fox News, the *Pittsburgh News-Tribune,* and the *Wall Street Journal* reported the exonerations, though the *WSJ* editorial pages predictably cried "white wash."[13] Especially gratifying for me personally was the widespread coverage by various regional and local papers and TV news stations. PennFuture demanded that the Commonwealth Foundation now publicly apologize to me.[14] While that didn't happen, a cloud that had been hanging over me for the better part of a year had nonetheless finally lifted. There was true cause for family celebration on the Fourth of July holiday.

There were nonetheless a few road bumps along the way as the various investigations unfolded. Climate change deniers, for example, successfully hijacked a small working group (the Energy Sub-Group) within a respected European scientific organization known as the Institute of Physics (IOP).[15] Using the imprimatur of the IOP, the group made a formal submission to the Muir Russell Commission filled with standard denier talking points, claiming among other things that the content of the hacked CRU e-mails revealed "worrying implications" for "scientific integrity in this field." There was an immediate outcry from prominent institute members who were deeply troubled by the development,[16] and the IOP disbanded the Energy Sub-Group altogether later that summer.[17]

The statistician on the Oxburgh panel, David Hand, caused a bit of trouble with offhand remarks he chose to make at the press conference announcing the panel's findings. Though our own work did not fall within the remit of the committee and the hockey stick was not mentioned at all in the report, Hand commented that "The particular technique [Mann et al.] used exaggerated the size of the blade at the end of the hockey stick." This was instant fodder for the denial mill, with papers such as the *Telegraph* happily reporting the claim that the "hockey stick was exaggerated."[18] The statement was nonsensical, however. The end of the blade of the hockey stick was simply the instrumental temperature record; there was no way that our reconstruction, or any reconstruction for that matter, could either underestimate it or overestimate it. Within a few days, a retraction of the Hand's claim was effectively issued by the committee.[19]

Witch Hunt Redux

By spring 2010, climate change deniers had lost some ground. With the result of each new exculpatory climategate investigation report, their narrative of alleged scientific malfeasance had become increasingly untenable.[20] Attempts by deniers to find yet more e-mails to mine proved fruitless. An attempt to break into the laboratory of prominent Canadian climate scientist Andrew Weaver at the University of Victoria in British Columbia in order to steal e-mails and other materials was foiled, for example.[21] And none of the efforts by the Scaife-linked Southeastern Legal Foundation and Landmark Legal Foundation to obtain additional climate scientist e-mails, discussed in the previous chapter, yielded anything useful.[22] Meanwhile, the Massey West Virginia coal mine disaster in February and the Deepwater Horizon Gulf oil spill disaster in April served as vivid reminders of the extreme hidden costs of dirty fossil fuel energy. The dominant media narrative was now shifting to the tactics of the climate change denial movement, its funding by big oil, and the harassment of climate scientists.[23] The denial machine needed a new distraction to try to regain control of the narrative. They got it—and perhaps far more than they bargained for—in the form of the newly elected Virginia attorney general named Ken Cuccinelli.

In his first few months as Virginia's attorney general, Cuccinelli had pursued an eyebrow-raising agenda that included questioning president Barack Obama's citizenship, instructing state colleges to stop protecting gay students from discrimination, issuing a legal challenge declaring federal health care reform as unconstitutional (which was lampooned on Comedy Central's *Jon Stewart Show*), and attacking the U.S. EPA's right to regulate greenhouse gas emissions, claiming the underlying science was fundamentally unsound. He had then sought to alter the state seal of Virginia, designed by George Wythe, a signer of the Declaration of Independence, which had been adopted in 1776. The original seal showed the Roman goddess Virtus, holding a spear, with her left breast exposed. Cuccinelli had begun circulating among his staff a new "family-friendly" version of the seal with the offending body part covered by a breastplate. He dropped the plan abruptly when it was picked up by the media and brought him widespread ridicule.[24]

Following up on his petition of the EPA, Cuccinelli decided to turn his attention a little closer to home. He demanded that the University of Virginia turn over essentially every e-mail, record, or document it had that related to me from my time on the faculty there from 1999 to 2005. Cuccinelli used a relatively new legal maneuver available to the attorney general known as a civil investigative demand (CID). With a CID, the attorney general could seek to subpoena materials from any state agency, including a university, by claiming that fraud concerning state funds might be involved.[25] This must have seemed a clever way to seize materials that other groups, such as Competitive Enterprise Institute (CEI), had been unable to obtain using standard state FOIA laws. Initially, the University of Virginia indicated that it would comply; it perceived it had little choice, given state law.[26] That would change, however, as Cuccinelli's attack became increasingly recognized as an abuse of power that threatened the very bedrock principles of academic freedom that the University of Virginia was founded upon.

The initial outcry came from politicians within Virginia. State delegate Mark Herring (D-Loudoun) was among the first to speak out: "On its face this seems to be a serious abuse of the power of the Attorney General. This is Virginia, since when do we investigate professors when we disagree with them?"[27] He was followed by U.S. House Representative Jim Moran (D-VA), who told Cuccinelli: "History will neither reflect kindly on those who reject science in the pursuit of short-term economic and political gain, nor will it look kindly on your attempt to tarnish the good name of Professor Michael Mann."[28]

Word soon spread through the mass media. *USA Today* labeled Cuccinelli's actions a witch hunt against climate scientists.[29] *Slate* magazine quoted H. L. Mencken from the 1920s Scopes Monkey Trial: "This old buzzard, having failed to raise the mob against its rulers, now prepares to raise it against its teachers."[30] The *Washington Post*, in the first of five editorials[31] on the subject over the next six months, wrote: "Ken Cuccinelli II (R) had declared war on reality. Now he has declared

Figure 15.1: Cuccinelli's Witch Hunt
Cuccinelli is *twice* lampooned by the *Washington Post's* award-winning cartoonist Tom Toles. [*Washington Post*, May 14, 2010, and October 6, 2010. TOLES © 2010 The Washington Post. Reprinted with permission of UNIVERSAL UCLICK. All rights reserved.]

war on the freedom of academic inquiry as well." The editorial implored the University of Virginia to "have the spine to repudiate Mr. Cuccinelli's abuse of the legal code."

Numerous newspaper editorial boards across Virginia, including some of the state's most conservative papers that had endorsed Cuccinelli's run for office, also denounced Cuccinelli's attack.[32] The editorial board of one such paper summarized Cuccinelli's message to Virginia scientists as "Don't venture down this path of inquiry or you'll pay with your reputation and career."[33] Perhaps it is simply because of the deeply chilling nature of that message that Cuccinelli's actions were so widely denounced by public policy groups, including the American Civil Liberties Union (ACLU), the American Association of University Professors (AAUP), and even the conservative Foundation for Individual Rights in Education (FIRE). The University of Virginia soon began to show signs of fight itself. On May 5, the Faculty Senate issued a formal denunciation of Cuccinelli's investigation: "His action and the potential threat of legal prosecution of scientific endeavor that has satisfied peer-review standards send a chilling message to scientists engaged in basic research involving Earth's climate and indeed to scholars in any discipline. Such actions directly threaten academic freedom and, thus, our ability to generate the knowledge upon which informed public policy relies."[34] Soon enough, the scientific community would fight back too.

Wegman Revisited

In late December 2009, near the height of the climategate feeding frenzy, and just as climate scientists were most on the defensive, a dramatic new development instead put climate change deniers on the spot. Thanks to the sleuthing of computer scientist John Mashey[35] and an anonymous blogger known as Deep Climate, some dramatic new revelations came to light that shattered the myth that the now-three-year-old Wegman Report (WR) had been the independent, nonpartisan scientific assessment climate change deniers continued to claim it to be. Through a series of posts at Deep Climate's site[36] and Mashey's own full-length report,[37] the WR was put in a new, very different light. Mashey showed, first of all, that a network of high-level partisan political operatives and professional climate change deniers had been involved in the planning of the report.[38] Mashey and Deep Climate moreover unearthed compelling evidence of a disturbing pattern of conduct by Wegman and coauthors in their preparation of the WR. The developments would be reported in a series of articles over the next year and a half by *USA Today* reporter Dan Vergano.[39]

Deep Climate and Mashey detailed striking similarities between large sections of the WR and other previously published works. Most remarkably, the background discussion of paleoclimate proxy data appeared largely taken without attribution from MBH hockey stick coauthor Ray Bradley's well-known *Paleoclimatology* textbook.[40] Where Wegman and coauthors did tweak Bradley's original text, the changes were often far from innocuous. Bradley's words were systematically altered in a way that downplayed the reliability of the science and, in a perverse twist of irony, made them appear to undermine the conclusions of Bradley's own work.[41]

Consider one of the report's "key findings": "As mentioned earlier in our background section, tree ring proxies are typically calibrated to remove low frequency variations. The cycle of Medieval Warm Period and Little Ice Age that was widely recognized in 1990 has disappeared from the MBH98/99 analyses." Let us neglect for the moment the resuscitation of the "Lamb" curve of the 1990 IPCC report to attack more recent work, the false assertion that the Little Ice Age and medieval warm period "disappeared" in MBH98/MBH99, and the false implication that the MBH hockey stick was based exclusively on tree ring data. Most significant is how the WR background discussion of low-frequency information in tree ring data was altered relative to Bradley, in what appears to have been a cynical effort to call our work into question.

A key passage in the Bradley account explained how data derived from a regional set of tree ring cores are combined to form a composite chronology that can preserve long-term ("low-frequency") variations even if the individual cores cannot.[42] In the WR, that passage was replaced[43] with a shorter statement that blurred the key distinction between a single tree core and a composite, providing a straw man that was then used to call into question the reliability of "low frequency, longer-term" information in tree ring data (and by direct implication, the MBH hockey stick).[44] Mashey and Deep Climate documented many other examples where small changes made in Bradley's original wording inevitably favored the WR's hockey stick-bashing narrative. The irony was not lost on Deep Climate: "That such a shoddy misrepresentation of another author's work has been used as part of a baseless, politically motivated attack on that author is beyond shameful."[45]

Further suspicions about plagiarism were raised about the WR's discussion of social networks, which appears[46] to have been lifted from two books, one by Wasserman and Faust,[47] the other by de Nooy, Mrvar, and Batagelj,[48] and from Wikipedia. Many of the small changes seemed to serve no other purpose than to introduce some difference—perhaps to make plagiarism more difficult to detect—while at the same time in many cases making sensible passages incomprehensible.[49] This produced some amusing results. For example, Wasserman and Faust refer to a particular type of social network link as "movement between places or *statuses* (migration, social or physical mobility)." In the WR this became "movement between places or *statues*."[50]

The summaries of "key papers" in the WR, too, turn out to have been mostly cut and pasted from the abstracts of the original papers, yet with subtle but important changes that served to weaken the apparent strength of the science.[51] In the end, John Mashey estimates that nearly a third of the WR was made up of plagiarized material.[52] Wegman apparently even lifted materials for lectures he was giving on climate change, including slides he took, without attribution, from my very own public lecture on climate change.[53] But plagiarism may be the least of the alleged issues that came to light; the fundamental claims advanced by the report—both that the peer review process was broken in the field of paleoclimate studies and that the hockey stick was an artifact—were demonstrably based on distorted evidence and arguments.

Recall Wegman et al.'s use of social network analysis (described in chapter 11) to portray paleoclimatologists as an inbred clique, unable to independently review each other's work. It was a charge whose origins lay with Stephen McIntyre, but Wegman et al. sought to grant it the imprimatur of mathematical rigor. The WR social network analysis—allegedly plagiarized material intact[54]—was published by Said et al. in the journal *Computational Statistics and Data Analysis* in July 2007, a year after the WR report was published.[55] The Said et al. paper—essentially a *post hoc* attempt to justify the WR's social network attacks against me[56]—purported to compare and contrast my supposed "entrepreneurial" style of authorship[57] with Wegman's "mentor" style of authorship.[58] The ostensible purpose was to indicate that Wegman's pattern of publication was somehow less susceptible to peer review bias than that of paleoclimatologists such as me, with our supposed "entrepreneurial" approach. According to the authors, "the mentor style of co-authorship . . . does suggest that younger co-authors are generally not editors or associate editors. . . . they are not in a position to become referees, so that the possibility of bias is much reduced." The assertion could not have dripped with more irony: Wegman was on the advisory board of the journal publishing this very paper, and Said—his graduate student—had somehow been appointed an associate editor of the journal prior to having earned a Ph.D!

To defend their conclusions, the authors made a quite misleading comparison between one individual (me) who was at the time only a junior faculty member and who had graduated very few students (thus forced by definition into the "entrepreneurial" classification), and a senior academic (Wegman) who had graduated many students over the years (thus forced by definition into the "mentor" classification). Even worse, Wegman and company had generated their clique diagrams differently for me and for Wegman, leaving out in Wegman's case the diagonal elements of the matrix, making it look, as one commenter put it, less "cliquey."[59]

Summarizing the problems with the WR social network analysis, Deep Climate noted: "[E]ven a cursory examination of the social network material betrays a shocking lack of understanding of social network analysis, accompanied by a complete failure to tie the background material to any meaningful analysis."[60] Reporting on the matter in *USA Today*, Dan Vergano noted that the editor of the journal—a close friend of Wegman's—had accepted the paper within five days of submission, suggesting there was no peer review of the paper at all.[61] Vergano also interviewed an independent expert in the field of social network analysis, who dismissed the article as an "opinion piece," offering the further critique: "The authors speculate that the entrepreneurial style leads to peer review abuse. No data is provided to support this argument."[62] In response to the overwhelming evidence that the article suffered not only from plagiarism, but also from shoddy scholarship that appears to have slipped through the review process by virtue of questionable editorial oversight, the publisher (Elsevier) officially retracted the article in May 2011.[63]

Let us consider the WR's attempt at supporting the claim that the hockey stick was an artifact of the statistical conventions used. Evidence now brought to light suggested that Wegman and coauthors, rather than having independently assessed the underlying statistical issues, had simply repackaged materials of Stephen McIntyre's. Not only had there apparently been[64] substantial undisclosed collaboration between the WR authors and Stephen McIntyre, as hinted at earlier[65]—something Wegman had denied in his testimony under oath in Congress[66]—but Wegman also appeared to have presented McIntyre's work as his own. Deep Climate showed that Wegman's support for the McIntyre claim that the MBH PCA convention "mined" simple red noise for "hockey sticks" (see chapter 9) rested on a kind of "bait-and-switch" McIntyre himself had introduced.[67]

It all had to do with an unsourced figure shown in the WR.[68] Deep Climate was able to trace the figure back to McIntyre. The results shown were not based on the appropriate model of standard red noise as Wegman had claimed. Rather, they were based on McIntyre's so-called "trendless persistent noise,"[69] which Wahl and Ammann had already demonstrated to be entirely inappropriate. The results appeared to be taken from an archive produced by McIntyre himself.[70]

Ironically, if there had been any mining for a particular result (an allegation that had often falsely been made against us), it was Wegman and McIntyre who had done it. While the WR figure purported to show twelve series randomly selected from twelve thousand realizations of simple red noise, what it in fact showed was a massive cherry-pick: twelve series selected from a ranking of the hundred most hockey stick–like series generated from twelve thousand realizations of McIntyre's unrealistic noise model.[71] As Mashey put it, it was "like declaring the average male height to be 6'6", without bothering to mention the sample was taken on an NBA basketball court."[72] Wegman apparently had simply repackaged McIntyre's flawed work as his own, while excluding any discussion of published refutations of the work. According to Mashey, "the legal term is culpable ignorance."[73]

In summary, then, the supposed independent review by Wegman et al. turned out to be a partisan hatchet job from the start.[74] Wegman was handpicked by Republican party operatives working for Joe Barton. Barton and his staff rejected the National Academy of Science's offer of an impartial review so they could manufacture a report whose content they could control. Wegman had accepted a Faustian bargain when he agreed to author the report.

In April 2010, the office of vice president for research at George Mason University initiated a formal investigation into allegations of misconduct.[75] As of November 2011, no official finding had been issued. Just days before Thanksgiving 2010, however, the scandal finally broke nationally when news of the plagiarism charges received prominent coverage in a feature article in *USA Today*.[76] The news quickly spread as other high-profile media outlets covered it.[77] Part of me genuinely felt bad for Wegman. Precisely one year earlier, I'd endured a Thanksgiving from Hell dealing with the aftermath of the hacked CRU e-mails. I knew as well as anyone what sort of Thanksgiving Wegman was in store for this year.

Wegman appears to have misled Congress in support of an agenda of policy inaction.[78] Where were the calls for investigations by those who'd been so quick to demand them over the faux climategate scandal? Here was a real scandal involving the use of taxpayer funds to support agenda-driven pseudoscience at a public institution in Virginia. Why no interest on Cuccinelli's part? The silence was deafening. In any case, the scandal was not going away any time soon[79]—something most inconvenient for climate change deniers[80] at a time they'd been looking to press their own attacks.

A Turn in the Tide

The climategate and Cuccinelli affairs might have had the desired short-term effect of generating further controversy over climate change. Both appear to have been long-term tactical errors by the climate change de-nial machine, however. Relying on stolen e-mails and the questionable use of political office to achieve their ends, these twin assaults were such an atrocity that they'd finally, to quote one colleague, awakened a "sleeping bear." No longer would scientists "stand by watching one of their own being attacked."[81] Perhaps in part recognizing that they were potentially vulnerable themselves, but—more important, I suspect—sensing the responsibility of their role in what was emerging as a monu-mental battle in which the stakes—the health of our planet—could not be higher, my fellow scientists would no longer sit silently by as the Seren-geti strategy of picking off scientists one by one was deployed against their colleagues.

I first sensed a shift in the wind from the outpouring of support I re-ceived from colleagues, fellow scientists, and academics—many of whom were pillars of the academic and scientific communities—following the various exonerations over the East Anglia e-mail hacking affair. There was the supportive phone call I received from Penn State President Gra-ham Spanier, and the kind notes of support from, among others, former *Science* editor in chief and Stanford University President Donald Ken-nedy; science celebrity Bill Nye "The Science Guy"; Paul Ehrlich, a per-sonal hero; and Steven Soter, the astrophysicist who cowrote the classic PBS series *Cosmos* with Carl Sagan. More than balancing the occasional coordinated bursts of hate mail were the thoughtful and supportive messages from members of the public of all walks of life: Penn State alumni, friends from my distant past, colleagues, ex-colleagues, and other citizens from all over the country who provided moral support.

Some in the scientific community had, of course, shown real signs of fight in the past. There were, for example, the numerous scientists and groups that came out to support my colleagues and me during the height of the 2005 Joe Barton attacks and that had spoken out earlier in defense of James Hansen, Ben Santer, and others subjected to similar attacks in the past. But now the community was showing a new willingness to engage in full-out battle against the attackers of our science. Scientific institutions such as AAAS, the American Meteorological Society, and *Nature* took uncharacteristically aggressive stands against the attacks, alongside various nongovernmental organizations like the Union of Concerned Scientists, the Natural Resources Defense Council, and Pennsylvania groups such as the Clean Air Council and Penn-Future. These various organizations rallied to defend not just me or a handful of other climate scientists, but the very endeavor of science itself, which they saw as under attack. Leading scientific figures such as Gerald North, Donald Kennedy, and many others took action by writing open letters and op-eds, organizing and signing petitions, and encouraging scientific and professional organizations to take principled stands against the attacks.

Perhaps the most dramatic move occurred in May 2010, when a group of more than 250 members of the National Academy of Sciences, including 11 Nobel laureates, banded together to publish an open letter in *Science*.[82] The letter began with a note of concern over the increasingly hostile attacks against climate science, stating that the group was "deeply disturbed by the recent escalation of political assaults on scientists in general and on climate scientists in particular." It went on to stress that the science of human-caused climate change—like all science—can never be "certain"; the demand for absolute certainty before taking action is not only illogical, but that "for a problem as potentially catastrophic as climate change, taking no action poses a dangerous risk for our planet." The letter blasted the tactics of the climate change denial movement, employing surprisingly frank language for a consensus document of hundreds of scientists: "We . . . call for an end to McCarthy-like threats of criminal prosecution against our colleagues based on innuendo and guilt by association, the harassment of scientists by politicians seeking distractions to avoid taking action, and the outright lies being spread about them. Society has two choices: We can ignore the science and hide our heads in the sand and hope we are lucky, or we can act in the public interest to reduce the threat of global climate change quickly and substantively."

The open letter could not have been better timed from my perspective; it happened to appear the very week that Ken Cuccinelli had launched his attack against me and the University of Virginia. It also roughly coincided in timing with an *ABC Evening News* piece on the harassment of climate scientists. Meanwhile, Congressman Edwin Markey (D-MA) had just held two hearings of the House Subcommittee on Climate and Energy focused, in large part, on the harassment and vilification of climate scientists. It was a convergence of independent events that collectively helped to reshape the popular narrative at the time. Rather than talking about the irrelevant content of purloined e-mails, or minor errors in the IPCC reports, the discussion was at least now focused on the McCarthyist attacks against climate scientists, if still not so much on the dangers of unchecked greenhouse gas emissions. Our detractors had helped accomplish something that nobody else had yet been able to: shifting the debate away from climategate and its aftermath. Indeed, this may help to explain why many climate change deniers were distancing themselves from Cuccinelli's inquisition and, in some cases, even denouncing it.[83]

The Cuccinelli attack in particular seemed to have helped mobilize the scientific community. The Union of Concerned Scientists, on May 7, 2010, issued a warning that Cuccinelli's "seemingly unprecedented action could set a dangerous precedent and stymie communication among scientists in many disciplines, preventing them from doing their best work."[84] *Nature* issued a bluntly worded editorial[85] stating that "Given the lack of any evidence of wrongdoing, it's "hard to see Cuccinelli's subpoena—and similar threats of legal action against climate scientists . . . as anything more than an ideologically motivated inquisition that harasses and intimidates climate scientists."[86]

Official denouncements of Cuccinelli's investigation were issued jointly by the American Meteorological Society and the University Corporation for Atmospheric Research[87] and by the American Association for the Advancement of Science.[88] And more than nine hundred Virginia scientists and academics signed a petition demanding that Cuccinelli withdraw his CID, noting that "The request is unfounded and could undermine the effectiveness of not only climate scientists but also thousands of other Virginia researchers."[89] Shortly thereafter, on May 27, the University of Virginia filed papers in court challenging Cuccinelli's CID. Its statement of justification[90] was both compelling and eloquent, invoking the name of the university's founder, Thomas Jefferson (incidentally, one of the first to collect climate observations in

America[91]). It stressed Jeffersonian principles of the "illimitable freedom of the human mind" and the "tradition of limited government framed by enumerated powers, which Jefferson ardently believed was necessary for a civil society to endure," and suggested that Cuccinelli's actions "threaten these bedrock principles."

In fall 2010, as Republicans vied to take back majority control of the U.S. House of Representatives, there were renewed threats. Several of the usual suspects, including Darrell Issa and James Sensenbrenner, promised they would once again pursue inquisitions against climate scientists were Republicans to regain control of the House.[92] The *Washington Post* granted me an op-ed in which I urged my colleagues to "be ready to stand up to the blatant abuse of politicians who seek to mislead and distract the public."[93] Other voices soon answered the call. Sherwood Boehlert in a subsequent *Washington Post* op-ed implored his "fellow Republicans to open their minds," explaining, "I can understand arguments over proposed policy approaches . . . what I find incomprehensible is the dogged determination by some to discredit distinguished scientists and their findings."[94] Later that month,[95] a group of scientists announced a Climate Change Rapid Response Team to combat climate change misinformation and disinformation, through a network of more than a hundred experts who would respond quickly to questions from journalists.[96] Similar efforts by other scientific groups were soon underway.[97]

Following the election, congressional deniers seemed surprisingly reticent to follow through with their campaign threats, despite having seized control of the House of Representatives. They collectively backed away from holding hostile show trials, diverting their attention instead to related issues, such as the supposed job losses that would arise from the EPA's endangerment finding.[98] I like to think that their hesitance was a result of the degree of mobilization demonstrated by the scientific community in recent months, and the worry they'd stir up a hornet's nest if they actually pursued McCarthyist attacks against scientists. But the overriding consideration, I suspect, was that fossil fuel special interests were actually quite happy with the current status quo. Thanks to an assist from a lagging economy, environmental matters had been pushed way down in the hierarchy of public concern. Any meaningful climate legislation had almost certainly been scuttled for the foreseeable future under partial Republican control of the government.

It simply wasn't worth risky McCarthyesque hearings that held out the distinct threat of backfiring.[99]

The years ahead will still present a great challenge for climate scientists—and, indeed, all of us whose future is threatened by the prospect of powerful vested interests dictating environmental policy. With key committees of the House now controlled by politicians with close ties to special interests such as Koch Industries,[100] perhaps for some time to come, attacks against the science of climate change are certain to continue.[101] While Cuccinelli was dealt a major setback in late August 2010 by a Virginia judge who found that the attorney general had not managed in a forty-page filing with the court to even explain "the nature of the conduct" supposedly constituting fraud[102]— and though the *Washington Post* has called his rationale for continuing "so tenuous as to leave only one plausible explanation: that he is on a fishing expedition designed to intimidate and suppress honest research . . . because he does not like what science says about climate change"—as of November 2011 he still hadn't given up his crusade.[103] Fossil fuel interests are working hard to assist him.[104]

Meanwhile, climate change contrarians continue to launch hand grenades,[105] the denial machine persists in churning out disinformation,[106] and the attacks against climate scientists[107]—myself included[108]— continue. Yet something is different now. The forces of climate change denial *have*, I believe, awakened a "sleeping bear." My fellow scientists will be fighting back, and I look forward to joining them in this battle.

Epilogue

All that is necessary for the triumph of evil is that good
men [and women] do nothing.

—Edmund Burke (attributed)

We look back now with revulsion at the corporate CEOs, representatives, lobbyists, and scientists-for-hire who knowingly ensured the suffering and mortality of millions by hiding their knowledge of tobacco smoking's ill effects for the sake of short-term corporate profits. Will we hold those who have funded or otherwise participated in the fraudulent denial of climate change similarly accountable—those individuals and groups who both made and took corporate payoffs for knowingly lying about the threat climate change posed to humanity, those who willfully have led the public and policy makers astray, and those politicians and media figures who have sought to intimidate climate scientists using McCarthyite tactics? Though the impact is more subtle and difficult to measure, their recalcitrance may end up costing more lives than cigarette smoking ever has. And, unlike the case of tobacco and cigarette smoking, many of those who will suffer the worst impacts of climate change will have played little or no role in creating the problem.

The decades of delay in reducing carbon emissions have already incurred a very real cost to humanity and our environment. Each year that emissions reductions are delayed, it becomes increasingly difficult to stabilize CO_2 concentrations below safe levels. This is the so-called "procrastination penalty" of delayed action. Simply from our emissions thus far, there is likely at least an additional 0.6°C (1°F) warming of Earth's surface and at least a foot or so of sea level rise already in the pipeline for the next century or so. That means that we have already, in all likelihood, ensured the obliteration of some low-lying island nations and the extinction of many animal species, among other impacts.

While the professional climate change denial machine is not the only reason for the delay in action, it is a major contributor. That delay has already committed many to irreparable harm. Some critics have gone so far as to pronounce those individuals behind the corporate-funded denial machine guilty of "crimes against humanity" because of the devastating effects on humanity to which we are already committed, especially among the world's poor and disadvantaged.[1]

As of late 2011, CO_2 concentrations are at roughly 392 ppm.[2] A year and a half earlier, when I began writing this book, they were at 388 ppm. When we reach concentrations of 450 ppm (about 2030, extrapolating from current trends), we will likely have locked in at least 2°C (3.5°F) warming of the climate relative to preindustrial levels, a level experts generally agree constitutes dangerous human interference with the climate system, far beyond what we have thus far witnessed. The likely impacts will be devastating sea level rise, more powerful hurricanes, more widespread drought, and increased weather extremes, with adverse impacts on human life and health, animal species, and our environment. To avoid reaching that level, we would need, as discussed in chapter 2, to bring our annual emissions to a peak in less than a decade and reduce them by nearly 80 percent (relative to 1990 levels) by mid-century.

Will human civilization have the will to confront this challenge before it's too late? Prominent commentators such as Jared Diamond have pondered the question. In *Collapse*, Diamond points to past examples of both success and failure when societies and cultures have been similarly challenged, whether by nature or by the consequences of their own behavior.[3] The demise of Easter Island's society, in Diamond's view, is a cautionary tale of how environmental exploitation can bring upon the downfall of an entire civilization. There, the inhabitants exploited the island's natural resources to the point of uninhabitability. That, of course, was a small island; the footprint of its "commons" was tangibly limited. The problem is less intuitive for global environmental problems, as the global commons illusively appears infinite. But as is increasingly being recognized, it is not. Not only are our land and seas finite, but so too is our atmosphere, and our practices are changing its composition detectably and profoundly. Ending those practices will require extraordinary cooperation across the governments and peoples of the world. It won't be easy. But like Diamond, I'm cautiously optimistic that we will meet the challenge.

We can look to history for cautionary tales, but can we look to it for hope? When lakes in eastern North America were being destroyed by acid rain in the 1970s, we as a people acted. Americans recognized the problem and passed legislation, the Clean Air Act, to scrub the culprit—sulfate emissions—from smoke stacks. In the 1980s, scientists recognized that the ozone-depleting properties of chlorofluorocarbons (CFCs), used at the time in spray cans, were responsible for the growing ozone hole, with its threat of increased skin cancer and other adverse impacts on us and other living things. Disregarding the naysayers,[4] countries around the world signed the Montreal Protocol, eventually banning the production of CFCs.

By the standard of what faces us now, this was a simpler problem to solve; only a handful of major industrialized nations were contributing to the problem, and cost-effective industrial alternatives were reasonably available, making concerted action less contentious. Carbon emissions, by contrast, are fundamental to the prevailing world economy. Ending our addiction to carbon-based power requires a fundamental revision of our energy infrastructure and a substantial shift from our current lifestyle. Climate change requires solutions of a far greater scope than other global environmental challenges we have faced, but successful past efforts to confront acid rain and ozone depletion suggest we can meet this greater challenge.

The failure to reach a binding international agreement in Copenhagen in December 2009 was a cause for pessimism among those concerned about the climate change threat. The forces of inaction gained the upper hand in part by engaging in a successful smear campaign, using the University of East Anglia hacking to attempt to call the fundamental science of climate change into question at the most opportune time. That the contrarian barrage had made serious inroads into the public's concern over climate change seemed tacitly acknowledged by President Barack Obama in his State of the Union address on January 21, 2010: "I know that there are those who disagree. . . . But even if you doubt the evidence, providing incentives for energy efficiency and clean energy are the right thing to do for our future." How far we had regressed in the ten years since President Bill Clinton's 2000 State of the

Union address, in which he issued a call to arms in confronting the climate change threat *based* on the strength of the science, not *in spite of* a conceded potential weakness in it.

In reality, it is unlikely that the climategate attacks were a primary factor behind the limited progress in Copenhagen. Most policy experts have acknowledged that the main problem remained the nagging political complications and competing economic interests of different nations. Chief among the challenges is getting developing and developed nations to agree on basic principles of how carbon credits should be allocated and to what extent contributions to global carbon emissions should be measured on a net or per capita basis (the former penalizes countries such as China far more, and countries like the United States far less, than does the latter), and how to factor in historical versus current and future emissions (countries like the United States have benefited economically from more than a century of past fossil fuel burning). These political and economic obstacles are considerable at present, but they are not insurmountable. They can, and hopefully will, be overcome through diplomacy and negotiation.

The legacy of the manufactured climategate scandal may in fact be the opposite of what climate change deniers had hoped. While the campaign did have the immediate impact of casting doubt over climate science, it also marked a critical juncture, and indeed potentially a turning point, in the climate change debate.[5] Perhaps "climategate" was the moment when the climate change denial movement conceded the legitimate debate, choosing instead to double down on smear and disinformation, a tacit acceptance that an honest, science-based case for denying the reality of human-caused climate change and the threat it presents could no longer be made. Maybe it was the moment when the seamy underbelly of the climate change denial movement became exposed for all to see. Some of the once stealth funding by Koch Industries and other funders of climate change denial was suddenly out in the open, thanks to the renewed scrutiny that climate change denial was now receiving. The attacks against climate scientists by politicians like Senator James Inhofe and Virginia Attorney General Ken Cuccinelli were now being identified by prominent media outlets for the witch hunts they were. Finally, I believe that the climategate attacks represented a turning point for my fellow climate scientist colleagues and how they viewed their role in the public debate. These latest attacks will fade from memory, and new ones will undoubtedly be launched to take their place. But I suspect that the change in heart among climate scientists regarding their role in the debate will be enduring.

The Role of the Scientist

What is the proper role for scientists in the societal discourse surrounding climate change? Should they remain ensconced in their labs, with their heads buried in their laptops? Or should they engage in vigorous efforts to communicate their findings and speak out about the implications? I once subscribed to the former point of view. As a graduate student and then a beginning postdoctoral researcher in the mid-1990s, I wanted nothing more than to be left alone analyzing data, constructing and running theoretical climate models, and pursuing curiosity-driven science. When we first published our hockey stick work in the late 1990s, I was of the belief that the role of a scientist was, simply put, to do science. Others, I felt, should be left to assess and publicize any implications of the science. Taking anything even remotely resembling a position regarding climate change policy was, to me, anathema. Doing so, I felt, would compromise the authority of my science. I felt that scientists should take an entirely dispassionate view when discussing matters of science—that we should do our best to divorce ourselves from all of our typically human inclinations—emotion, empathy, concern. In the interviews I conducted with reporters, I was careful not to wade into the dangerous waters of expressing a personal opinion and to avoid entirely the subject of policy implications.[6]

Everything I have experienced since then has gradually convinced me that my former viewpoint was misguided. I became a public figure involuntarily when our work was thrust into the public spotlight in the late 1990s. I have remained a public figure since, but I have come to embrace, rather than eschew, that role. Despite the battle scars I've suffered from having served on the front lines in the climate wars—and they are numerous—I remain convinced that there is nothing more noble than striving to communicate, in terms that are simultaneously accurate and accessible, the societal implications of our scientific knowledge. Indeed, much of my time and effort over the past decade has been dedicated to doing so.

I can continue to live with the cynical assaults against my integrity and character by the corporate-funded denial machine. What I could not live with is knowing that I stood by silently as my fellow human beings, confused and misled by industry-funded propaganda, were unwittingly led down a tragic path that would mortgage future generations. How could we explain to our grandchildren that we saw the threat coming, but did not do all we could to ensure that humankind took the necessary precautions? Scientists who study climate change and its potential impacts understand better than anyone the nature of the climate change threat. It would, in my view, be irresponsible for us to silently stand by while industry-funded climate change deniers succeed in confusing and distracting the public and dissuading our policy makers from taking appropriate actions. If climategate and the other related attacks against climate science have served no purpose other than to awaken the scientific community to the reality that we are in a war and to move some of my colleagues off the fence, then they will have served a purpose.

The challenge climate scientists face remains monumental. Scientists understand the processes that lead to scientific consensus because these processes are intrinsic to the culture of science. They involve a good faith give-and-take of scientific ideas, through publication in the peer reviewed literature, the exchanges that take place at scientific meetings, and the participation in scientific assessments that attempt to characterize our collective knowledge. The processes that lead to a public consensus, however, are different, and by contrast are generally foreign to most scientists.

Scientific truth alone is not enough to carry the day in the court of public opinion. The effectiveness of one's messaging and the resources available to support and amplify it play a far greater, perhaps even dominant role. And here, as we have seen, the scientific community and those seeking to communicate its message are greatly outmatched by a massive disinformation campaign funded by powerful vested interests driven by a single goal. That goal is to thwart efforts to regulate carbon emissions—a necessary step if we are to stabilize greenhouse gas concentrations below dangerous levels.

Foes of emissions limits have been able to exploit deficiencies in the public's understanding of the problem: the reality of the problem, the risks it presents, an honest accounting of the economic and environmental costs, and the ethical quandaries of inaction. These deficiencies are inextricably linked to the willingness—indeed, the inclination—of our mass media to present the views of a small number of mavericks, iconoclasts, and even crackpots as being on a par with those of the world's leading scientists. While there are mavericks in every field, only in areas like climate science—where they are funded heavily and there are both personal and economic interests in denying mainstream science—are plain-old charlatans so readily presented as credible voices in the public debate, used to advance the policy agenda of inaction favored by powerful economic interests and the politicians in their pocket. As atmospheric scientist Kerry Emanuel eloquently put it during testimony he gave at a March 2011 congressional hearing, "mavericks are indispensible to the progress of science, but politicians who make mascots out of mavericks are invariably engaging in advocacy."[7]

Looking Ahead

In the end, what are the larger lessons to be taken away from my story? One lesson, I suppose, is a familiar one: Change is not easy. Rightly or wrongly, the hockey stick was a game-changer in the climate change debate. It was a powerful, easily digestible icon, enough so that the forces of climate change denial saw the need to employ their full arsenal of resources in an attempt to destroy it—and, indeed, my collaborators and me. They supported scientific hired guns to assail us with a constant barrage of specious criticisms and attacks. They found ideologically sympathetic politicians, including powerful Senate and House committee chairs and state attorneys general, who were willing to sponsor inquisitions of climate science and climate scientists. They used stolen private e-mails to find material to misrepresent and smear us and other climate scientists. And they made use of their access to an array of media outlets to issue attack after vitriolic attack against us, the hockey stick, and the work of other scientists around the world. The assaults were in a sense the logical continuation of corporate-funded campaigns to attack inconvenient scientific findings that had begun decades ago.

There are other sobering lessons that might be taken away as well. It is clear that the scientific community is, at present, ill-equipped to deal with direct assaults upon its integrity. A fundamental principle of scientific inquiry is the honest exchange of ideas, the communication of caveats and uncertainty. Without a science-literate and politically aware populace, there can be no match against well-funded, well-organized groups that place little value on honesty or integrity, that cleverly masquerade denialism as skepticism, and that are more than willing to state their own positions in the most absolute of terms, while exploiting and indeed misrepresenting the frank admissions of uncertainty by those they view as their opponents.

As *Nature* put it in editorial in the wake of the CRU e-mail hack in March 2010, climate scientists must acknowledge that "they are in a street fight."[8] *Nature* went on to assess the predicament:

> Climate scientists are on the defensive, knocked off balance by a re-energized community of global-warming deniers who, by dominating the media agenda, are sowing doubts about the fundamental science. Most researchers find themselves completely out of their league in this kind of battle because it's only superficially about the science. The real goal is to stoke the angry fires of talk radio, cable news, the blogosphere and the like, all of which feed off of contrarian story lines and seldom make the time to assess facts and weigh evidence. Civility, honesty, fact and perspective are irrelevant.

All is not lost, however. When scientists are willing to fight for their cause—in this case, communicating the potential climate change threat—there are many good men and women who will not simply stand by and do nothing. There were politicians of principle on both sides of the aisle, such as Democrat Henry Waxman and Republican Sherwood Boehlert, who were willing to stand up to McCarthyist attacks against me and my coauthors. There were reporters and editors who displayed thoughtfulness and sanity in their coverage of events even during the height of the media feeding frenzy. Dan Vergano and Brian Winter of *USA Today*, Seth Borenstein of the AP, and the BBC's Richard Black stand out as shining examples, as do the editorial boards of the *Washington Post*, the *New York Times*, and numerous smaller newspapers around the country. Nongovernmental organizations such as the Union of Concerned Scientists, the Natural Resources Defense Council, and the oft-criticized but ever-gutsy Greenpeace all rose to the occasion. Many fellow scientists, academics, and other citizens did as well.

Yes, the public discourse has been polluted now for decades by corporate-funded disinformation—not just with climate change, but with a host of health, environmental, and societal threats. Public perception is fickle. Things could change quickly with a concerted effort to improve the public's understanding of climate change risks and what's likely to be lost by not addressing them. Such an effort would of necessity require scientists to work closely with social scientists and with public policy and communication experts. It would require the financial support of foundations and private sector interests that genuinely care about Earth and its future. It would need to take full advantage of Internet-age organizing opportunities, using social media and online networking tools, to build a true grassroots movement that can go toe-to-toe with the massively-funded "Astroturf" campaigns. And it would certainly need to work toward a dramatic improvement in the accuracy and objectivity of mainstream media coverage of the climate change issue.[9] Only an informed electorate can hold our policy makers accountable to represent our interests and values and insist on the development of a sensible climate change strategy. Fortunately, there is evidence that some prominent media outlets are awakening to this reality and are willing to do their part.[10]

So there is reason for hope. I will end with a personal story that conveys my cautious optimism. At the height of the climategate assault, just after New Year's Day 2010, I was on a brief family vacation in the Florida Keys, a good opportunity, I felt, to get away from it all, at least for a short while. It was just my luck that Key West was experiencing its most severe cold snap in decades. I was subjected to more than one local resident who, when informed that I was an atmospheric scientist, skeptically inquired something akin to, "So what about this *supposed* global warming?" I was happy to explain the difference between weather and climate, the fact that El Niño years tend to be cold in the southeastern United States, that other regions such as the tropics, the Arctic, and the Pacific Northwest were unusually warm, that 2009 had just gone down as one of the warmest years globally, and that the decade that had just ended was the warmest we had yet witnessed. Nonetheless, it was all a bit dispiriting, as if Mother Nature herself had decided to kick me while I was down.

But there was something more here. My four-year-old daughter was entranced by the Keys—the mangrove forests, the sonorous birds, the leaping dolphins, the coral reefs with their exotic and colorful fish. It was unlike anything she had ever seen. In fact, three generations of my family—my parents, my wife and me, and our daughter—were all sharing this mutual opportunity to enjoy one of Earth's true wonders—and authentic key lime pie. I didn't have the heart to tell our daughter that this island paradise was under assault—by us. That the warming and increasingly acidic ocean was slowly killing the reefs, that increasingly destructive hurricanes would subject them to further insult, and that projected sea level rise over the next few decades under "business as usual" emissions would literally submerge vast regions of the Florida Keys, including the wildlife refuges home to so many of its unique species. Nor do I have the heart to tell her now that the majestic scene of giraffes and elephants looming in the foreground of Ernest Hemingway's *Snows of Kilimanjaro* may soon become a casualty of our warming of the planet.

I am determined to do whatever I can to make sure that it will be possible for us to return decades from now—my wife and me, our daughter, her children, and perhaps theirs—to again marvel at these natural wonders. While slowly slipping away, that future is still within the realm of possibility. It is a matter of what path we choose to follow. I hope that my fellow scientists—and concerned individuals everywhere—will join me in the effort to make sure we follow the right one.

Glossary

Aerosols: Aerosols are small particles suspended in the atmosphere that can reflect and/or absorb various types of radiation, like dust or soot from smoke. Natural sulfate aerosol particles are injected into the stratosphere by explosive volcanic eruptions and may reside there for several years. Their primary effect is to cool Earth's surface by reflecting away incoming solar radiation that would otherwise be available to warm Earth's surface. Human industrial activity has also led to the production of sulfate and other aerosol particles that reside for a short period of time in industrial and downwind regions. This human aerosol production has led to a cooling effect in some regions, particularly in the Northern Hemisphere, which has offset some of the human-caused greenhouse warming.

Atlantic Multidecadal Oscillation (AMO): The AMO is a putative multidecadal oscillation of the climate system with a periodicity between fifty and eighty years. It is associated with changes in surface temperatures, wind patterns, and ocean circulation in the North Atlantic Ocean, with an apparent influence on North American and European climate, and potentially other phenomena such as Atlantic hurricanes and drought in sub-Saharan Africa. Substantial uncertainty remains over the precise character of the AMO and its wider influences.

Climate Proxy: A climate proxy is a natural archive of information, such as the growth history recorded by tree rings or the isotopic composition of ice in an ice core, that reflects at least partially some aspect of climate. Such a record can thus serve as a proxy for key climate variables (e.g., temperature or rainfall) prior to the modern instrumental period.

Divergence Problem: The divergence problem applies primarily to tree ring density data (in contrast to ring width data), especially those from high northern latitudes. These particular tree ring data suffer from an enigmatic decline in their response to warming temperatures after about 1960. Various explanations for the phenomenon have been offered, including the possible impacts of pollution in the latter part of the twentieth century. Due to this problem, scientists who work this precise type of tree ring data have advised against using them as proxies for temperature changes after about 1960. This problem is observed to a much lesser extent with tree ring width data and does not apply to other types of proxy data, such as from corals, ice cores, and sediment records.

El Niño: El Niño is a well-known natural climate phenomenon associated with a periodic warming of sea surface temperatures in the eastern and central equatorial Pacific Ocean that takes place every few years. This pattern of warming in the equatorial Pacific in turn perturbs worldwide weather patterns. During El Niño events, winter temperatures over much of the northern half of the United States are typically warmer than usual, and rainfall in the desert Southwest of the United States is enhanced. The flip side of El Niño is known as La Niña (see La Niña).

El Niño/Southern Oscillation (ENSO): The southern oscillation is the change in sea level pressure across the equatorial Pacific associated with the alternation between El Niño and La Niña events. While El Niño and La Niña measure the oceanic expression of the phenomenon, the southern oscillation measures the atmospheric changes. The oscillation in the combined oceanic/atmosphere system is known as the El Niño/southern oscillation or ENSO.

Feedback: A feedback, in the context of climate studies, is some response that can act to either enhance or diminish the initial response to an impulse, for example, an increase in greenhouse gas concentrations. Positive (or amplifying) feedbacks lead to more warming than would be expected based on the elevation of greenhouse gases alone in the atmosphere, because that elevation leads to other warming effects such as decreased snow and ice cover (which causes Earth's surface to absorb more sunlight) or increased evaporation of water into the atmosphere (water vapor itself is an important greenhouse gas). Negative (or dampening) feedbacks, by contrast, lead to less warming than would be expected. One possible negative feedback is that a warmer Earth might have more low clouds, which would reflect more of the incoming sunlight out to space.

Greenhouse Effect: The balance between the loss of emitted infrared radiation from Earth's surface and the incoming heating by solar radiation establishes Earth's equilibrium temperature. By absorbing some fraction of the outgoing infrared radiation and sending half of it back down toward Earth's surface, greenhouse gases force Earth's surface to warm. This warming effect on the surface (and lower atmosphere) is known as the greenhouse effect.

Greenhouse Gases: Greenhouse gases are gases, such as water vapor, carbon dioxide, methane, ozone, and various other trace gases in the atmosphere, that absorb infrared radiation such as that emitted from Earth's surface (see Greenhouse Effect).

La Niña: La Niña is essentially the mirror opposite of El Niño. During La Niña events, which occur every few years on average, sea surface temperatures in the eastern and central equatorial Pacific Ocean are cooler than usual. The impacts on worldwide weather patterns are broadly the opposite of those associated with El Niño, e.g., the desert Southwest of the United States is generally dry (see El Niño).

Little Ice Age: The Little Ice Age generally refers to an interval spanning roughly the fifteenth to the nineteenth centuries when evidence from the expansion of mountain glaciers and other data suggests substantially cooler conditions in some regions of the world such as Europe and parts of North America. While temperatures during this time interval may have been as much as 2°C cooler than present in those regions, assessments of more widespread evidence indicate that conditions were not especially cold in many other regions, and that globally temperatures were probably at most about 1°C cooler than present (see Medieval Warm Period).

Medieval Climate Anomaly (MCA): The term *medieval climate anomaly* is a more recently favored alternative to the term *medieval warm period*. This alternative terminology emphasizes that evidence of warmth during the period A.D. 800 to 1400 was highly variable regionally, and not as widespread as originally believed. Indeed, some regions, such as the tropical Pacific, appear to have been unusually cold. Moreover, many of the most significant alterations in climate at this time had little to do with temperature, but instead corresponded to shifts in rainfall and drought patterns (see Medieval Warm Period).

Medieval Warm Period (MWP): The so-called medieval warm period refers to an interval, generally thought to lie somewhere between A.D. 800 and 1400, when conditions in some regions of the Northern Hemisphere (e.g., Europe and Greenland) appear to have been relatively mild. The term was originally coined by British paleoclimatologist Hubert Lamb in characterizing the evidence for a past period of relatively mild weather in Europe. Paleoclimate evidence suggests, however, that other regions, such as the tropical Pacific, may have been relatively cold at this time. Recent evidence suggests that globally, peak warmth during this period was similar to that of the mid-twentieth century, but cooler than present, and that warmth was highly regional in nature (see Medieval Climate Anomaly).

North Atlantic Oscillation (NAO): The NAO measures the surface pressure difference between the subpolar North Atlantic (e.g., Iceland) and the subtropical North Atlantic (e.g., the Azores), which is a measure of the strength and direction taken by the Northern Hemisphere storm track. The positive phase of the NAO is associated with a stronger, more northerly jet stream and relatively warm and wet winters in Europe, relatively dry winters in the Middle East, and relatively warm winters in the eastern United States. The negative phase of the NAO is associated with a more easterly tracking jet stream and roughly the opposite impacts on these regions. The NAO is closely related to another pattern sometimes referred to as the Arctic oscillation (AO).

Principal Component Analysis (PCA): PCA is a statistical method to represent a large dataset in terms of a smaller number of components that explain most of the variation in the data. Typically in atmospheric science and climate studies, PCA expresses the main patterns of variation in the data in terms of a time series characterizing the temporal evolution of the pattern (the principal component or PC), and a spatial map characterizing how the amplitude and sign of the signal varies with region (the empirical orthogonal function or EOF).

Red Noise: A type of random noise considered appropriate to describe natural, internally generated climate variability. Unlike white noise, which is associated with random fluctuations of uniform magnitude over all timescales, red noise exhibits greater amplitude fluctuations on longer timescales. In the climate system, such noise is a result of the way that the sluggish components of the climate system, such as the oceans, respond to the impact of shorter-term weather fluctuations.

Stratosphere: The stratosphere is the region of the atmosphere typically found between a lower altitude of between 8 and 17 km (higher near the equator) and an upper altitude of approximately 50 km. Unlike the lower part of the atmosphere known as the troposphere, temperatures actually warm with increasing altitude within the stratosphere. This makes the stratosphere highly stable in its properties, inhibiting the sorts of vertical motions that are responsible for the weather phenomena of the troposphere. The boundary between the troposphere and stratosphere is known as the tropopause. It is typically at this height that the jet stream currents of the atmosphere are strongest (see Troposphere).

Troposphere: The troposphere is the region of the atmosphere, confined to the lower 8 to 17 km (extending higher near the equator). Within this lowest layer of the atmosphere, temperatures tend to decline with altitude, which leads to a tendency for instability and the familiar phenomenon of weather. This temperature trend reverses once one reaches the next layer up in the atmosphere, known as the stratosphere (see Stratosphere).

Notes

Prologue

1. John M. Broder, "Climate Change Doubt Is Tea Party Article of Faith," *New York Times*, October 20, 2010.

2. This paleothermometer is imprecise, however, as other factors, such as the source region of the moist air that produced the snow, can also have an impact on these ratios.

3. M. E. Mann, R. S. Bradley, and, M. K. Hughes, "Global-Scale Temperature Patterns and Climate Forcing Over the Past Six Centuries," *Nature*, 392 (1998): 779–787.

4. M. E. Mann, R. S. Bradley, and M. K. Hughes, "Northern Hemisphere Temperatures During the Past Millennium: Inferences, Uncertainties, and Limitations," *Geophysical Research Letters*, 26 (1999): 759–762.

5. The colleague was Jerry Mahlman, director of NOAA's Geophysical Fluid Dynamics Laboratory (GFDL) in Princeton, New Jersey.

1. Born in a War

1. The paper, submitted November 7, 1995, was published the following year as M. Mann and J. Park, "Greenhouse Warming and Changes in the Seasonal Cycle of Temperature: Model Versus Observations," *Geophysical Research Letters*, 23 (1966): 1111–1114. It focused on a discrepancy between observations and theoretical climate model predictions—the sort of thing that climate change deniers love to take out of context and hype. The conservative

organization Accuracy in Media took note of the study, citing lack of media coverage of it as some sort of evidence of media bias in coverage of climate change—something that I, to this day, find puzzling as the paper actually dealt with a relatively obscure technical detail of climate models and hardly challenged the mainstream view that human activity was leading to the warming of the globe. An interesting follow-up to the study was published more recently: A. R. Stine et al., "Changes in the Phase of the Annual Cycle of Surface Temperature," *Nature,* 457 (2009): 435–440.

2. M. E. Mann, J. Park, and R. S. Bradley, "Global Interdecadal and Century-Scale Climate Oscillations During the Past Five Centuries," *Nature,* 378 (1995): 266–270.

3. An excellent account of these events can be found in chapter 6 of Naomi Oreskes and Eric Conway, *Merchants of Doubt* (New York, Bloomsbury Press, 2010).

4. P. J. Michaels, P. C. Knappenberger, and D. A. Gay, "General Circulation Models: Testing the Forecast," *Journal of the Franklin Institute,* 331A (1994): 123–133.

5. It was originally called the *World Climate Review* and later changed to the *World Climate Report*; see, e.g., Ross Gelbspan, *The Heat Is On* (New York, Basic Books, 1997), 36–39.

6. In *The Heat Is On,* Gelbspan quotes directly from a Western Fuels Association annual report from the early 1990s: "Western Fuels approached Pat Michaels about writing a quarterly publication designed to provide its readers critical insight concerning the global climatic change and greenhouse effect controversy. . . . Western Fuels agreed to finance publication and distribution of *World Climate Review* magazine" (36).

7. See, e.g., Gelbspan, *The Heat Is On,* 46, citing a segment that ran on ABC News's *Nightline* ("Is Environmental Science for Sale," no. 3329, February 24, 1994), noting that Singer had admitted to receiving "funding from Exxon, Shell, ARCO, Unocal, and Sun Oil." In a separate piece that ran on March 23, 2008 ("Global Warming Denier: Fraud or 'Realist'?"), ABC News noted that Singer "admits he once accepted an unsolicited check from Exxon for $10,000." According to *Mother Jones* magazine, Singer's organization (SEPP) had received money from Philip Morris, Texaco, and Monsanto. Keith Hammond, "Wingnuts in Sheep's Clothing," *Mother Jones,* December 4, 1997.

8. S. Fred Singer, "Climate Change and Consensus," *Science,* 271 (February 2, 1996): 581.

9. See, e.g., Gelbspan, *The Heat Is On,* 55–56.

10. Benjamin D. Santer, Testimony for House Select Committee on Energy Independence and Global Warming, May 20, 2010.

11. While the group no longer lists its funders, in 2000 it acknowledged present or past funding from Exxon as well as the Scaife Foundations. See George C. Marshall Institute, "George C. Marshall Institute: Recent Funders," August 2000, http://web.archive.org/web/20000823170917/www.marshall.org/funders.htm.

12. Frederick Seitz, "A Major Deception on 'Global Warming,'" *Wall Street Journal*, June 12, 1996; Santer's edited response ran as a letter to the editor on June 25, 1996.

13. In reality, the problem is a bit more complicated than described and requires additional programming, since it is possible for the computer to put itself in a position where there are no non-losing moves left.

14. My first scientific paper described this work: M. E. Mann, C. H. Marshall, and A. D. J. Haymet, "Nematic Liquid Crystals: A Monte Carlo Simulation Study in Higher Dimensions," *Molecular Physics*, 66 (1989): 493.

15. This work was described in G. Ceder, M. Asta, W. C. Carter, M. Kraitchman, D. de Fontaine, M. E. Mann, and M. Sluiter, "Phase Diagram and Low-Temperature Behavior of Oxygen Ordering in YBa2Cu3Oz," *Physical Review B*, 41 (1990): 13.

16. Some years ago, MIT hurricane researcher Kerry Emanuel and colleagues suggested the possibility that a warmer climate, by favoring greater tropical cyclone activity, could enhance the mixing of the upper ocean and hence the transport of heat by the ocean from low to high latitudes: R. L. Korty and K. A. Emanuel, "The Dynamic Response of the Winter Stratosphere to an Equable Climate Surface Temperature Gradient," *Journal of Climate*, 20 (2007): 5213–5228. My collaborators and I have recently done research on this problem as well: R. L. Sriver, M. Goes, M. E. Mann, and K. Keller, "Climate Response to Tropical Cyclone-Induced Ocean Mixing in an Earth System Model of Intermediate Complexity," *Journal of Geophysical Research*, 115, C10042.

2. Climate Science Comes of Age

1. See James Hansen et al., "Global Temperature Change," *Proceedings of the National Academy of Sciences*, 103 (2006): 14288–14293.

2. See James Hansen et al., "Potential Climate Impact of the Mount Pinatubo Eruption," *Geophysical Research Letters*, 19 (1992): 215–218.

3. The irregularity of the surface pattern arises primarily from amplified warming in the Arctic due to disappearing ice and snow (which allows the surface to absorb more incoming sunlight and thus warm more) and the fact

that continents tend to warm faster than oceans. The vertical pattern of a warming lower atmosphere (troposphere) and cooling upper atmosphere (stratosphere) is consistent with the effect of increased greenhouse gas concentrations, but inconsistent with natural causes of surface warming.

4. This warming marker was originally chosen by the European Union in 1996 as constituting the level of "dangerous anthropogenic interference" with the climate; see, e.g., M. E. Mann, "Defining Dangerous Anthropogenic Interference," *Proceedings of the National Academy of Sciences*, 106 (2009): 4065–4066.

5. Because of the slow uptake of CO_2 by long-term geological processes, emissions need not quite reach zero but must instead stay below a rate of about 0.2 billion tons of carbon per year, amounting to roughly 2 percent of current emissions.

6. James Hansen, "Is There Still Time to Avoid 'Dangerous Anthropogenic Interference' with Global Climate? The Importance of the Work of Charles David Keeling," lecture at the American Geophysical Union meeting, December 6, 2005, San Francisco, California.

7. See, e.g., Bill Blakemore, "Micronesia: A Third Kind of Nation, Written Off? Island Nations Slowly Drowning; Ambassador Says World Has 'Written Us Off,'" *ABC Evening News*, December 9, 2009.

8. In 2011, a team of scientists and I published a paper in the *Proceedings of the National Academy of Sciences* that used paleodata to better constrain the parameters used in empirical models of future sea level rise: A. C. Kemp, B. P. Horton, J. P. Donnelly, M. E. Mann, M. Vermeer, and S. Rahmstorf, "Climate Related Sea-Level Variations Over the Past Two Millennia," *Proceedings of the National Academy of Sciences* (early edition), doi:10.1073/pnas.1015619108, 2011. Our findings supported previous work suggesting the possibility of as much as two meters of sea level rise by 2100, given business as usual carbon emissions.

9. See, e.g., J. S. Smith et al., "Assessing Dangerous Climate Change Through an Update of the Intergovernmental Panel on Climate Change (IPCC) 'Reasons for Concern,'" *Proceedings of the National Academy of Sciences*, 106 (2009): 4133–4137.

10. Hansen, "Is There Still Time to Avoid 'Dangerous Anthropogenic Interference' with Global Climate?"

11. Bill McKibben, *Eaarth: Making a Life on a Tough New Planet* (New York, Times Books, 2010).

12. Schneider used this analogy as early as 1989. He was quoted in a written statement for his testimony to a hearing of the House Energy and Commerce Committee: "I am a believer in insurance against potentially

catastrophic loss at both personal and national levels"; see, e.g., "Experts Blow Hot and Cold on Severity of Global Warming," *Pittsburgh Press*, February 21, 1989. I can't recall for certain, but I believe that the article I had read was S. H. Schneider, "The Greenhouse Effect: Science and Policy," *Science*, 243 (1989): 771–781.

13. The memo, "The Environment: A Cleaner, Safer, Healthier America," was leaked by the Environmental Working Group, a nongovernmental organization.

14. See, e.g., R. W. Spencer and J. R. Christy, "Precision and Radiosonde Validation of Satellite Gridpoint Temperature Anomalies, Part I: MSU Channel 2," *Journal of Climate*, 5 (1992): 847–857; R. W. Spencer and J. R. Christy, "Precision and Radiosonde Validation of Satellite Gridpoint Temperature Anomalies, Part II: A Tropospheric Retrieval and Trends 1979–90," *Journal of Climate*, 5 (1992): 858–866.

15. See, e.g., C. Prabhakara, R. Iacovassi Jr., and J.-M. Yoo, "Global Warming Deduced from MSU," *Geophysical Research Letters*, 25 (1998): 1927–1930.

16. See Roy Spencer, "Increasing Atmospheric CO_2: Manmade . . . or Natural," www.drroyspencer.com/2009/01/increasing-atmospheric-co2-manmade%E2%80%A6or-natural/.

17. David Perlman, "Earth Warming at Faster Pace, Say Top Science Group's Leaders," *San Francisco Chronicle*, December 18, 2003, A-6.

18. S. Fred Singer and Dennis T. Avery, *Unstoppable Global Warming: Every 1,500 Years* (Lantham, MD, Rowman and Littlefield Publishers, 2007); see also the commentary "Avery and Singer: Unstoppable Hot Air," www.realclimate.org/index.php/archives/2006/11/avery-and-singer-unstoppable-hot-air/.

19. Ernst-Georg Beck, "180 Years of Atmospheric CO_2 Gas Analysis by Chemical Methods," *Energy and Environment*, 18 (2007): 259–282.

20. Keeling originally published his response online on May 14, 2007, and it was later published as H. A. J. Meijer and R. F. Keeling, "Comment on '180 Years of Atmospheric CO_2 Gas Analysis by Chemical Methods,'" *Energy and Environment*, 18 (2007): 635–641.

21. See Georg Hoffmann, "Beck to the Future," www.realclimate.org/index.php/archives/2007/05/beck-to-the-future/.

3. Signals in the Noise

1. Technically speaking, the behavior isn't random in the pure sense of the word, but is instead what is termed *chaotic*. In other words, the behavior is governed by deterministic physics, but the nature of the physics is such that

the tiniest perturbation can cause the system to veer off into a different trajectory. Such chaos is sometimes referred to as the "butterfly effect." For our purposes here, it can be characterized as essentially random.

2. M. E. Schlesinger and N. Ramankutty, "An Oscillation in the Global Climate System of Period 65–70 Years," *Nature*, 367 (1994): 723–726.

3. The method was referred to as "MTM-SVD." *MTM* refers to a method of spectral analysis (a fancy term for an attempt to identify oscillations) called the multi-taper method, while *SVD* refers to singular value decomposition, a set of mathematical operations that can be used to identify the leading patterns in a spatial array of data.

4. M. E. Mann and J. Park, "Global-Scale Modes of Surface Temperature Variability on Interannual to Century Timescales," *Journal of Geophysical Research*, 99 (1994): 819–825, 833.

5. T. Delworth, S. Manabe, and R. J. Stouffer, "Interdecadal Variations of the Thermohaline Circulation in a Coupled Ocean-Atmosphere Model," *Journal of Climate*, 6 (1993): 1993–2011.

6. See the news summary in R. A. Kerr, "A North Atlantic Climate Pacemaker for the Centuries," *Science*, 288 (2000): 1984–1985, which in part reflected my conversation with Kerr, including the AMO terminology I coined in the interview.

7. In 2006 I published a paper with MIT hurricane researcher Kerry Emanuel making the case that the tropical Atlantic warming associated with increased hurricane activity in recent decades is more closely associated with climate change than with the AMO. M. E. Mann and K. A. Emanuel, "Atlantic Hurricane Trends Linked to Climate Change," *Eos*, 87 (2006): 233, 238, 241.

8. For further details about climate proxy data, including how they are recovered, analyzed, and interpreted, see R. S. Bradley, *Paleoclimatology: Reconstructing Climates of the Quaternary*, 2nd ed. (San Diego, Academic Press, 1999).

9. We published our findings in M. E. Mann, J. Park, and R. S. Bradley, "Global Interdecadal and Century-Scale Climate Oscillations During the Past Five Centuries," *Nature*, 378 (November 1995): 266–270.

10. The estimates were first published in H. H. Lamb, "The Earlier Medieval Warm Epoch and Its Sequel," *Palaeogeography, Palaeoclimatology, Palaeoecology*, 1 (1965): 13–37. Various revised versions of the series were published by Lamb in 1977 and 1982.

11. P. D. Jones, K. R. Briffa, T. J. Osborn, J. M. Lough, T. D. van Ommen, B. M. Vinther, J. Luterbacher, E. R. Wahl, F. W. Zwiers, M. E. Mann, G. A.

Schmidt, C. M. Ammann, B. M. Buckley, K. M. Cobb, J. Esper, H. Goosse, N. Graham, E. Jansen, T. Kiefer, C. Kull, M. Kuttel, E. Mosely-Thompson, J. T. Overpeck, N. Riedwyl, M. Schulz, A. W. Tudhope, R. Villalba, H. Wanner, E. Wolff, and E. Xoplaki, "High-Resolution Paleoclimatology of the Last Millennium: A Review of Current Status and Future Prospects," *Holocene*, 19 (2009): 3–49.

12. See, e.g., M. K. Hughes and H. F. Diaz, "Was There a 'Medieval Warm Period' and If So, When and Where?" *Climate Change*, 26 (1994): 109–142.

13. The MCA terminology was first introduced by paleoclimatologist Scott Stine, "Extreme and Persistent Drought in California and Patagonia During Mediaeval Time," *Nature*, 369 (1994): 546–549. Stine had been studying relict tree stumps found today in seasonally flooded lake beds and riverbeds throughout the Sierra Nevada. Using radiocarbon dating to estimate the age of the stumps in places like California's Mono Lake, he found that they invariably dated to medieval times, suggesting an absence of the flooding that today makes these same environments inhospitable for tree growth. In other words, the tree stumps told an important climate story: that medieval climate was characterized by unusually dry conditions in this region of western North America.

4. The Making of the Hockey Stick

1. R. S. Bradley and P. D. Jones, "'Little Ice Age' Summer Temperature Variations: Their Nature and Relevance to Recent Global Warming Trends," *Holocene*, 3 (1993): 367–376.

2. This approach is often referred to as "composite-plus-scale" (CPS), as it involves first compositing the data and then scaling the result so it matches the instrumental data during the interval of common overlap.

3. J. Lean, J. Beer, et al., "Reconstruction of Solar Irradiance Since 1610: Implications for Climate Change," *Geophysical Research Letters* 22 (1995): 3195–3198.

4. J. A. Eddy, "The Maunder Minimum," *Science*, 192 (1976): 1189–1202.

5. Direct observations of the Sun via telescope have been possible since the time of Galileo, allowing for reasonably accurate estimates of sunspot activity, believed to be a proxy—albeit an indirect one—for solar output changes. Records of longer-term, albeit less reliable, naked-eye observations of sunspot activity are available further back in time based on observations from Asia. Cosmogenic isotopes of carbon (^{14}C) and beryllium (^{10}Be) found on Earth are

produced by interactions with solar radiation in the upper atmosphere and, when deposited in ice cores or assimilated into the organic matter of trees, can also be used to estimate, somewhat indirectly, changes over time in solar output.

6. J. Overpeck, K. Hughen, et al., "Arctic Environmental Change of the Last Four Centuries," *Science*, 278 (1997): 1251–1256. Overpeck and collaborators examined the various factors that could explain trends in Arctic summer temperatures over the past four centuries. While natural factors appear to have been important, they concluded, only human influences appear to be able to explain modern warming.

7. I was supported on an Alexander Hollaender postdoctoral fellowship from the Department of Energy. I completed and defended my Ph.D. work in 1996, which is when the postdoctoral fellowship began. I did not actually turn in the final copy of the dissertation and receive my Ph.D. from Yale until 1998 owing to a combination of procrastination and a hard disk failure of the research group's main computer that required me to regenerate from scratch substantial parts of my thesis.

8. A particularly instructive review of the methods of climate field reconstruction is E. R. Cook, K. R. Briffa, et al., "Spatial Regression Methods in Dendroclimatology: A Review and Comparison of Two Techniques," *International Journal of Climatology*, 14 (1994): 379–402.

9. Its EOF would be represented by a spatial pattern that is uniformly positive across the globe (i.e., west and east).

10. Its EOF would be represented by a spatial pattern that is positive across the west and negative across the east.

11. Using this representation of the data yielded 112 distinct series representing a much larger number of 415 proxy records. The data became increasingly sparse as we went further back in time. Only twenty such series were available all the way back to A.D. 1400.

12. By convention, instrumental surface temperature datasets are often represented in terms of regional averages of stations, gridded at a resolution of, e.g., 5° x 5° latitude/longitude.

13. EOF#1 was a global map with positive values nearly everywhere, while PC#1 showed a long-term increasing trend—that is, it was much like the uniform global warming pattern in the simple PCA example provided earlier. The next most important pattern was related to the global influence of the El Nino/ Southern Oscillation (ENSO) phenomenon, with EOF#2 looking much like the classic El Niño pattern of a horseshoe-shaped pattern of warmth straddling the equatorial Pacific, cool temperatures in the central North Pacific

ocean, and various patches of warming and cooling in various other regions of the globe influenced by ENSO. The associated PC#2 (together with PC#4) primarily showed the interannual oscillations of ENSO, but there were also decadal variations and a long-term trend. The third most important pattern showed features reminiscent of, e.g., the North Atlantic oscillation (NAO) phenomenon discussed earlier, while the fifth pattern, characterized by large amplitude variations in the North Atlantic, appeared to relate to the AMO climate signal.

14. Various forms of this sort of validation test have been used in past work. In some cases, the dataset is split in two, with half used to calibrate and the other half to validate. In other cases, a single year is left out for validation purposes, and all other data are used to calibrate, with the process repeated for all possible choices of a missing year. Another variant is to leave out a block of data, say ten years rather than a single year at a time. Each of these approaches has advantages and disadvantages. One limitation of all such validation tests is that they cannot test for the impact of possible degradation of the proxy records back in time, since the proxy data are only being evaluated based on their performance in the modern period. Furthermore, such exercises cannot provide a rigorous test of how well the reconstruction method works on timescales longer than are resolved by the relatively short validation interval. Other tests using so-called pseudoproxy data derived from climate model simulations, as discussed elsewhere in the book, can be used to address this latter issue. A good review of this topic is provided by E. R. Wahl and C. M. Ammann, "Robustness of the Mann, Bradley, Hughes Reconstruction of Surface Temperatures: Examination of Criticisms Based on the Nature and Processing of Proxy Climate Evidence," *Climatic Change*, 85 (2007): 33–69, and various references cited therein.

15. We used a version of the CRU instrumental surface temperature record that was available in the mid-1990s, which spanned the interval 1854–1993. Certain key temperature series (e.g., the Northern Hemisphere mean) were updated through the late 1990s using supplementary CRU data. There were 1,082 temperature grid boxes in the CRU dataset for which there were nearly continuous monthly data from 1902 to 1993, but only 219 of them had nearly continuous data back to 1854. For this reason, we chose to use the data-rich post-1902 interval for calibration, reserving the sparser pre-1902 instrumental data for validation.

16. To say a statistical prediction or estimate is "skillful" is to say that it performs better than would be expected from random chance alone. One can generate such random predictions and evaluate the performance of the actual statistical model (i.e., the prediction using the proxy data) relative to the

benchmarks obtained from the random data. Typically, if the model outperforms, say, 95 percent or more of the random model predictions, we conclude that it is in fact skillful.

17. M. E. Mann, R. S. Bradley, and M. K. Hughes, "Global-Scale Temperature Patterns and Climate Forcing Over the Past Six Centuries," *Nature*, 392 (1998): 779–787.

18. There are two primary competing estimates of global mean temperature: (1) the "HadCRUT" product of CRU/UK Met Office in the UK and (2) the "GISSTemp" product of the NASA Goddard Institute for Space Studies. Both groups treat the issue of missing data differently, leading to small differences in their estimates. HadCRUT still yields 1998 as the warmest year, while GISS-Temp has 2005 and 2010 in a statistical tie for the warmest year.

19. The "blade" of the hockey stick—the modern temperature record including the twentieth-century spike—is determined by the instrumental record, against which the proxy data were calibrated to yield a long-term reconstruction. Despite claims that have been made to the contrary, it is not a product—let alone an artifact—of our proxy reconstruction of past temperatures—i.e., the "handle" of the hockey stick.

20. The George C. Marshall Institute, for example, issued a press release on June 30, 1998—"1997: Warmest Year Since 1400?"—in which it commented on our finding: "But why stop at 1400? Because that is just about the farthest back in the recorded past for which this statement is true. Go back just a few hundred years more to the period 1000–1200 A.D. and you find that the climate was considerably warmer than now. This era is known as the Medieval Warm Period."

21. P. D. Jones et al., "High-Resolution Paleoclimatic Records for the Last Millennium: Interpretation, Integration and Comparison with Circulation Model Control-Run Temperatures," *Holocene*, 8 (1998): 455–471.

22. They hadn't performed statistical validation exercises, so it remained unclear whether their reconstruction was actually skillful a thousand years back.

23. The original (and prescient) article presenting these data was G. C. Jacoby and R. D'Arrigo, "Reconstructed Northern Hemisphere Annual Temperature Since 1671 Based on High-Latitude Tree-Ring Data from North America," *Climatic Change*, 14 (1989): 39–59. As early as the late 1980s, these authors interpreted their boreal North American tree ring data as providing a plausible representation of Northern Hemisphere average temperature, though they only attempted to go back a little bit more than three hundred years (for which they had estimates from eleven distinct regions). The authors went further than this, however. They concluded that the recent warming was

anomalous in the context of their three-hundred-plus-year record, and even speculated that their results supported the hypothesis of anthropogenic climate change.

24. See R. S. Bradley, "Are There Optimum Sites for Global Paleotemperature Reconstruction?" in P. D. Jones, R. S. Bradley, and J. Jouzel, eds., NATO ASI Series, vol. 41, series 1, *Climatic Variations and Forcing Mechanisms of the Last 2000 Years* (Berlin and Heidelberg, Springer-Verlag, 1996), 603–624. Bradley analyzed a long-term climate model simulation to determine the sweet spots for estimating the average temperature of the Northern Hemisphere given only a modest sample of sites, and found North America to be a key region. The model simulation was primitive by today's standards, as it lacked many of the relevant external forcings that we know to be important (e.g., solar and volcanic) over this time frame. The basic finding of the study nonetheless still appears to be valid.

25. D. A. Graybill and S. B. Idso, "Detecting the Aerial Fertilization of Atmospheric CO_2 Enrichment in Tree-Ring Chronologies," *Global Biogeochemical Cycles*, 7 (1993): 81–95.

26. Since the extra component of growth since roughly 1800 in the high-elevation western U.S. trees appeared to be due to CO_2, the necessary correction could simply be estimated from the post-1800 difference between the high-elevation western U.S. tree ring data and the low-elevation North American boreal tree line data.

27. We used *likely* in the same sense that the IPCC had used it: a conclusion that has a moderately high, roughly two-thirds (67 percent) likelihood of being correct. This is far lower than the "very likely" conclusion (a finding with a 90 percent likelihood, in IPCC parlance) that was attached to, for example, the 2007 AR4 report conclusion that most of the warming of the past fifty years could be attributed to human impacts.

28. M. E. Mann, R. S. Bradley, and M. K. Hughes, "Northern Hemisphere Temperatures During the Past Millennium: Inferences, Uncertainties, and Limitations," *Geophysical Research Letters*, 26 (1999): 759–762.

29. H. Pollack, S. Huang, and P. Y. Shen, "Climate Change Revealed by Subsurface Temperatures: A Global Perspective," *Science*, 282 (1998): 279–281.

30. Despite recent high-profile criticisms regarding errors or alleged errors in parts of the impacts report of the IPCC fourth assessment of 2007, to date no errors have been found in any of the scientific assessment (working group 1) reports or in any of the reports of the IPCC third assessment.

31. While I contributed to all aspects of our chapter, my main responsibilities related to the section entitled "Is the Recent Warming Unusual?" that reviewed the paleoclimate record of past centuries. The subsections included

a background that reviewed the conclusions of the previous IPCC (Second Assessment) report in this area, and the main overall developments of the science since, while a subsection "Temperature of the Past 1000 Years," characterized the relative strengths and weaknesses of each of the main sources of proxy temperature information available and described the recent multi-proxy syntheses of past temperature changes derived from combinations of various proxy sources. The chapter assessed the range of evidence regarding the Little Ice Age and medieval warm period and examined the role of natural and human impacts on temperature trends in past centuries.

32. K. R. Briffa, F. H. Schweingruber, P. D. Jones, T. J. Osborn, S. G. Shiyatov, and E. A. Vaganov, "Reduced Sensitivity of Recent Tree-Growth to Temperature at High Northern Latitudes," *Nature*, 391 (1998): 678–682.

33. T. J. Crowley and T. Lowery, "How Warm Was the Medieval Warm Period?" *Ambio*, 29 (2000): 51–54.

34. Examples of this false claim about our reconstruction are common in the public discourse, e.g., a letter to the editor of *The Day* of New London, Connecticut (March 12, 2005): "In a stroke, this graph eliminated the Medieval Warm Period (from about A.D. 1000 to 1400) and the Little Ice Age (from about 1400 to 1850) from history," and more recently the article "A Disgrace to Science" in the conservative *American Spectator* by Tom Bethell (February 2010): "[Mann] manufactured the misleading 'hockey stick' temperature graph that eliminated the Medieval Warm Period."

35. T. J. Crowley, "Causes of Climate Change Over the Past 1000 Years," *Science*, 289 (2000): 270–277.

5. The Origins of Denial

1. David Michaels, *Doubt Is Their Product* (New York, Oxford University Press, 2008).

2. Chris Mooney, *The Republican War on Science* (New York, Basic Books, 2005).

3. Naomi Oreskes and Erik M. Conway, *Merchants of Doubt* (New York, Bloomsbury Press, 2010).

4. This tactic is described by Philip Mirowski in "The Rise of the Dedicated Natural Science Think Tank," *Social Science Research Council*, July 2008. According to Mirowski, "The key tenets were to promote otherwise isolated scientific spokespersons (from gold plated universities, if possible) who would take the industry side in the debate, manufacture uncertainty

about the existing scientific literature, launder information through seemingly neutral third party fronts, and wherever possible recast the debate by moving it away from aspects of the science which it would seem otherwise impossible to challenge" (3).

5. Paul R. Ehrlich and Anne H. Ehrlich, *Betrayal of Science and Reason: How Anti-environmental Rhetoric Threatens Our Future* (Washington, DC, Island Press, 1996), 1.

6. See Mooney, *The Republican War on Science*, 6.

7. This particular phrase was emphasized by the Discovery Institute, a conservative think tank, in its campaign to convince educators to offer the pseudoscientific concept of intelligent design as a credible scientific alternative to the theory of evolution.

8. See, e.g., Elizabeth Lopatto, Jef Feeley, and Margaret Cronin Fisk, "Eli Lilly 'Ghostwrote' Articles to Market Zyprexa, Files Show," *Bloomberg News*, June 12, 2009; Stephanie Saul, "Merck Wrote Drug Studies for Doctors," *New York Times*, April 16, 2008.

9. See, e.g., the discussion of Knoll Pharmaceuticals in Sheldon Krimsky, "Threats to the Integrity of Biomedical Research," in Wendy Wagner and Rena Steinzor, eds., *Rescuing Science from Politics* (New York, Cambridge University Press, 2006), 77. In this example, a peer-reviewed publication presenting damaging findings with regard to one of Knoll's new drugs was retracted by the author after threats of legal action by Knoll.

10. See, e.g., SourceWatch.org and exxonsecrets.org.

11. See, e.g., Chris Mooney, "Some Like It Hot," *Mother Jones*, May/June 2005; Sharon Begley, "The Truth About Denial," *Newsweek*, August 13, 2007; Jane Mayer, "Covert Operations: The Billionaire Brothers Who Are Waging a War Against Obama," *New Yorker*, August 30, 2010.

12. Ehrlich and Ehrlich, *Betrayal of Science and Reason*; Ross Gelbspan, *The Heat Is On* (New York, Basic Books, 1997) and *Boiling Point* (New York, Basic Books, 2004); and more recently James Hoggan and Richard Littlemore, *Climate Cover-Up* (Vancouver, BC, Greystone Books, 2009); Stephen Schneider, *Science as a Contact Sport* (Washington DC, National Geographic, 2009); Oreskes and Conway, *Merchants of Doubt*.

13. Oreskes and Conway in *Merchants of Doubt* provide an especially lucid account of the origins of the climate change denial campaign. They describe how the campaign can be traced back to Cold War hawks' distrust of scientists who questioned the efficacy and appropriateness of developing missile defense systems like the Reagan administrations' Strategic Defense Initiative (popularly known as "Star Wars") in the 1980s.

14. John H. Cushman Jr., "Industrial Group Plans to Battle Climate Treaty," *New York Times,* April 26, 1998.

15. Prior to the April 2010 Deepwater Horizon spill in the Gulf of Mexico—by most standards the worst oil spill disaster in history, BP had cultivated a reputation as one of the more forward-thinking fossil fuel companies, its advertising motto being *"Beyond* Petroleum."

16. See Greenpeace USA, "Koch Industries: Secretly Funding the Climate Denial Machine," March 2010. The report outlines how brothers Charles G. Koch and David H. Koch, who own and control Koch Industries, an oil corporation that is the second largest privately held company in the United States, have, along with family members and their employees, "directed a web of financing that supports conservative special interest groups and think-tanks, with a strong focus on fighting environmental regulation, opposing clean energy legislation, and easing limits on industrial pollution." Greenpeace notes that the money "is typically funneled through one of three 'charitable' foundations the Kochs have set up: the Claude R. Lambe Foundation; the Charles G. Koch Foundation; and the David H. Koch Foundation." See also Mayer, "Covert Operations."

17. The Scaife Foundations comprise the Sarah Mellon Scaife Foundation, the Carthage Foundation, the Allegheny Foundation, and the Scaife Family Foundation. Through these foundations, Richard Mellon Scaife, whose wealth was inherited from the Mellon industrial, oil, mining, and banking fortune, has financed numerous right-wing groups involved in the climate change denial campaign. According to Sourcewatch.org (accessed April 13, 2011), between 1985 and 2001 he donated $15,860,000 to the Heritage Foundation, $4,411,000 to the American Enterprise Institute, $2,575,000 to the Manhattan Institute for Policy Research, $1,855,000 to the George C. Marshall Institute, $1,808,000 to the Hudson Institute, and $1,697,000 to the Cato Institute.

18. In October 2006, Senators Olympia Snowe (R-ME) and Jay Rockefeller (D-WV) publicly demanded that ExxonMobil "end any further financial assistance" to groups "whose public advocacy has contributed to the small but unfortunately effective climate change denial myth." A July 19, 2010, investigative report in the *Times* of London revealed that despite the company's promise in 2007 that "In 2008, we will discontinue contributions to several public policy groups, whose position on climate change could divert attention from the important discussion on how the world will secure energy required for economic growth in a responsible manner," ExxonMobil continued to fund climate change deniers through at least 2009. The *Times* article quotes Bob Ward, policy director at the London School of Economics Grantham Research Institute: "Exxon has engaged in a public relations campaign to con-

vince the world that it has stopped funding climate sceptic groups. But this has turned out to be pure greenwash. Exxon has continued to provide financial support for many groups that are engaged in activities to persuade the public and policy-makers into wrongly believing that climate change is a hoax." The *Guardian* more recently reported that ExxonMobil continued to fund at least one climate change contrarian through at least 2010. John Vidal, "Climate Sceptic Willie Soon Received $1m from Oil Companies, Papers Show," *Guardian*, June 28, 2011.

19. Evidence of fossil fuel industry funding of climate change denial by each of these (in particular, by Koch Industries and ExxonMobil) and numerous other groups can be found, for example, in the following sources: Greenpeace USA, "Koch Industries Secret Funding the Climate Denial Machine," March 2010; Gelbspan, *Boiling Point*; Oreskes and Conway, *Merchants of Doubt*; James Powell, *The Inquisition of Climate Science* (New York, Columbia University Press, 2011).

20. In *Merchants of Doubt*, Oreskes and Conway give a detailed account of these individuals, including their backgrounds and the history of how they became involved in industry-funded public relations campaigns including climate change denial.

21. The physics faculty included Edward Teller, commonly considered the Father of the Hydrogen Bomb for his role in the Manhattan Project in the 1940s, who was one of the most prominent advocates for the Star Wars program, and Charles Schwartz, who staunchly opposed the program. An excellent account of the conflict between the two is provided by Darwin Bondgraham, Nicholas Robinson, and Will Parrish, "California's Nuclear Nexus: A Faux Disarmament Plan Has Roots in the Golden State's Pro-Nuclear Lobby," *Z Magazine*, December 2009.

22. Begley, "The Truth About Denial."

23. The song included the lines "Hey there mighty brontosaurus, / Don't you have a lesson for us. / You thought your rule would always last. / There were no lessons in your past" (Walking in Your Footsteps," Music and Lyrics by Sting, © 1983 G. M. SUMNER, administered by EMI MUSIC PUBLISHING LIMITED. All Rights Reserved, International Copyright Secured. Used by Permission. *Reprinted by Permission of Hal Leonard Corporation*).

24. See Mark Hertsgaard, "While Washington Slept," *Vanity Fair*, May 2006; PBS Frontline, "Hot Politics," April 3, 2006, www.pbs.org/wgbh/pages /frontline/hotpolitics/interviews/seitz.html; Union of Concerned Scientists, "Smoke, Mirrors and Hot Air: How ExxonMobil Uses Big Tobacco's Tactics to 'Manufacture Uncertainty' on Climate Change," January 2007.

25. See, e.g., Hertsgaard, "While Washington Slept."

26. "Statement by the Council of the National Academy of Sciences Regarding Global Change Petition," April 20, 1998, www.nationalacademies .org/onpinews/newsitem.aspx?RecordID=s04201998.

27. "Skepticism About Skeptics," *Scientific American*, October 16, 2001.

28. See "Oregon Institute of Science and Malarkey," RealClimate.org, October 10 2007.

29. The story was reported by Ian Sample, "Scientists Offered Cash to Dispute Climate Study," *Guardian*, February 2, 2007. However, the issue came to light earlier still, in July 2006 when a colleague, Andrew Dressler, scanned and posted on his blog the letter two climate scientists at Texas A&M University had received. The letter begins "The American Enterprise Institute is launching a major project to produce a review and policy critique of the forthcoming Fourth Assessment Report (FAR)."

30. Amusingly, some token mainstream climate scientists, myself included, have been invited by the organizers to participate in such meetings, presumably in the hope that our attendance might grant these PR events scientific credibility.

31. Singer took leave as a faculty member of the Department of Environmental Sciences at the University of Virginia in the early 1990s to spend full time on advocacy activities, formally retiring from his university position in 1994.

32. *Rolling Stone* magazine had this to say about Singer: "A former mouthpiece for the tobacco industry, the 85-year-old Singer is the granddaddy of fake 'science' designed to debunk global warming. The retired physicist—who also tried to downplay the danger of the hole in the ozone layer—is still wheeled out as an authority by big polluters determined to kill climate legislation. For years, Singer steadfastly denied that the world is heating up: Citing satellite data that has since been discredited, he even made the unhinged claim that 'the climate has been cooling just slightly.' Last year, Singer served as a lead author of 'Climate Change Reconsidered'— an 880-page report by the right-wing Heartland Institute that was laughably presented as a counterweight to the Intergovernmental Panel on Climate Change, the world's scientific authority on global warming. Singer concludes that the unchecked growth of climate-cooking pollution is 'unequivocally good news.' Why? Because 'rising CO2 levels increase plant growth and make plants more resistant to drought and pests.' Small wonder that Heartland's climate work has long been funded by the likes of Exxon and reactionary energy barons like Charles Koch and Richard Mellon Scaife." Tim Dickinson, "The Climate Killers: Meet the 17 Polluters and Deniers Who Are Derailing Efforts to Curb Global Warming," *Rolling Stone*, January 6, 2010.

33. Oreskes and Conway, *Merchants of Doubt*, 129–130.

34. For further information about Singer's funding by industry special interest groups, see Gelbspan, *The Heat Is On*, 46–47; Hoggan and Littlemore, *Climate Cover-Up*, 30, 80, 138–140, 156–157.

35. Dan Harris, Felicia Biberica, Elizabeth Stuart, and Nils Kongshaug, "Global Warming Denier: Fraud or 'Realist'?" ABC News, March 23, 2008.

36. The episode in question is detailed in Hoggan and Littlemore, *Climate Cover-Up*, 135–138.

37. According to Revelle's former colleague, the distinguished oceanographer Walter Munk, "Singer wrote the paper and, as a courtesy, added Roger as a co-author based upon his willingness to review the manuscript and advise on aspects relating to sea-level rise." See Oreskes and Conway, *Merchants of Doubt*, 195.

38. Richard Littlemore, "The Deniers? The World Renowned Scientist Who Got Al Gore Started," DeSmogBlog, April 16, 2008, www.desmogblog.com /the-deniers-the-world-renowned-scientist-who-got-al-gore-started.

39. The widespread funding of climate change contrarians by industry and industry-funded front groups is detailed in books such as Oreskes and Conway, *Merchants of Doubt*; Hoggan and Littlemore, *Climate Cover-Up*; Gelbspan, *Boiling Point*; and others. Useful online sources are the Web sites Sourcewatch.org and ExxonSecrets.org, which provide details regarding the ties between various professional climate change deniers and fossil fuel industry interests and their front groups.

40. In *The Heat Is On*, Ross Gelbspan reports an interview he conducted with Lindzen in which Lindzen admits to receiving (as of 1995) roughly $10,000 per year from fossil fuel industry consulting alone. In "The Heat Is On: The Warming of the World's Climate Sparks a Blaze of Denial," *Harper's*, December 1995, Ross Gelbspan noted, again based on his own interview, that Lindzen "charges oil and coal interests $2,500 a day for his consulting services; his 1991 trip to testify before a Senate committee was paid for by Western Fuels, and a speech he wrote, entitled 'Global Warming: The Origin and Nature of Alleged Scientific Consensus,' was underwritten by OPEC." Lindzen currently lists himself as a member of the Science, Health, and Economic Advisory Council of the Annapolis Center on his curriculum vitae (www -eaps.mit.edu/faculty/lindzen/CV.pdf, July 3, 2011). This organization has been funded by ExxonMobil at least as recently as 2009 (Exxon Mobil Corporation 2009 Worldwide Contributions and Community Investments, www .exxonmobil.com/Corporate/files/gcr_contributions_pub-policy09.pdf).

41. The negative feedback in question relates to the so-called tropical Pacific thermostat mechanism. This mechanism implies a counterintuitive La

Niña-like response to increased greenhouse gas concentrations that leads to less warming or even cooling of sea surface temperatures in the eastern and central tropical Pacific.

42. See, e.g., "Why So Gloomy," *Newsweek*, April 16, 2007; Richard Lindzen, "Climate of Fear," *Wall Street Journal*, April 12, 2006; Lindzen's UK House of Lords testimony on January 25, 2005, www.publications.parliament .uk/pa/ld200506/ldselect/ldeconaf/12/5012501.htm.

43. See, e.g., the discussion in James Hansen, Makiko Sato, Reto Ruedy, Ken Lo, David W. Lea, and Martin Medina-Elizade, "Global Temperature Change," *Proceedings of the National Academy of Science*, 103 (2006): 14288–14293.

44. Richard S. Lindzen, "Some Coolness Concerning Global Warming," *Bulletin of the American Meteorology Society*, 71 (March 1990): 288–299.

45. In Global Climate Coalition, "Primer on Climate Change Science," January 18, 1996, obtained as part of a court action against the automobile industry, the following is stated: "Lindzen's hypothesis that any warming would create more rain which would cool and dry the upper troposphere did offer a mechanism for balancing the effect of increased greenhouse gases. However, the data supporting this hypothesis is weak, and even Lindzen has stopped presenting it as an alternative to the conventional model of climate change."

46. R. S. Lindzen, M.-D. Chou, and A. Y. Hou, "Does the Earth Have an Adaptive Infrared Iris?" *Bulletin of the American Meteorological Society*, 82 (2001): 417–432.

47. Unlike low clouds, high clouds such as cirrus clouds typically have a warming influence on the surface as they block the escape of infrared radiation to space.

48. As summarized in the IPCC's Fourth Assessment Report (2007), "numerous objections have been raised about various aspects of the observational evidence provided so far [for Lindzen's iris hypothesis]." D. A. Randall, R. A. Wood, S. Bony, R. Colman, T. Fichefet, J. Fyfe, V. Kattsov, A. Pitman, J. Shukla, J. Srinivasan, R. J. Stouffer, A. Sumi, and K. E. Taylor, "Climate Models and Their Evaluation," in S. Solomon, D. Qin, M. Manning, Z. Chen, M. Marquis, K. B. Averyt, M. Tignor, and H. L. Miller, eds., *Climate Change 2007: The Physical Science Basis: Contribution of Working Group I to the Fourth Assessment Report of the Intergovernmental Panel on Climate Change* (Cambridge, UK, and New York, Cambridge University Press, 2007). Some authors found no evidence that warming ocean surface temperatures would influence cirrus clouds; D. L. Hartman and M. L. Michelsen, "No Evidence for Iris," *Bulletin of*

the American Meteorology Society, 83 (2002): 249–254; while others argued there would actually be more cirrus clouds, therefore favoring instead a positive feedback; Q. Fu, M. Baker, and D. L. Hartman, "Tropical Cirrus and Water Vapor: An Effective Earth Infrared Iris Feedback?" *Atmospheric Chemistry and Physics*, 2 (2002): 31–37; B. Lin, B. Wielicki, L. Chambers, Y. Hu, and K.-M. Xu, "The Iris Hypothesis: A Negative or Positive Cloud Feedback?" *Journal of Climate*, 15 (2002): 3–7.

49. R. S. Lindzen and Y.-S. Choi, "On the Determination of Climate Feedbacks from ERBE Data," *Geophysical Research Letters*, 36 (2009): L16705, doi:10.1029/2009GL039628, www.agu.org/pubs/crossref/2009/2009GL039628.shtml.

50. K. E. Trenberth, J. T. Fasullo, Chris O'Dell, and T. Wong, "Relationships Between Tropical Sea Surface Temperature and Top-of-Atmosphere Radiation," *Geophysical Research Letters*, 37 (2010): L03702, doi:10.1029/2009GL042314. See also the commentary by Fasullo, Trenberth, and O'Dell, "Lindzen and Choi Unraveled," RealClimate, January 8, 2010, www.realclimate.org/index.php/archives/2010/01/lindzen-and-choi-unraveled/.

51. Trenberth and colleagues pointed out that the study also fundamentally misinterpreted the observations by assuming that any changes in cloud distribution automatically reflected a response to changes in sea surface temperature.

52. A. E. Dessler, "A Determination of the Cloud Feedback from Climate Variations Over the Past Decade," *Science*, 330 (2010): 1523–1527.

53. Journalist Paul Thacker reported in the January 2006 issue of the *New Republic* that Milloy had received substantial payments from big tobacco (Phillip Morris, specifically) for nearly two decades, as well as substantial money from ExxonMobil, while presenting himself as an independent expert. (According to Thacker's article, Fox News claimed to be unaware of Milloy's financial ties to Philip Morris, admitting that "any affiliation he had should have been disclosed." Industry funding of Milloy's organization, the Advancement of Sound Science Coalition, is discussed, for example, by Mooney in *Republican War Against Science*, 67–68.

54. The title was "awarded" in George Monbiot, "The Patron Saint of Charlatans," *Guardian*, September 23, 2008.

55. Richard Black, "Climate Documentary 'Broke Rules,'" British Broadcasting Corporation, July 21, 2008, http://news.bbc.co.uk/2/hi/7517509.stm.

56. Wunsch's expression of these opinions is described in the BBC piece cited in the preceding note. Wunsch is a skeptic, as all practicing scientists should be. It was Wunsch, for example, who correctly anticipated—as we will

see later—that the claims by one research team of a dramatically weakening conveyor belt ocean circulation were premature. He is not, as the film seemed to imply, a denialist, however.

57. See, e.g., Richard Littlemore, "Pompous Prat Alert!" DeSmogBlog, April 4, 2009, www.desmogblog.com/pompous-prat-alert-viscount-monckton-tour.

58. In an open letter from Christopher Monckton to Senator John Mc-Cain (R-AZ), Monckton states in his biographical sketch that "his contribution to the IPCC's Fourth Assessment Report in 2007 . . . earned him the status of Nobel Peace Laureate" (http://scienceandpublicpolicy.org/images/stories/pa pers/reprint/Letter_to_McCain.pdf, accessed August 11, 2011). As Littlemore notes in "Pompous Prat Alert!" ibid., "Monckton claimed to also be a Nobel winner because he had done such good work trying to undermine their effort. He even got a friend to melt down an old science experiment so they could fashion a little Nobel Prize pin, later presented to Monckton in a highly unofficial ceremony. (For the record, Monckton claims he deserves the accolade because he was a 'reviewer' of the IPCC report. The IPCC accepts reviews, unsolicited, from all parties . . .)."

59. See Leo Hickman, "Christopher Monckton Told to Stop Claiming He Is a Member of the Lords," *Guardian*, August 11, 2010, http://scienceblogs.com /deltoid/2008/10/monckton_has_a_gold_nobel_priz.php.

60. Stephen McIntyre, "Is Gavin Schmidt Honest?" climateaudit, October 29, 2005.

61. The report, entitled "Is the Surface Temperature Record Reliable," was published by the Heartland Institute in March 2009 and is available electronically at http://wattsupwiththat.files.wordpress.com/2009/05/surfacesta tionsreport_spring09.pdf.

62. Matthew J. Menne, Claude N. Williams Jr., and Michael A. Palecki, "On the Reliability of the U.S. Surface Temperature Record," *Journal of Geophysical Research*, 115 (2010): D11108, doi:10.1029/2009JD013094, www1.ncdc.noaa.gov /pub/data/ushcn/v2/monthly/menne-etal2010.pdf. The summary conclusion by the authors was "we find no evidence that the CONUS [Continental U.S.] average temperature trends are inflated due to poor station siting."

63. Brad Stone, "The Hand That Controls the Sock Puppet Could Get Slapped," *New York Times*, July 16, 2007.

64. See, e.g., Tim Lambert, "Sock Puppet Guide," Deltoid, http://science blogs.com/deltoid/2005/08/sockpuppets.php; with regard to Fumento, see his "Fumento's Sidekick," http://scienceblogs.com/deltoid/2005/12/fumentos-sidekick.php.

65. See, e.g., Hoggan and Littlemore, *Climate Cover-Up*, 96. Cybercast News Service is a project of the Media Research Center (MRC), which received

over $400,000 from ExxonMobil between 1998 and 2009 (www.exxonse crets.org/html/orgfactsheet.php?id=110, accessed July 16, 2011). According to *Media Transparency*, it has received over $3 million from the Sarah Scaife foundation between 1998 and 2009 (http://mediamattersaction.org/ transparency/organization/Media_Research_Center/funders, accessed July 16, 2011).

66. See "Kerry 'Unfit to Be Commander-in-Chief,' Say Former Military Colleagues," CNS, May 3, 2004.

67. ClimateDepot.com, July 13, 2010, "Climate Depot's Morano calls NASA's Hansen a 'wannabe Unabomber' for endorsing book urging 'ridding the world of industrial civilization' and 'razing cities' and 'blowing up dams,'" www.climatedepot.com/a/7355/a/4993/Time-for-Meds-NASA-scientist-James -Hansen-endorses-book-which-calls-for-ridding-the-world-of-Industrial -Civilization-ndash-Hansen-declares-author-has-it-rightthe-system-is-the -problem.

68. Clive Hamilton, "Silencing the Scientists: The Rise of Right-Wing Populism," *Our World*, March 2, 2011, http://ourworld.unu.edu/en/silencing-the -scientists-the-rise-of-right-wing-populism/#authordata.

69. Morano's position on the EPW staff was terminated in spring 2009 for reasons that have not been made public.

70. See the Sourcewatch.org pages on Committee for a Constructive Tomor- row, archived May 3, 2011, www.sourcewatch.org/index.php?title=Committee _for_a_Constructive_Tomorrow; and Climate Depot, archived May 3, 2011, www.sourcewatch.org/index.php?title=ClimateDepot.com.

71. Marc Morano, "ClimateDepot.com Launch Aims to Redefine Global Warming Reporting: Climate Clearinghouse to Challenge Mainstream Media's Eco-Reporting," ClimateDepot.com, April 6, 2009.

72. Douglas Fisher and the Daily Climate, "Cyber Bullying Intensifies as Climate Data Questioned," *Scientific American*, March 1, 2010.

73. Rachel Carson, *Silent Spring* (New York, Houghton Mifflin, 1962).

74. Christopher J. Bosso, *Pesticides and Politics: The Life Cycle of a Public Issue* (Pittsburgh, University of Pittsburgh Press, 1987), 116.

75. Ed Regis, "The Doomslayer," *Wired*, February 1997.

76. World Scientists' Warning to Humanity, Union of Concerned Scien- tists, 1993.

77. For example, a joint statement of fifty-eight of the world's academies of science, including the United States and United Kingdom, concluded that "As human numbers further increase, the potential for irreversible changes of far- reaching magnitude also increases." "A Joint Statement by Fifty-Eight of the World's Scientific Academies," *Population Summit of the World's Scientific*

Academies, 1993, available through National Academies Press, www.nap.edu /openbook.php?record_id=9148&page=R2.

78. Joseph Palca, "Get-the-Lead-Out Guru Challenged," *Science*, 253 (1991): 842–844.

79. A personal favorite of mine was his quip with regard to the plethora of climate change deniers in the petroleum geology industry: A "petroleum geologist's opinion on climate science is as good as a climate scientist's opinion on oil reserves."

80. U.S. National Academy of Sciences, "Understanding Climate Change: A Program for Action," The 1975 US National Academy of Sciences/National Research Council Report; Thomas Peterson, William Connolley, and John Fleck, "The Myth of the 1970s Global Cooling Scientific Consensus," Bulletin of the American Meteorological Society, 89 (2008): 1325–1337.

81. S. I. Rasool and S. H. Schneider, "Atmospheric Carbon Dioxide and Aerosols: Effects of Large Increases on Global Climate," *Science*, 173 (1971): 138–141, doi:10.1126/science.173.3992.138.

82. *Discover*, October 1989, 45–48.

83. Philip Shabecoff, "Global Warming Has Begun, Expert Tells Senate," *New York Times*, June 24, 1988.

84. See Mark Bowen, *Censoring Science: Inside the Political Attack on Dr. James Hansen and the Truth of Global Warming* (New York, Dutton, 2007).

85. Ibid.

86. Conservative Southern California congressman Dana Rohrabacher tried to pressure his employer to fire him based on the specious allegations made by the GCC. He was at the receiving end of death threats credible enough that law enforcement was forced to step in.

87. Fred Pearce, "Climate Change Special: State of Denial," *New Scientist*, November 2006.

6. A Candle in the Dark

1. Carl Sagan provides a particularly cogent discussion of the role of skepticism in science in his classic *The Demon-Haunted World* (New York, Random House, 1996). This chapter pays homage to that work, drawing directly from its subtitle, *Science as a Candle in the Dark*.

2. The senator was James Inhofe (R-OK), the occasion was an October 30, 2003, vote on a key bipartisan Senate climate change bill, and the study was coauthored by energy industry consultant Stephen McIntyre and climate change contrarian Ross McKitrick.

3. Neither quantum mechanics nor special relativity required the rejection of the prevailing Newtonian understanding of the laws of physics. Instead, they led to corrections to Newtonian mechanics at very small spatial scales or very high velocities, conditions that do not apply to Newton's proverbial apple.

4. Several such examples are discussed at RealClimate.org. See, e.g., M. Mann and G. Schmidt, "Peer Review: A Necessary but Not Sufficient Condition," January 20, 2005.

5. See, e.g., examples in G. Schmidt, "Science at the Bleeding Edge," RealClimate.org, July 2, 2009; G. Schmidt and M. Mann, "Decrease in Atlantic Circulation," RealClimate.org, November 30, 2005.

6. N. Shaviv and J. Veizer, "Celestial Driver of Phanerozoic Climate?" *GSA Today*, 13 (2003): 4–10.

7. In the press release, Shaviv, the lead author, claimed that "The operative significance of our research is that a significant reduction of the release of greenhouse gases will not significantly lower the global temperature, since only about a third of the warming over the past century should be attributed to man."

8. Lorne Gunter, "Here Comes the Sun," TechCentralStation, July 23, 2003.

9. S. Rahmstorf, D. Archer, D. S. Ebel, O. Eugster, J. Jouzel, D. Maraun, G. A. Schmidt, J. Severinghaus, A. J. Weaver, and J. Zachos, "Cosmic Rays, Carbon Dioxide, and Climate," *Eos*, 85 (2004): 38, 41. The authors showed that the claimed relationship was an artifact of unreliable and poorly replicated estimates, and selective alteration of the chronology (in one case, by 40 million years), among other problems.

10. Kate Ravilious, "Blame It on the Supernova," *Independent*, August 13, 2003.

11. See James Hoggan and Richard Littlemore, *Climate Cover-Up* (Vancouver, BC, Greystone Books, 2009), 76.

12. R. McKitrick and P. J. Michaels, "A Test of Corrections for Extraneous Signals in Gridded Surface Temperature Data," *Climate Research*, 26 (2004): 159–173.

13. Patrick J. Michaels, S. Fred Singer and David H. Douglass, "Settling Global Warming Science," TechCentralStation, August 12, 2004.

14. Tim Lambert, "McKitrick Screws Up Yet Again," Deltoid, scienceblogs.com, August 26, 2004.

15. Norwegian climate researcher Rasmus Benestad demonstrated that the authors had not accounted for an important statistical property known as autocorrelation (a function of the fact that neighboring samples in a dataset

are often not independent of each other), which led them to dramatically over-estimate their sample sizes and, as a consequence also the statistical signifi-cances of supposed relationships between economic factors and temperature trends. R. E. Benestad, "Are Temperature Trends Affected by Economic Activ-ity? Comment on McKitrick and Michaels (2004)," *Climate Research*, 27 (2004): 171–173; see also: G. A. Schmidt, "Spurious Correlations Between Recent Warming and Indices of Local Economic Activity," *International Journal of Climatology*, 29 (2009): 2041–2048.

16. H. L. Bryden et al., "Slowing of the Atlantic Meridional Overturning Circulation at 25°," *Nature*, 438 (December 1, 2005): 655–657.

17. Tom Delworth (from the GFDL group in Princeton) and I, for example, detected an AMO-like signal in the MBH98 temperature reconstructions that bore a striking resemblance to the AMO signal seen in his long control simu-lations of the GFDL coupled model; T. L. Delworth and M. E. Mann, "Ob-served and Simulated Multidecadal Variability in the Northern Hemisphere," *Climate Dynamics*, 16 (2000): 661–676. Similar AMO oscillations have more recently been found in simulations of the independent United Kingdom Met Office coupled model: e.g., J. R. Knight, R. J. Allan, C. K. Folland, M. Vellinga, and M. E. Mann, "A Signature of Persistent Natural Thermohaline Circulation Cycles in Observed Climate," *Geophysical Research Letters*, 32 (2005): L20708, doi:10.1029/2005GL02423.

18. D. T. Shindell, D. Rind, N. Balachandran, J. Lean, and P. Lonergan, "Solar Cycle Variability, Ozone, and Climate," *Science*, 284 (1999): 305–308, doi:10.1126/science.284.5412.305.

19. As part of her Ph.D. thesis at U. Mass., supervised by Ray Bradley and me, Anne Waple estimated the pattern of response of surface temperatures to solar output changes over the past four centuries using the correlation be-tween reconstructions of solar output and our surface temperature recon-structions. A. Waple, M. E. Mann, and R. S. Bradley, "Long-Term Patterns of Solar Irradiance Forcing in Model Experiments and Proxy-Based Surface Temperature Reconstructions," *Climate Dynamics*, 18 (2002): 563–578.

20. The pattern of the Northern Hemisphere jet stream is influenced by latitudinal and vertical variations in atmospheric temperature. Drew Shindell and his GISS collaborators performed a simulation with the GISS climate model in which solar output was lowered by the estimated drop of the Maunder minimum. In part as a result of the enhanced temperature response in the up-per atmosphere due to changes in stratospheric ozone concentrations, the model yielded a large change in atmospheric circulation consistent with the negative phase of the NAO, and a resulting temperature pattern that was similar

to that we had established empirically; see D. T. Shindell, G. A. Schmidt, M. E. Mann, D. Rind, and A. Waple, "Solar Forcing of Regional Climate Change During the Maunder Minimum," *Science*, 294 (2001): 2149–2152.

21. P. Handler, "Possible Association of Stratospheric Aerosols and El Niño Type Events," *Geophysical Research Letters*, 11 (1984): 1121–1124; P. Handler, and K. Andsager, "Possible Association Between the Climatic Effects of Stratospheric Aerosols and Sea Surface Temperatures in the Eastern Tropical Pacific Ocean," *International Journal of Climatology*, 10 (1990): 413–424.

22. For criticisms of Handler's findings, see A. Robock, "Volcanic Eruptions and Climate," *Reviews of Geophysics*, 38 (2000): 191–219; N. Nicholls, "Low-Latitude Volcanic Eruptions and the El Niño/Southern Oscillation: A Reply," *International Journal of Climatology*, 10 (1990): 425–429.

23. M. A. Cane et al., "Twentieth-Century Sea Surface Temperature Trends," *Science*, 275 (1997): 957–960.

24. The wind currents in the stratosphere—the atmospheric layer where aerosols are deposited by particularly explosive volcanic eruptions—are poleward. So if an eruption takes place outside the tropics, the volcanic aerosol cloud will never reach the tropics.

25. J. B. Adams, M. E. Mann, and C. M. Ammann, "Proxy Evidence for an El Niño-Like Response to Volcanic Forcing," *Nature*, 426 (2003): 274–278.

26. M. E. Mann, M. A. Cane, S. E. Zebiak, and A. Clement, "Volcanic and Solar Forcing of the Tropical Pacific Over the Past 1000 Years," *Journal of Climate*, 18 (2005): 447–456.

27. Handler was in many ways a kindred spirit. He too had started out in the field of condensed matter physics, having worked with the physicist John Bardeen, a corecipient of the Nobel Prize for his theoretical explanation of conventional superconductivity. Handler, too, had wandered into the field of climate dynamics through a circuitous career path. An obituary captured something of his scientific spirit: "Paul Handler possessed one of the greatest gifts a physicist can have—eternal curiosity, the pressing need to know how things work" (http://physics.illinois.edu/people/memorials/handler.asp).

28. We published a new set of spatial temperature reconstructions spanning the past 1,500 years in M .E. Mann, Z. Zhang, S. Rutherford, R. S. Bradley, M. K. Hughes, D. Shindell, C. Ammann, G. Falugevi, and F. Ni, "Global Signatures and Dynamical Origins of the 'Little Ice Age' and 'Medieval Climate Anomaly,'" *Science*, 326 (2009): 1256–1260.

29. With the more extensive climate proxy data available now relative to our original 1990s work, we were able to obtain a more reliable estimate of the pattern of the MCA (a term we favored over MWP), defined as the

three-century-long interval from A.D. 950 to 1250. We found evidence of warmth to be most pronounced over North America and northern Eurasia. In fact, our analysis indicated that, within the uncertainties, these regions might have been as warm then as they are today. On the other hand, the existence of cool conditions elsewhere—for example, in the tropical Pacific—meant that when averaged over the entire Northern Hemisphere or globe, medieval warmth did not quite approach the levels of warmth seen in the most recent decades, consistent with the conclusions of the hockey stick and many other confirmatory studies published since. Using a simulation of an updated version of the NASA GISS climate model, we showed that aspects of the MCA pattern were reproduced when the model was subject to the conditions estimated for medieval times—in particular, relatively high levels of solar output. This was the flip side of what we had found for the Little Ice Age in our 2001 *Science* article. The jet stream was now wiggling in the opposite way (it was in the positive phase of the NAO pattern) in response to the higher solar output of medieval times.

30. Andrew C. Revkin, "Climate Experts Tussle Over Details, Public Gets Whiplash," *New York Times*, July 29, 2008.

31. In a 2004 article entitled "The Scientific Consensus on Climate Change," *Science*, 306: 1386, UCSD geoscientist Naomi Oreskes surveyed all peer reviewed articles corresponding to the search term "global climate change" over the past decade present in the Institute for Scientific Information's database. In the 928 articles returned by the search, she did not find a single paper that rejected the scientific consensus position that "most of the observed warming of the last 50 years is likely to have been due to the increase in greenhouse gas concentrations."

32. For further details, see Appendix A to the Principles Governing IPCC Work, "Procedures for the Preparation, Review, Acceptance, Adoption, Approval, and Publication of IPCC Reports, Adopted at the Fifteenth Session (San Jose, April 15–18, 1999), amended at the Twentieth Session (Paris, February 19–21, 2003) and Twenty-First Session (Vienna, November 3 and 6–7, 2003).

33. Reviewers chosen in this way are not in fact expert reviewers, though there are a number of amusing instances of confusion on this point, wherein individuals with no expertise or training have publicly boasted of their status as expert reviewers of the IPCC report. See, e.g., Tim Lambert, "You Too Can Be a Leading Climate Scientist," Deltoid, May 9, 2006, http://scienceblogs.com /deltoid/2006/05/you_too_can_be_a_leading_clima.php.

34. To be precise, the IPCC is charged by the United Nations Environmental Program (UNEP) to assess the risk of "dangerous anthropogenic

interference with the climate." This mission is in turn tied to the United Nations Framework Convention on Climate Change (UNFCCC), which requires signatory nations (which includes all major nations) to stabilize greenhouse gas concentrations at levels short of DAI. For further discussion, see, e.g., M. E. Mann, "Defining Dangerous Anthropogenic Interference," *Proceedings of the National Academy of Science*, 106 (2009): 4065–4066, and the various other references given within.

35. An unintentionally amusing piece attacking climate change science was Barry Napier, "One World Government the Real Aim of Environmentalism," *Canada Free Press*, November 10, 2009.

36. U.S. National Academy of Sciences, *Climate Change Science: An Analysis of Some Key Questions*, 2001.

37. Royal Society, "Joint Science Academies' Statement: Global Response to Climate Change," 2005.

38. See Mark Townsend and Paul Harris, "Now the Pentagon Tells Bush: Climate Change Will Destroy Us," *Observer*, February 22, 2004.

39. S. Rahmstorf et al., "A Semi-Empirical Approach to Projecting Future," *Science*, 315 (2007): 368.

40. Thomas J. Kuhn, *The Structure of Scientific Revolutions* (Chicago, University of Chicago Press, 1962).

41. While Copernicus introduced the concept of heliocentrism nearly a century earlier, he was unwilling to abandon the Ptolemaic model consisting of so-called cycles and epicycles of planetary motion and attempted without success to fit an Earth-centric model into the Ptolemaic framework. Only Galileo, with a more physically driven model that broke with the Ptolemaic view, was able to provide convincing support for this new paradigm.

42. For detailed treatments of the history, see, e.g., Stillman Drake, *Galileo at Work: His Scientific Biography* (New York, Dover Publications, 1995); Giorgio de Santillana, *The Crime of Galileo* (Chicago, University of Chicago Press, 1978).

43. See, e.g., "Professor Barry Saltzman, a Pioneer in the Study of the Atmosphere and Climate, Dies," *Yale Bulletin and Calendar*, February 9, 2001. Lorenz's original paper contains the acknowledgement "The writer is indebted to Dr. Barry Saltzman for bringing to his attention the existence of nonperiodic solutions of the convection equations." E. N. Lorenz, "Deterministic Non-Periodic Flow," *Journal of Atmospheric Science*, 20 (1963): 130–141.

44. The term appears to have arisen in the blogosphere around 2004. One of the earliest references I could find is http://oracknows.blogspot.com/2005/03/galileo-gambit.html.

45. S. L. Myers, "A Growing and Expanding Earth Is No Longer Questionable," abstract submitted to 2008 AGU fall meeting in San Francisco.

46. This characterization is by Gavin Schmidt, www.realclimate.org/index .php/archives/2010/11/science-narrative-and-heresy/, November 3, 2010.

47. John P. Holdren, Comments on "The Shaky Science Behind the Climate Change Sense of the Congress Resolution," U.S. Senate Republican Policy Committee, June 9, 2003.

48. An August 12, 2006, article by Charles Montgomery in the *Globe and Mail* of Canada has characterized the group as "oil-patch geologists, Tory insiders, anonymous donors, and oil PR professionals."

49. See Hoggan and Littlemore, *Climate Cover-Up*, 49–52.

50. Listed as Independent Advisers at www.galileomovement.com.au /who_we_are.php#G (archived July 3, 2011). Listed advisers also include Patrick Michaels, Christopher Monckton, Fred Singer, and Richard Lindzen.

51. See www.galileomovement.com.au/who_we_are.php#B (archived July 3, 2011).

52. Hoggan and Littlemore note that, over the course of his career, Ball published a grand total of just four journal articles in the peer reviewed literature (many scientists publish more papers in a single year), none of which deal with the topic of atmospheric science. *Climate Cover-Up*, 144.

53. See ibid., 141–144.

54. Stephen Hawking, "Galileo and the Birth of Modern Science," *American Heritage's Invention and Technology*, 24 (Spring 2009): 36.

7. In the Line of Fire

1. These included, at this point in time, the various reconstructions discussed earlier: Briffa et al., Jones et al., Crowley and Lowery, and MBH98/ MBH99

2. Phil Jones published a commentary when our original 1998 *Nature* article was published: P. Jones, "It Was the Best of Times, It Was the Worst of Times," *Science*, 280 (1998): 544–545. We published a response contesting and/or clarifying some of the statements in Jones's piece, with Jones joining us as a coauthor: M. E. Mann et al., "Global Temperature Patterns," *Science*, 280 (1998): 2029–2030. Jones and his collaborator Tim Barnett and their colleagues expressed some additional reservations about the reliability of proxy records over past centuries in another article published in 1999: T. P. Barnett et al., "Detection and Attribution of Recent Climate Change: A Status Report," *Bulletin of the American Meteorological Society*, 80 (1999): 2631–2637. Here

they argued that theoretical climate models didn't support the information in the proxy data. We felt they had it backwards; one doesn't validate data with models, but rather models are validated by data, and we said as much in our response: R. S. Bradley, M. K. Hughes, and M. E. Mann, "Comments on 'Detection and Attribution of Recent Climate Change: A Status Report,'" *Bulletin of the American Meteorological Society*, 81 (2000): 2987–2990.

3. S. Huang, H. N. Pollack, et al., "Temperature Trends Over the Past Five Centuries Reconstructed from Borehole Temperatures," *Nature*, 403 (2000): 756–758.

4. E. R. Cook, K. R. Briffa, et al., "The 'Segment Length Curse' in Long Tree-Ring Chronology Development for Palaeoclimatic Studies," *Holocene*, 5 (1995): 229–237.

5. In M. E. Mann, S. Rutherford, R. S. Bradley, M. K. Hughes, and F. T. Keimig, "Optimal Surface Temperature Reconstructions Using Terrestrial Borehole Data," *Journal of Geophysical Research*, 108 (D7), doi:10.1029/2002 JD002532,2003, we argued that the admittedly sparse instrumental records that stretched back several centuries did not support the very cold temperatures suggested by the borehole evidence from the eighteenth and early nineteenth centuries, agreeing better with the proxy-based temperature reconstructions.

6. In M. E. Mann and G. A. Schmidt, "Ground vs. Surface Air Temperature Trends: Implications for Borehole Surface Temperature Reconstructions," *Geophysical Research Letters*, 30 (2003): 1607, doi:10.1029/2003GL017170, 2003, using climate model simulations, we argued that changes in seasonal snow cover over time could unduly influence borehole-based ground surface temperatures, since snow cover would change the degree of insulation of the ground surface from cold air during winter. In Mann et al., "Optimal Surface Temperature Reconstructions Using Terrestrial Borehole Data," we attempted to back out an alternative estimate of past temperatures from the boreholes by emphasizing the component of information in the borehole data most closely correlated with instrumental surface air temperature estimates. Our alternative approach yielded an estimate of surface temperatures in past centuries more in line with what the proxy-based studies showed.

7. Our interpretation of the simulations was challenged by D. S. Chapman, M. G. Bartlett, and R. N. Harris, "Comment on 'Ground vs. Surface Air Temperature Trends: Implications for Borehole Surface Temperature Reconstructions' by M. E. Mann and G. Schmidt," *Geophysical Research Letters*, 31 (2004): L07205, doi:10.1029/2003GL019054, which we rebutted in G. A. Schmidt and M. E. Mann, "Reply to 'Comment on "Ground vs. Surface Air Temperature Trends: Implications for Borehole Surface Temperature Reconstructions" by

D. Chapman et al.,'" *Geophysical Research Letters*, 31 (2004): L07206, doi:10.1029/2003GL0119144, 2004. Other borehole researchers argued for a smaller bias due to snow cover based on simple conceptual models and analysis of other climate model simulations. See, e.g., J. E. Smerdon, H. N. Pollack, V. Cermak, J. W. Enz, M. Kresl, J. Safanda, and J. F. Wehmiller, "Air-Ground Temperature Coupling and Subsurface Propagation of Annual Temperature Signals," *Journal of Geophysical Research*, 109 (2004): D21107, doi:10.1029/2004JD005056; F. Gonzalez-Rouco, H. von Storch, and E. Zorita, "Deep Soil Temperature as Proxy for Surface Air-Temperature in a Coupled Model Simulation of the Last Thousand Years," *Geophysical Research Letters*, 30 (2003): 2116, doi:10.1029/2003GL018264. The debate continues, with us most recently using simulations of climate changes over the course of the Holocene period to demonstrate the potential for a large snow cover influence on ground temperatures. See, e.g., M. E. Mann, G. A. Schmidt, S. K. Miller, and A. N. LeGrande, "Potential Biases in Inferring Holocene Temperature Trends from Long-Term Borehole Information," *Geophysical Research Letters*, 36 (2009): L05708, doi:10.1029/2008GL036354, and other references cited within.

8. W. S. Broecker, "Paleoclimate: Was the Medieval Warm Period Global?" *Science*, 291 (February 23, 2001): 1497–1499. "Perspective" pieces in *Science* are opinion pieces that are edited, but not, in a strict sense, peer reviewed.

9. R. S. Bradley, K. R. Briffa, et al., "The Scope of Medieval Warming," *Science*, 292 (2001): 2011–2012.

10. M. E. Mann, Z. Zhang, S. Rutherford, R. S. Bradley, M. K. Hughes, D. Shindell, C. Ammann, G. Falugevi, and F. Ni, "Global Signatures and Dynamical Origins of the 'Little Ice Age' and 'Medieval Climate Anomaly,'" *Science*, 326 (2009): 1256–1260.

11. While it was originally believed that these oscillations were largely restricted to the past glacial period, Bond and collaborators argued for the persistence of these roughly 1,500-year "Bond cycles": G. Bond, W. Showers, et al., "A Pervasive Millennial-Scale Cycle in North Atlantic Holocene and Glacial Climates," *Science*, 278 (1997): 1257–1266.

12. J. Esper, E. R. Cook, et al., "Low-Frequency Signals in Long Tree-Ring Chronologies for Reconstructing Past Temperature Variability," *Science*, 295 (March 22, 2002): 2250–2253.

13. The method is known as regional curve standardization (RCS). RCS is especially aggressive, methodologically speaking, in attempting to retain long-term trends, but this can prove problematic. Too tight a retention criterion and real climate trends can be eliminated, but too loose a criterion and artificial long-term trends (e.g., changes in tree growth related to non-climatic factors

like forest disturbance and competition between neighboring trees) can creep in. This is increasingly problematic with RCS when sample sizes become small.

14. If they had used the full instrumental record back through the mid-nineteenth century—rather than the twentieth-century portion only—to scale their reconstruction, the amplitude of the centennial fluctuations in temperature in their reconstruction would have been substantially reduced.

15. Malcolm Hughes and I made these points in a comment we published later that year: M. E. Mann, M. K. Hughes, et al., "Tree-Ring Chronologies and Climate Variability," *Science,* 296 (May 3, 2002): 848–849. We specifically noted Cook's caution in previously published work that "Successful use of the RCS method generally requires a large number of tree ring series" and questioned whether this condition was met in the small samples (twenty-five to thirty-five trees) used in the early part (900–1300) of their record.

16. E. R. Cook, J. Esper, et al., "Extra-Tropical Northern Hemisphere Land Temperature Variability Over the Past 1000 Years," *Quaternary Science Reviews,* 23 (2004): 2063–2074.

17. See, e.g., the comparison of a dozen reconstructions published in chapter 6 of the 2007 IPCC Fourth Assessment Report: E. Jansen, J. Overpeck, K. R. Briffa, J.-C. Duplessy, F. Joos, V. Masson-Delmotte, D. Olago, B. Otto-Bliesner, W. R. Peltier, S. Rahmstorf, R. Ramesh, D. Raynaud, D. Rind, O. Solomina, R. Villalba, and D. Zhang, "Palaeoclimate," in *Climate Change 2007: The Physical Science Basis. Contribution of Working Group I to the Fourth Assessment Report of the Intergovernmental Panel on Climate Change,* ed. S. Solomon, D. Qin, M. Manning, Z. Chen, M. Marquis, K. B. Averyt, M. Tignor, and H. L. Miller (Cambridge, UK, and New York, Cambridge University Press, 2007).

18. M. E. Mann and S. Rutherford, "Climate Reconstruction Using 'Pseudoproxies,'" *Geophysical Research Letters,* 29 (2002): 1501, doi:10.1029/2001GL014554. The use of climate model simulations to test climate reconstruction approaches was described in S. Rutherford, M. E. Mann, T. L. Delworth, and R. Stouffer, "Climate Field Reconstruction Under Stationary and Nonstationary Forcing," *Journal of Climate,* 16 (2003): 462–479.

19. H. Von Storch, E. Zorita, J. M. Jones, Y. Dimitriev, F. Gonzalez-Rouco, and S. F. B. Tett, "Reconstructing Past Climate from Noisy Data," *Science,* 306 (2004): 679–682.

20. M. E. Mann and S. Rutherford, "Climate Reconstruction Using 'Pseudoproxies,'" *Geophysical Research Letters,* 29 (2002): 1501, doi:10.1029/2001GL014554.

21. The step involved the removal of trends (detrending) from the data prior to estimating statistical relationships. It had the net effect of removing the main pattern of variation from the data. The detrending error was pointed

out by E. R. Wahl, D. M. Ritson, and C. M. Ammann, "Comment on 'Reconstructing Past Climate from Noisy Data,'" *Science*, 312 (2006): 529b. A separate criticism by S. Rahmstorf, "Testing Climate Reconstructions," *Science*, 312 (2006): 1872, argued that results contradicting the author's conclusions were buried in the "Supplementary Information" of the paper, while only the results more consistent with their claims were shown in the main article. Others have gone on to repeat the detrending error in subsequent publications—e.g., G. Burger and U. Cubasch, "Are Multiproxy Climate Reconstructions Robust?" *Geophysical Research Letters*, 32 (2005): L23711, doi:10.1029/2005GL024155; G., Burger, I. Fast, and U. Cubasch, "Climate Reconstruction by Regression," *Tellus*, Ser. A, 58 (2006): 227–235—perpetuating a misleading criticism of proxy-based temperature reconstructions.

22. See, e.g., Ross Gelbspan, *The Heat Is On* (New York, Basic Books, 1998), 40–44; James Hoggan and Richard Littlemore, *Climate Cover-Up* (Vancouver, BC, Greystone Books, 2009), 105–108. Citing rebuttal testimony of Patrick J. Michaels in *Before the Minnesota Public Utilities Commission: In the Matter of the Quantification of Environmental Costs Pursuant to Laws of Minnesota*, 1993, chap. 356, sec. 3, March 15, 1995, Gelbspan notes that Michaels disclosed under oath that he had received a $63,000 research grant from the Western Fuels Association. Michaels also received $49,000 from the German Coal Mining Electric Institute and a grant of $40,000 from the Cyprus Minerals mining company.

23. Michaels has disclosed under oath that the Western Fuels $63,000 research grant had funded his publications, including the World Climate Report (Gelbspan, ibid.).

24. The divergence problem applies primarily to tree ring density (as opposed to ring width) data such as those used by K. R. Briffa, F. H. Schweingruber, P. D. Jones, T. J. Osborn, S. G. Shiyatov, and E. A. Vaganov, "Reduced Sensitivity of Recent Tree-Growth to Temperature at High Northern Latitudes," *Nature*, 391 (1998): 678–682. These particular proxy data suffer from an enigmatic decline in their response to warming temperatures after about 1960. MBH98 and MBH99 used very few tree ring density records, and the divergence problem has not been demonstrated to be a problem with the hockey stick.

25. Daniel Grossman, "Dissent in the Maelstrom," *Scientific American*, November 2001.

26. Bruce Sullivan, "Scientists Say Global Warming Theory Is a Lot of Hot Air," CNS News, May 30, 2000.

8. Hockey Stick Goes to Washington

1. "A Comprehensive National Energy Policy," Saginaw, Michigan, Friday, September 29, 2000.

2. Chris Mooney, *The Republican War on Science* (New York, Basic Books, 2005).

3. Televised interview, June 26, 2006, http://thinkprogress.org/2006/06/26/bush-debate-climate/.

4. "Bush Pulls U.S. Out of Kyoto Talks," *Boston Globe*, March 29, 2001, A01.

5. Michael Abramowitz and Steven Mufson, "Papers Detail Industry's Role in Cheney's Energy Report," *Washington Post*, July 18, 2007.

6. Through at least 2003, CEI received funding from ExxonMobil; see Jennifer Lee, "Exxon Backs Groups That Question Global Warming," *New York Times*, May 28, 2003. According to *Media Transparency*, CEI has also received funding over the years from the Scaife and Koch Foundations, among other conservative/free market foundations; see http://mediamattersaction.org/transparency/organization/Competitive_Enterprise_Institute/funders.

7. This information was revealed in a memo from Ebell to Cooney obtained through the Freedom of Information Act (FOIA) by Greenpeace on September 9, 2003: "Greenpeace Obtains Smoking-Gun Memo: White House/Exxon Link," www.greenpeace.org/international/en/news/features/investigation-of-exxon-front-g/.

8. Jake Tapper, "Bush's EPA Chief Seeks Greener Pastures," Salon.com, May 22, 2003. Vice President Dick Cheney's role has been reported elsewhere, e.g., in an interview on NPR's *Frontline* on January 9, 2007, in which Whitman revealed that the opposition to carbon caps was coming directly from his office: "The vice president's fingerprints were all over this," she commented.

9. See "Bush White House Charged with Conspiracy with Private Think Tank," Office of Maine Attorney General, August 11, 2003, www.heatisonline.org/contentserver/objecthandlers/index.cfm?id=4391&method=full; Ross Gelbspan, *Boiling Point* (New York, Basic Books, 2004), 57–58.

10. *Competitive Enterprise Institute v. George Bush*, Complaint for Declarative Relief, No. 1:03CV1670 RJL (D.D.C. August 6, 2003); *Competitive Enterprise Institute v. George Bush*, Stipulation of Dismissal with Prejudice, No. 1:03CV1670 RJL (D.D.C. November 4, 2003).

11. Press Release from the Office of Senator Joseph Liebermann, "Lieberman Seeks White House Contacts Regarding Global Warming Lawsuit," September 24, 2003.

12. Andrew C. Revkin, "Bush Aide Softened Greenhouse Gas Links to Global Warming," *New York Times*, June 8, 2005.

13. Ibid.

14. Andrew Revkin, "Former Bush Aide Who Edited Reports Is Hired by Exxon," *New York Times*, June 15, 2005.

15. Andrew C. Revkin with Katharine Q. Seelye, "Report by the E.P.A. Leaves Out Data on Climate Change," *New York Times*, June 19, 2003. The Soon and Baliunas paper mentioned the API funding in the acknowledgements section; W. Soon and S. Baliunas, "Proxy Climatic and Environmental Changes Over the Past 1000 Years," *Climate Research*, 23 (2003): 89–110.

16. This information was documented in a publicly released report by the U.S. House of Representatives Committee on Oversight and Government Reform, "Political Interference with Climate Change Science Under the Bush Administration," December 12, 2007.

17. M. E. Mann and J. Park, "Spatial Correlations of Interdecadal Variation in Global Surface Temperatures," *Geophysical Research Letters*, 20 (1993): 1055–1058.

18. S. Baliunas and R. Jastrow, "Evidence for Long-Term Brightness Changes of Solar-Type Stars," *Nature*, 348 (1990): 520–522.

19. P. Forster, V. Ramaswamy, P. Artaxo, T. Berntsen, R. Betts, D. W. Fahey, J. Haywood, J. Lean, D. C. Lowe, G. Myhre, J. Nganga, R. Prinn, G. Raga, M. Schulz, and R. Van Dorland, "Changes in Atmospheric Constituents and in Radiative Forcing," in S. Solomon, D. Qin, M. Manning, Z. Chen, M. Marquis, K. B. Averyt, M. Tignor, and H. L. Miller, eds., *Climate Change 2007: The Physical Science Basis*, Contribution of Working Group I to the Fourth Assessment Report of the Intergovernmental Panel on Climate Change (Cambridge, UK, and New York, Cambridge University Press, 2007).

20. W. Soon and S. Baliunas, "Proxy Climatic and Environmental Changes Over the Past 1000 Years," *Climate Research*, 23 (2003): 89–110; W. Soon, W., S. Baliunas, C. Idso, S. Idso, and D. R. Legates, "Reconstructing Climatic and Environmental Changes of the Past 1000 Years: A Reappraisal," *Energy and Environment*, 14 (2003): 233–296.

21. See, e.g., James Hoggan and Richard Littlemore, *Climate Cover-Up* (Vancouver BC, Greystone Books, 2009), 105; Mooney, *The Republican War on Science*, 87; Jeff Nesmith, "Foes of Global Warming Theory Have Energy Ties," *Seattle Post-Intelligencer*, June 2, 2003.

22. An e-mail from editorial board member Mike Hulme of the University of East Anglia on June 12, 2003, documents past concerns on his part and on the part of Dr. Wolfgang Cramer (a former editorial board member who had

quit the journal in part over his concern with editorial practices) dating back to November 1999.

23. De Freitas participated, for example, in the preparation of an amicus brief by the fossil fuel industry-funded Competitive Enterprise Institute challenging the right of the U.S. EPA to regulate greenhouse gas emissions; http://cei.org/pdf/5572.pdf.

24. Richard Monastersky, "Storm Brews Over Global Warming," *Chronicle of Higher Education*, September 5, 2003.

25. Boehmer-Christiansen has further elaborated on this point in recent years. See the online comment thread of the news article "Institute Denies Censoring 'Global Cooling' Article," *Times Higher Education*, August 13, 2009, comment by Boehmer-Christiansen, September 3, 2009, www.timeshighereducation.co.uk/story.asp?storycode=407763 (accessed July 6, 2011): "By the way, E&E is not a science journal and has published IPCC critiques to give a platform [to] critical voices and 'paradigms' because of the enormous implications for energy policy, the energy industries and their employees and investors, and for research. We do not claim to be right . . ."

26. Monastersky, "Storm Brews Over Global Warming." Black told Monastersky that "winds don't meet their definition of warm, wet, or dry" and that, according to Monastersky, he had found no fifty-year period of medieval extremes in his record, contradicting claims to the contrary made about his record by Soon and Baliunas.

27. Quoted in Irene Sanchez, "Warming Study Draws Fire," *Harvard Crimson*, September 12, 2003.

28. Daniel Schrag, a paleoclimatologist in the Department of Earth and Planetary Sciences at Harvard who went on to win a MacArthur "genius" award a couple years later, was also quoted: "this paper is suggesting that the unusually warm weather we've been having for the last 100 years is part of natural variability. We have observations to show that that's not the case."

29. M. E. Mann, C. M. Ammann, R. S. Bradley, K. R. Briffa, T. J. Crowley, M. K. Hughes, P. D. Jones,. M. Oppenheimer, T. J. Osborn, J. T. Overpeck, S. Rutherford, K. E. Trenberth, and T. M. L. Wigley, "On Past Temperatures and Anomalous Late 20th Century Warmth," *Eos*, 84 (2003): 256–258; a comment by Soon and Baliunas and our reply were published later in M. E. Mann, C. M. Ammann, R. S. Bradley, K. R. Briffa, T. J. Crowley, M. K. Hughes, P. D. Jones, M. Oppenheimer, T. J. Osborn, J. T. Overpeck, S. Rutherford, K. E. Trenberth, and T. M. L. Wigley, "Response to Comment on 'On Past Temperatures and Anomalous Late 20th Century Warmth,'" *Eos*, 84 (2003): 473.

30. Climate Change Update, Senate Floor Statement by U.S. Senator James M. Inhofe (R-OK), January 4, 2005.

31. Hoggan and Littlemore, *Climate Cover-Up*, 96.

32. This hearing would be the first of two back-to-back hearings, to be followed by a second hearing chaired by Inhofe on the threats of environmental mercury contamination. I expected Soon to testify for Inhofe in that hearing, too, given his claimed expertise on that topic. TechCentralStation listed Soon as a science director between 2003 and 2007, with expertise including mercury and global warming. His only apparent writing on this topic, however, was a press release, "Is EPA Ignoring the Science on Mercury?" put out on April 18, 2003, by the Center for Science and Public Policy, a project of the industry-funded Frontiers of Freedom (a group that has received hundreds of thousands of dollars from ExxonMobil; see Jennifer Lee, "Exxon Backs Groups That Question Global Warming," *New York Times*, May 28, 2003). Soon did not in fact testify in the mercury hearing.

33. Mooney, *The Republican War on Science*, 87; Nesmith, "Foes of Global Warming Theory Have Energy Ties."

34. See Chris Mooney, "Earth Last," *American Prospect*, April 13, 2004, for an account. Both Stephen Schneider and Tom Wigley followed up with formal letters protesting Inhofe's misrepresentation of their work. Inhofe, for example, had characterized Schneider as a critic of the IPCC whose work "cast serious doubt" on IPCC's projections of warming. Schneider responded: "It is misrepresenting my views to characterize them as even implying that IPCC has exaggerated or failed to describe the state of the science fairly."

35. I referred to the so-called FACE (free-air CO_2 enhancement) experiments at Duke Forest, in which CO_2 was actively pumped into the atmosphere below the forest canopy, and detailed measurements over time of the growth and productivity response of the trees and net carbon uptake were made. The experiments indicated that any increases in uptake tend to saturate quickly with further warming.

36. The group in question is WEFA, the Wharton Econometric Forecasting Association, an economics forecasting and consulting organization founded as a spinoff of the University of Pennsylvania's Wharton School of Business by faculty member Lawrence Klein.

37. Antonio Regalado, "Global Warming Skeptics Are Facing Storm Clouds," *Wall Street Journal*, July 31, 2003.

38. See, e.g., Hoggan and Littlemore, *Climate Cover-Up*, 104–105.

39. The chair of my department at the time, Bruce Hayden (a close friend of controversial University of Virginia climate change contrarian Patrick

Michaels) initially wrote letters apologizing to both Christy and de Freitas for my Senate testimony, copying President Casteen. He did so without consulting either me or the dean of our college—a rather unconventional action for a department chair to take in such a situation. Following my complaint to Casteen, where I pointed out the falsehood of the criticisms made by both Christy and de Freitas, and my dissatisfaction with how Hayden had acted, Hayden wrote a follow-up letter to Casteen retracting his original apologies. "It is clear in my mind," Hayden wrote, "that Professor Mann has provided additional information and clarification of the content and the context of his Senate testimony. In this way he has diminished, in my mind, the substance of the accusations by Drs. Christy and de Frietas."

40. National Research Council, *Reconciling Observations of Global Temperature Change* (Washington, DC, National Academy Press, 2000). John M. Wallace of the University of Washington chaired the committee, and John Christy was one of the coauthors. The report concluded: "In the opinion of the panel, the warming trend in global-mean surface temperature observations during the past 20 years is undoubtedly real and is substantially greater than the average rate of warming during the twentieth century. The disparity between surface and upper air trends in no way invalidates the conclusion that surface temperature has been rising."

41. See, e.g., Monastersky, "Storm Brews Over Global Warming."

42. Gavin Schmidt and I wrote "Peer Review: A Necessary but Not Sufficient Condition," RealClimate.org, January 20, 2005, www.realclimate.org/index.php/archives/2005/01/peer-review-a-necessary-but-not-sufficient-condition/, explaining this principle, and citing the Soon and Baliunas study as an example.

43. S. McIntyre and R. McKitrick, "Corrections to the Mann et al. [1998] Proxy Database and Northern Hemisphere Average Temperature Series," *Energy and Environment*, 14 (2003): 751–771.

44. See Paul D. Thacker, "How the Wall Street Journal and Rep. Barton Celebrated a Global Warming Skeptic," *Environmental Science and Technology*, August 31, 2005.

45. To be specific, they claimed that the hockey stick was an artifact of four supposed "categories of errors": "collation errors," "unjustified truncation and extrapolation," "obsolete data," and "calculation mistakes." As we noted in a reply to a McIntyre and McKitrick comment on MBH98 that had been submitted to and rejected by *Nature* (because their comment was rejected anyway, our reply would not appear there either), those claims were false, resulting from their misunderstanding of the format of a spreadsheet version of the

dataset they had specifically requested from my associate, Scott Rutherford. None of the problems they cited were present in the raw, publicly available version of our dataset, which was available at that time at ftp://holocene.evsc .virginia.edu/pub/MBH98/.

46. Because the journal that published the McIntyre and McKitrick article—*Energy and Environment*—was not a recognized scientific journal, we chose not to submit a comment to it. The errors in their claims, including their deletion of key records from our proxy network, were detailed in the RealClimate.org post "False Claims by McIntyre and McKitrick Regarding the Mann et al. (1998) Reconstruction," December 4, 2004, and in subsequent peer reviewed work published in other journals, as described in the next chapter.

47. I would go on to interact cordially with her when she became deputy communications director for California Governor Schwarzenegger (now going by the name Laura Braden) and, in a rather striking reversal of her previous work, became responsible for coordinating the governor's proactive climate change policy agenda.

48. TechCentralStation bills itself as a Web site "where free markets meet technology." According to SourceWatch.org (www.sourcewatch.org/index .php?title=Tech_Central_Station, accessed July 23, 2011), the site was established by the DCI Group, a Republican consulting and lobbying firm based in Washington, D.C., that derives its funding from a variety of corporate sponsors in the fossil fuel, pharmaceutical, fast food, and telecommunications industries, including fossil fuel giant ExxonMobil.

49. Nick Schulz, "Researchers Question Key Global Warming Study," *USA Today*, October 30, 2003.

50. The Sense of the Senate resolution, cosponsored by Senators Robert C. Byrd (D-WV) and Chuck Hagel (R-NE), stated that it was not the sense of the Senate that the United States should be a signatory to the Kyoto Protocol. The 95–0 vote was in fact a 95–0 defeat of support of climate action.

51. The correction, published in the Corrections and Clarifications box at the bottom of the op-ed page in the November 13, 2003, issue of *USA Today*, stated: "In an Oct. 29 Forum article about new research that challenges the findings of an earlier study on global warming, the writer said the data on the original study by University of Virginia assistant professor Michael Mann aren't available online. The data can be accessed at ftp://holocene.evsc.virginia .edu/pub/MBH98/."

52. Dan Vergano, "Global Warming Debate Heats Up Capitol Hill," *USA Today*, November 18, 2003. Vergano quoted climate researcher Tom Wigley of NCAR as calling McIntyre and McKitrick's attacks "seriously flawed" and

"silly," noting also that about a dozen independent studies suggest that the twentieth century was warmer than normal.

9. When You Get Your Picture on the Cover of . . .

1. Ironically, the song did end up landing Dr. Hook on the coveted cover.

2. Tim Dickinson, "The Climate Killers: Meet the 17 Polluters and Deniers Who Are Derailing Efforts to Curb Global Warming," *Rolling Stone*, January 6, 2010 (cover article).

3. Antonio Regalado, "In Climate Debate, the 'Hockey Stick' Leads to a Face-Off," *Wall Street Journal*, February 14, 2005.

4. McIntyre was included in a list of "climate villains" published around the same time by an alternative newspaper: Michael Roddy and Ian Murphy, "The 14 Most Heinous Climate Villains," *Buffalo Beast*, December 29, 2009. Its list included other familiar individuals: Patrick Michaels, Sallie Baliunas, Steve Milloy, Roy Spencer, Richard Lindzen, and Christopher Monckton, as well as Stephen McIntyre.

5. Recall from chapter 8 that promotion for the 2003 McIntyre and McKitrick paper was coordinated by the ExxonMobil-funded outlet TechCentralStation, a site that was singled out by Senators Olympia Snow (R-ME) and Jay Rockefeller (D-WV) in demanding that ExxonMobil "end any further financial assistance" to groups "whose public advocacy has contributed to the small but unfortunately effective climate change denial myth."

6. Paul D. Thacker, "How the *Wall Street Journal* and Rep. Barton Celebrated a Global Warming Skeptic," *Environmental Science and Technology*, August 31, 2005.

7. NOAA's Geophysical Fluid Dynamics Laboratory (GFDL) in Princeton, New Jersey.

8. Mahlman had explicitly pointed out to Regalado that "numerous other studies have confirmed Mann's original results."

9. This information was provided in e-mail correspondence between Zwiers and Phil Jones of the University of East Anglia that was made available to the author.

10. I was not given an opportunity to review the statements Regalado attributed to me. Had I, I would have pointed out the problematic nature of the attributed statement.

11. Dr. David J. Verardo, Director, Paleoclimate Program, Division of Atmospheric Sciences, National Science Foundation, wrote this in an e-mail to

Mr. Steve McIntyre (copied to me) on December 17, 2003: "Dear Mr. McIntyre, I apologize if my last electronic message was not clear but let me clarify the US NSF's view in this current message. Dr. Mann and his other US colleagues are under no obligation to provide you with any additional data beyond the extensive data sets they have already made available. He is not required to provide you with computer programs, codes, etc. His research is published in the peer-reviewed literature which has passed muster with the editors of those journals and other scientists who have reviewed his manuscripts. You are free to your analysis of climate data and he is free to his. The passing of time and evolving new knowledge about Earth's climate will eventually tell the full story of changing climate. I would expect that you would respect the views of the US NSF on the issue of data access and intellectual property for US investigators as articulated by me to you in my last message under the advisement of the US NSF's Office of General Counsel. Respectfully, David J. Verardo, Director, Paleoclimate Program, Division of Atmospheric Sciences."

12. Texas A&M climate scientist Gerald North later characterized Regalado's depiction of the hockey stick debate as "downright dishonest": Richard Monastersky, "Climate Science on Trial," *Chronicle of Higher Education*, September 8, 2006.

13. McIntyre argued on his Web site that his submission was rejected simply due to "lack of space." *Nature* makes its policy on such submissions quite clear, however: "The Brief Communications editor will decide how to proceed on the basis of whether the central conclusion of the earlier paper is brought into question; of the length of time since the original publication; and of whether a comment or exchange of views is likely to seem of interest to nonspecialist readers. Because *Nature* receives so many comments, those that do not meet these criteria are referred to the specialist literature." Since *Nature* chose to send the comment out for review in the first place, the "time since the original publication" was clearly not deemed a problematic factor. One is left to conclude that the grounds for rejection were the deficiencies in the authors' arguments explicitly noted by the reviewers. Although McIntyre and McKitrick's comment was rejected by *Nature*, it did bring to our attention some minor errors in the online supplement to our article in the locations or names of some of the proxy series used. *Nature* asked us to publish an online correction or "corrigendum" correcting those details (*Nature*, 430 [2004]: 105). All too aware that some climate change deniers would do their best to misrepresent the corrigendum, which in no way impacted any of the results of our study, we made sure that the corrigendum carried the unambiguous statement that "None of these errors affect our previously published results."

Predictably, some of the usual disinformation outlets nonetheless tried to promote the canard that we had issued some sort of "correction of MBH98" or that the corrigendum indicated that there were "errors in MBH98."

14. S. McIntyre and R. McKitrick, "Hockey Sticks, Principal Components, and Spurious Significance," *Geophysical Research Letters*, 32 (2005): L03710, doi: 10.1029/2004GL021750.

15. Stephen J. Gould, *The Mismeasure of Man* (New York, W.W. Norton, 1981); the book was updated and reissued in 1996 to address more recent claims made in Richard Herrnstein and Charles Murray, *The Bell Curve*.

16. It was another famous statistician, Karl Pearson, who had actually developed the statistical technique several years earlier, in 1901. Interestingly, both Pearson and Spearman are associated with the two prevailing ways of estimating correlation coefficients (the Pearson and the Spearman correlation coefficients).

17. Gould, *The Mismeasure of Man*, 1996 edition, 24.

18. In our case, we used a criterion known as the Preisdendorfer's Rule N— a fancy-sounding name for what is in fact a conceptually simple approach. To implement this criterion, one compares how much variation each PC explains with results from a parallel set of analyses in which the actual dataset has been replaced with a purely random dataset of the same size. One repeats this process numerous times, generating a family of random surrogates that characterize the range of results one would expect for purely random data. One ends up keeping as many PCs as there are in the original data that exceed, in how much variation they explain, the levels found for the purely random surrogates. By convention, a particular PC (say, PC#*N*) is considered statistically significant if it explains an amount of variation in the data that is greater than in 95 percent of the random PC#*N*s generated.

19. While the centering of the data is the primary consideration, there is the secondary issue of the normalization of the data. If one is dealing with, say, an instrumental thermometer-based surface temperature dataset, then each series in the dataset (the series of temperatures at each location) has dimensions of temperature, e.g., degrees Celsius. So no further adjustments are necessary. However, if one is analyzing proxy records, then one is potentially comparing apples and oranges. An ice core record, for example, might have dimensions of an isotopic composition ratio, while a sediment record might have dimensions of inches of annual sediment accumulation. This principle applies even to a set of tree ring records, since some tree ring data reflect annual ring widths, while other tree ring data reflect wood density. In such cases, the dimensions of the different series are not comparable, and the only meaningful way to analyze them simultaneously is to normalize each so they have

the same overall amplitude of fluctuation. The easiest way to do this is to divide each series by its standard deviation.

20. Using their long-term centering, the hockey stick pattern emphasizing the high-elevation western North American tree ring data was no longer PC#1, but was demoted to PC#4. However, the selection rule for the long-term centering indicates that a greater number of PCs (five to be precise) should be retained using that convention. So the hockey stick pattern is still kept. But McIntyre and McKitrick in their 2005 paper did not derive the selection rule appropriate for their convention. They simply assumed that our selection rule—keeping the first two PCS only—that had been derived based on a modern-centering convention, could be applied to their results.

21. We originally demonstrated these findings in a series of posts on Real-Climate.org, including "False Claims by McIntyre and McKitrick Regarding the Mann et al. (1998) Reconstruction," December 4, 2004; "On Yet Another False Claim by McIntyre and McKitrick," January 6, 2005. All of these findings would later be confirmed by independent peer reviewed work by Wahl, Ammann, and other researchers as detailed in subsequent discussion in this chapter.

22. See the RealClimate.org post "False Claims by McIntyre and McKitrick Regarding the Mann et al. (1998) Reconstruction."

23. For example, S. Rutherford, M. E. Mann, T. J. Osborn, R. S. Bradley, K. R. Briffa, M. K. Hughes, and P. D. Jones, "Proxy-Based Northern Hemisphere Surface Temperature Reconstructions: Sensitivity to Methodology, Predictor Network, Target Season and Target Domain," *Journal of Climate*, 18 (2005): 2308–2329, demonstrated that essentially the same MBH98 hockey stick reconstruction was obtained whether or not the PCA step was used to represent tree ring data networks. With the improved Climate Field Reconstruction (CFR) method we had been using since 2001 based on a statistical approach known as regularized expectation-maximization (RegEM), that step was no longer necessary. See M. E. Mann and S. Rutherford, "Climate Reconstruction Using 'Pseudoproxies,'" *Geophysical Research Letters*, 29 (2002): 1501, doi:10.1029/2001GL014554; S. Rutherford, M. E. Mann, T. L. Delworth, and R. Stouffer, "Climate Field Reconstruction Under Stationary and Nonstationary Forcing," *Journal of Climate*, 16 (2003): 462–479; Z. Zhang, M. E. Mann, and E. R. Cook, "Alternative Methods of Proxy-Based Climate Field Reconstruction: Application to the Reconstruction of Summer Drought Over the Conterminous United States Back to 1700 from Drought-Sensitive Tree Ring Data," *Holocene*, 14 (2004): 502–516.

24. The work was originally announced in a May 11, 2005, NCAR press release "The Hockey Stick Controversy: New Analysis Reproduces Graph of

Late 20th Century Temperature Rise," www.ucar.edu/news/releases/2005
/ammann.shtml. It was later published in two peer reviewed articles: E. R.
Wahl and C. M. Ammann, "Robustness of the Mann, Bradley, Hughes Reconstruction of Surface Temperatures: Examination of Criticisms Based on the
Nature and Processing of Proxy Climate Evidence," *Climatic Change*, 85
(2007): 33–69; C. M. Ammann and E. R. Wahl, "The Importance of the Geophysical Context in Statistical Evaluations of Climate Reconstruction Procedures," *Climatic Change*, 85 (2007): 71–88.

25. This was first shown by us in the RealClimate post "On Yet Another
False Claim by McIntyre and McKitrick," January 6, 2005, and later verified by
Wahl and Ammann.

26. McIntyre has variously referred to this as "persistent red noise" or
"trendless persistent red noise" on his climateaudit blog. It is a noise model
that allows for unrealistically large long-term swings. In the case at hand, Mc-
Intyre and McKitrick had taken time series that have a prominent twentieth-
century trend that is almost certainly related to human-caused climate change.
However, they produce random "surrogate" data that build the "persistence" in
the time series resulting from that trend into a statistical model for natural
variability. By pretending that natural, pre-anthropogenic variations can be
characterized by the same degree of persistence that results from human-
caused climate trends, the model dramatically overestimates the natural
swings that can be expected to be found in proxy data such as tree rings. Wahl
and Ammann (2007) demonstrated that McIntyre's estimates of natural variability and statistical significance were thoroughly compromised by this inappropriate approach.

27. H. Von Storch and E. Zorita, "Comment on 'Hockey Sticks, Principal
Components, and Spurious Significance' by S. McIntyre and R. McKitrick,"
Geophysical Research Letters, 32 (2005): L20701, doi:10.1029/2005GL022753.
They used a pseudoproxy analysis from a long-term climate model simulation
to demonstrate that the hockey sticks that McIntyre and McKitrick claimed
they could manufacture from noise alone had tiny amplitudes in comparison
with the actual hockey stick, and the PCA conventions had little if any impact
on the resulting reconstruction.

28. P. Huybers, "Comment on 'Hockey Sticks, Principal Components, and
Spurious Significance' by S. McIntyre and R. McKitrick," *Geophysical Research
Letters*, 32 (2005): L20705, doi:10.1029/2005GL023395.

29. Huybers demonstrated that McIntyre and McKitrick's claimed high
thresholds for statistical skill were based on a data handling error: They failed
to adjust their random surrogates to have the correct amplitude of variability.
Performing the procedure correctly, Huybers independently affirmed that

the MBH98 reconstruction did indeed pass the threshold for statistical skillfulness.

30. See, e.g., M. N. Juckes et al., "Millennial Temperature Reconstruction Intercomparison and Evaluation, *Climate of the Past*, 3 (2007): 591–609; G. C. Hegerl, T. J. Crowley, W. T. Hyde, and D. J. Frame, "Climate Sensitivity Constrained by Temperature Reconstructions Over the Past Seven Centuries," *Nature*, 440 (2006): 1029–1032; T. C. K. Lee, F. W. Zwiers, and M. Tsao, "Evaluation of Proxy-Based Millennial Reconstruction Methods," *Climate Dynamics*, 31 (2008): 263–281.

31. Climate policy expert Joe Romm of the Center for American Progress, for example, had this to say in a review of Muller's *Physics for Future Presidents* on his Climate Progress blog: "the book is full of opinions and misinformation, not science, and . . . what is being taught would certainly mislead Future Presidents on issues such as terrorism, climate, and electric cars." Joe Romm, "Confusing Future Presidents, Part 1," September 13, 2008, http://climateprogress.org/2008/09/13/confusing-future-presidents-part-1/.

32. Richard A. Muller, "Medieval Global Warming," *Technology Review*, December 17, 2003.

33. Richard A. Muller, "Global Warming Bombshell," *Technology Review*, October 14, 2004.

34. "Welcome Climate Bloggers," *Nature*, 432 (December 23, 2004): 933.

35. "Web Logs: Sifting for Truth About Global Warming," *Science*, 306 (December 24, 2004): 2167.

36. "Science and Technology Web Awards 2005," *Scientific American*, October 2005.

37. Gavin Schmidt, "Michael Crichton's State of Confusion," www.realclimate.org/index.php/archives/2004/12/michael-crichtons-state-of-confusion/.

38. U.S. Senator James M. Inhofe (R-OK), "Climate Change Update," Senate Floor Statement, January 4, 2005, http://inhofe.senate.gov/pressreleases/climateupdate.htm.

39. Michael Mann, Stefan Rahmstorf, Gavin Schmidt, Eric Steig, and William Connolley, "Senator Inhofe on Climate Change," www.realclimate.org/index.php/archives/2005/01/senator-inhofe/.

40. E.g., T. J. Crowley, "Causes of Climate Change Over the Past 1000 Years," *Science*, 289 (2000): 270–277, discussed in chapter 4.

41. Inhofe held a hearing of the Senate Committee on Environment and Public Works on "The Role of Science in Environmental Policy-Making" with Crichton as one of the witnesses on September 28, 2005.

42. Boxer' specific statement was, "A lot of people are being maligned here. I take great offense at that. They're not here, but they are being maligned.

One of them, Dr. Mann, who was the main subject of Dr. Crichton's testimony, and I'd like to place in the record a letter from Dr. Mann which was sent to congressional committee in which he shows how in fact his data were reproduced and used and studied."

43. For example, there were other scientist-run climate blogs like James Empty Blog, Stoat, Rabett Run, Atmos, and Head in a Cloud that were largely about climate science. *Nature* eventually created its own climate science blog, Climate Feedback.

44. For example, RealClimateEconomics, Climate Ethics, and Climate Policy.

45. Among the key initial articles I wrote on the topic and published when the site was launched in December 2004 were "Temperature Variations in Past Centuries and the So-Called 'Hockey Stick,'" "Myth vs. Fact Regarding the 'Hockey Stick,'" "False Claims by McIntyre and McKitrick Regarding the Mann et al. (1998) Reconstruction," and "On Yet Another False Claim by McIntyre and McKitrick." A number of other articles correcting falsehoods about the hockey stick, and paleoclimate reconstructions more generally were written by other RealClimate contributors, such as Stefan Rahmstorf, "What If . . . the 'Hockey Stick' Were Wrong?" January 27, 2005; Gavin Schmidt and Caspar Ammann, "Dummies Guide to the Latest 'Hockey Stick' Controversy," February 18, 2005. Dozens of later posts by various RealClimate authors would address the increasingly politically driven and highly charged attacks against the hockey stick.

10. Say It Ain't So, (Smokey) Joe!

1. Carl Sagan, *The Demon-Haunted World* (New York, Random House, 1996), chapter 14, "Antiscience."

2. Future President Richard Nixon, then a House member, had called for the revocation of Condon's security clearance.

3. Condon recounted the story to Sagan, who was briefly his student one summer.

4. Jasper Becker, *Hungry Ghosts: Mao's Secret Famine* (New York, Free Press, 1996).

5. Conventionally, a tenure case is put together in the sixth year by a committee of the individual's peers (the tenure committee), which is constituted of faculty both within and outside the individual's department. The committee's deliberations are informed by letters of reference solicited from leading scholars in the field. The case is then voted on by the

entire faculty of the individual's department. The result of that vote is typically advisory to the department chair, the dean, the university president, and potentially the board of visitors, trustees, or regents, who make a final determination.

6. C. Mooney, *The Republican War on Science* (New York, Basic Books, 2005), 7.

7. "Climate of Distrust," *Nature*, July 17, 2005.

8. Tim Dickinson, "The Climate Killers: Meet the 17 Polluters and Deniers Who Are Derailing Efforts to Curb Global Warming," *Rolling Stone*, January 6, 2010.

9. Dave Michaels, "U.S. Rep. Joe Barton's Gas Well Stake Raises Ethical Questions," *Dallas Morning News*, February 3, 2010. The article notes that Barton "earned nearly $100,000 from an interest in natural gas wells that he purchased from a longtime campaign donor who also advised the congressman on energy policy, according to interviews and records."

10. The apology came after the Obama administration had sought to hold BP accountable for its role in the Deepwater Horizon oil spill. The disaster killed eleven oil workers and spilled nearly 5 million barrels of oil into the Gulf of Mexico. See, e.g., Aaron Blake and Paul Kane, "GOP Leaders Forced Rep. Barton to Retract Apology to BP," *Washington Post*, June 17, 2010.

11. Copies of the letters are available at the House of Representatives Committee on Energy and Commerce Web site, http://energycommerce.house .gov/108/Letters/06232005_1570.htm (accessed July 7, 2011).

12. "Climate of Distrust."

13. On July 13, Alan Leshner, executive publisher of *Science*, wrote to Barton expressing "deep concern about letters recently sent by the Committee to several scientists . . . regarding their research in climate science," adding that "it would be unfortunate if Congress tried to become a participant in the scientific peer-review process itself" (www.aaas.org/spp/cstc/docs/05–7-13_climatebarton.pdf, accessed July 7, 2011). On July 15, National Academy of Sciences President Ralph Cicerone wrote to Barton and Whitfield that he had been "contacted by members of the National Academy of Sciences, including members of my council . . . [and] officers of various science organizations . . . expressing considerable concern regarding your letter" (www.realclimate.org /Cicerone_to_Barton.pdf, accessed July 7, 2011).

14. Www.realclimate.org/Scientists_to_Barton.pdf (accessed July 7, 2011).

15. Donald Kennedy, "Silly Season on the Hill," *Science*, 39 (August 25, 2005).

16. This followed an op-ed in the *Post* the day before (July 22, 2005) by David Ignatius, "A Bit to Chill Thinking: Behind Joe Barton's Assault on Climate Scientists."

17. Chris Mooney, for example, entitled his article in the July issue of *American Prospect* "Mann Hunt."

18. Larry Neal, Letter to editor, *Toledo Blade*, July 28, 2005.

19. Henry A. Waxman, Senior Democratic Member of Committee on Government Report, Letter to the Honorable Joe Barton, Chairman of the Committee on Energy and Commerce, copied to the Honorable John D. Dingell, Ranking Minority Member, July 1, 2005.

20. Jay Inslee and Jan Schakowsky, Members of Congress, Letter to the Honorable Joe Barton, Chairman of the Committee on Energy and Commerce and the Honorable Ed Whitfield, Chairman of the Subcommittee on Oversight and Investigations, copied to the Honorable John D. Dingell and the Honorable Bart Stupak, July 27, 2005.

21. Sherwood Boehlert, Chair of Committee on Science, Letter to the Honorable Joe Barton, Chairman of the Committee on Energy and Commerce, copied to the Honorable John D. Dingell, Ranking Minority Member, July 14, 2005.

22. The letter was coauthored with University of Arizona President Peter Likens. John McCain and Peter Likens, "Politics vs. the Integrity of Research," *Chronicle of Higher Education*, September 2, 2005.

23. McCain said, in addition, "asking for all of the data produced in a scientist's career is highly irregular. It represents a kind of intimidation, which threatens the relationship between science and public policy. That behavior must not be tolerated."

24. Our letters were posted publicly: "Scientists Respond to Barton," RealClimate.org, July 18, 2005.

25. Unique to the letter I received were demands for extensive information about my involvement with the IPCC and the following specific demand: "According to The Wall Street Journal, you have declined to release the exact computer code you used to generate your results. (a) Is this correct? (b) What policy on sharing research and methods do you follow? (c) What is the source of that policy? (d) Provide this exact computer code used to generate your results."

26. Vladeck had been involved in the legal actions against the tobacco industry in the 1990s. In 2008, he was honored by *Legal Times* as one of the leading Washington, D.C., lawyers of the past thirty years.

27. See discussion in the section "David Versus Goliath" in chapter 9.

28. Antonio Regalado, "Academy to Referee Climate-Change Fight," *Wall Street Journal*, February 10, 2006.

29. I was granted tenure at U. Va, but it was a formality by the time it happened (early summer), as I had already accepted the Penn State position, which came with tenure.

11. A Tale of Two Reports

1. G. R. North, F. Biondi, P. Bloomfield, J. R. Christy, K. M. Cuffey, R. E. Dickinson, E. R. M. Druffel, D. Nychka, B. Otto-Bliesner, N. Roberts, K. K. Turekian, J. M. Wallace, *Surface Temperature Reconstructions for the Last 2,000 Years* (Washington, DC, National Academies Press, 2006).

2. Edward J. Wegman, David W. Scott, and Yasmin H. Said, "Ad Hoc Committee Report on the 'Hockey Stick' Global Climate Reconstruction," July 14, 2006, available at: http://republicans.energycommerce.house.gov/108/home /07142006_Wegman_Report.pdf.

3. Wegman also acknowledged input from several of his other previous students.

4. Wegman had claimed that the report was subject to formal peer review similar to that used by the National Research Council/National Academy of Sciences. However, as shown by computer scientist John Mashey, this claim was untrue. John Mashey, "Strange Scholarship in the Wegman Report (SSWR): A Facade for the Climate Anti-science PR Campaign," http://deepcli mate.org/2010/09/26/strange-scholarship-wegman-report/, 56–61; also see chapter 15 for further discussion of this report. Mashey notes: "Wegman tried to describe this to Congress as being 'like the NRC' . . . whereas commenters were: 1) entirely chosen by the authors, 2) not anonymous, but not listed in the WR either as their identities were only supplied in response to questioning, 3) all represented as 'outside Wegman's network,' although only via strange definitions. Again, that is not the commenters' problem, nothing is wrong with sharing a network with Wegman. The problem is Wegman's misrepresentation to Congress, 4) in some cases asked with adequate time, in other cases not, 5) apparently not asked early in the process, in at least 2 cases, 6) and whose feedback was not incorporated, in at least 2 cases." Mashey reports that statistician Grace Wahba, for example, complained "they used my name and they said I was a referee. He sent it to me about 3 days beforehand and I sent him a bunch of criticisms which they didn't take into account."

5. Panel members clarified, in their press conference, the issue of whether the panel's use of *plausible* conflicted with our use of *likely*. John (Mike) Wallace, an atmospheric scientist from the University of Washington and a National

Academy member, provided clarification: The panel concurred with the fact that the conclusions were *likely* true. Peter Bloomfield clarified the meaning of the panel's conclusion that there is "less confidence" in conclusions prior to A.D. 1600: "Where we speak of 'less confidence,' we're more into a level of sort of 2 to 1 odds, which IPCC, they interpreted 'likely' as that level, roughly two to one odds or better." In other words, the panel's assessment was in fact consistent with the finding of both MBH99 and the IPCC TAR that recent warmth is *likely* anomalous in the context of the past millennium.

6. The report's conclusion that prior to A.D. 900 "annual data series are very few in number, and the non-annually resolved data used in reconstructions introduce additional uncertainties" was in fact the reason that the hockey stick didn't extend further back than A.D. 1000.

7. The report was also somewhat uncritical in citing the previously discussed Von Storch et al. claim that climate field reconstruction techniques dramatically underestimate long-term trends, when more recent work had found significant fault with that paper (see chapter 7).

8. "Surface Temperature Reconstructions for the Last 2,000 Years," June 22, 2006, www.nap.edu/webcast/webcast_detail.php?webcast_id=327; www.nationalacademies.org/podcast/20060622.mp3.

9. At the press conference, North stated that "there might have been things that maybe they could have done differently . . . it was after all the first paper on the subject," but he further expanded on his thoughts in a later interview (an online discussion forum on the Web site of the *Chronicle of Higher Education* that accompanied publication of Richard Monastersky, "Climate Science on Trial," *Chronicle of Higher Education*, September 8, 2006, stating: "There is a long history of making an inference from data using pretty crude methods and coming up with the right answer. Most of the great discoveries have been made this way. . . . It turns out that their choices led them to essentially the right answer."

10. R. Pielke Jr., "Quick Reaction to the NRC Hockey Stick Report," Prometheus (blog), June 22, 2006; http://cstpr.colorado.edu/prometheus/archives/climate_change/000859quick_reaction_to_th.html.

11. John Heilprin, "Study Says Earth's Temp at 400-Year High," Associated Press, June 22, 2006.

12. E.g., "Earth Warmest in at Least 400 Years," MSNBC, June 22, 2006.

13. Andrew Revkin, "Science Panel Backs Study on Warming Climate," *New York Times*, June 22, 2006.

14. Juliet Eilperin, "Study Confirms Past Few Decades Warmest on Record," *Washington Post*, June 23, 2006.

15. "Backing for 'Hockey Stick' Graph," BBC, June 23, 2006.

16. Geoff Brumfiel, "Academy Affirms Hockey-Stick Graph," *Nature*, June 28, 2006.

17. E.g., Beth Daley, "Report Backs Global Warming Claims," *Boston Globe*, June 22, 2006, reported that "A signature piece of evidence for global warming—a chart showing that a sharp rise in temperatures made the late 20th century the warmest period in at least 1,000 years—is most likely correct, a national panel of scientific experts concluded today." And the *Los Angeles Times* reported "U.S. Panel Backs Data on Global Warming" (Thomas H. Maugh II and Karen Kaplan, June 22, 2006). The *Guardian* summarized the panel's findings as "US Scientists Back Manmade Warming Claim" (David Adam, June 23, 2006), while *New Scientist* weighed in with "US Report Backs Study on Global Warming" (Roxanne Khamsi, June 23, 2006).

18. *Lou Dobbs's Report*, CNN, June 26, 2006.

19. Dobbs overstated our role in the larger climate change debate, though, by characterizing us as the individuals "who first raised the alarm a decade ago."

20. Antonio Regalado, "Panel Study Fails to Settle Debate on Past Climates," *Wall Street Journal*, June 23, 2006.

21. These comments were made in a conference call Gore did with reporters, as reported by the Web site Think Progress, June 23, 2006, http://think progress.org/politics/2006/06/23/5962/warming-skeptics/.

22. Sensenbrenner delivered this statement as ranking member of the U.S. House of Representatives Select Committee on Energy Independence and Global Warming in a hearing held on December 2, 2009. See, e.g., coverage by the AP: Seth Borenstein, "Obama Science Advisers Grilled Over Hacked E-mails," Associated Press, December 3, 2009.

23. See ibid. Borenstein contacted NAS panel chair Gerald North who, in Borenstein's words, "confirmed in an interview Wednesday that Holdren was right, not Sensenbrenner."

24. Revkin, "Science Panel Backs Study on Warming Climate."

25. I'm a coauthor with Gavin Schmidt, who has an Erdos number of three.

26. See "What's Your Mann Number?" Stoat, http://scienceblogs.com/stoat /2006/02/whats_your_mann_number.php.

27. Lauren Morello, "Scientists Clash with Barton, Others on Cause of Global Warming," *Environment and Energy Daily*, July 20, 2006, www.eenews .net. I was given only three days' notice, and the committee was informed that I had a commitment to take care of our infant daughter while my wife attended a conference in Vermont. Barton's cochair, Ed Whitfield (R-KY), nonetheless

mischaracterized the situation (emphasis added): "Let me add that we did invite Dr. Mann to this hearing, but his attorney explained that he was unavailable, *on family vacation.*"

28. I was on full-time infant duty, watching from a slow Internet connection, and with little or no cell phone reception.

29. The full transcript of this hearing is available at http://ftp.resource .org/gpo.gov/hearings/109h/31362.txt.

30. This finding had already been published by Von Storch in the peer reviewed literature, and indeed was supported by the findings of several independent groups (see chapter 9).

31. North's full statement was: "Now, do people collaborate and think similarly? Of course they do. But, you know, if you look back at the history of, say, quantum mechanics in the early 1920s, it was Einstein, Bohr, Heisenberg, all these people. I am sure if you did a similar analysis, you would probably find something very like that, but in fact these guys hated each other. I mean, they were very, very competitive. And if you look at the 43 authors, I am sure that not all of them like to go out and have a beer together. This is pretty competitive business, and I will tell you, if somebody can find a way to knock down someone else's theory, that is their road to recognition and fame. We all do that. That is part of the game and we really enjoy that part of the game."

32. Monastersky, "Climate Science on Trial."

33. Richard Harris, "Global Warming a Hot Topic in Congressional Hearing," NPR, July 20, 2006.

34. In his opening statement, http://ftp.resource.org/gpo.gov/hearings /109h/31362.txt, Whitfield states: "certainly the Kyoto arguments were primarily based on this new chart."

35. The full transcript of this hearing is available at http://ftp.resource .org/gpo.gov/hearings/109h/31362.txt.

36. The prelude to Stupak's statement was: "Today we are holding a very strange hearing. Originally scheduled to give Dr. Michael Mann a chance to respond to critics who provided testimony to this committee last week, this hearing has now expanded to allow these critics to attack the very science of global warming. Witnesses reappearing in the committee today . . . had attempted to discredit Dr. Mann's 8-year old study . . . and his conclusion that the earth is warming at an unprecedented rate. However, as Dr. North testified last week, a comprehensive review of temperature reconstruction research by the National Academy of Science at the request of the Science Committee found that there were numerous other studies concluding that the Earth is warming at an unprecedented rate. Now instead of allowing Dr. Mann to

respond to last week's allegations, two of our witnesses, apparently unhappy with the outcome of last week's hearing have decided to rewrite and expand their testimony to raise new issues, new complaints, and new questions. This re-written testimony . . . includes areas of climatology totally outside their expertise. As a result, it appears that these critics have lost interest in simply attacking Dr. Mann's work."

37. Waxman's statement was: "In effect, it is back to the tactics of the tobacco industry. I remember well when they would send their scientists to come in and just cast a little doubt about whether smoking cigarettes really do cause cancer. . . . These are bullying tactics and they drew highly unusual protests from the American Association for the Advancement of Science, the National Academy of Sciences, and the Republican Chairman of the House Science Committee, among others."

38. Stupak, in his opening statement, noted how Barton and company had curiously avoided the issue: "Dr. Wegman was not even asked if Dr. Mann's conclusions would change if the criticisms [by McIntyre/Wegman] were incorporated and the analysis were re-created, nor did he volunteer to do that."

39. Blackburn is one of the last few CO_2 deniers in Congress, as revealed in a CNN interview she gave on December 10, 2009. She was identified in 2010 by Greenpeace as the fourth leading U.S. House of Representatives recipient of Koch Industries funds (Greenpeace USA, "Koch Industries: Secretly Funding the Climate Denial Machine," March 2010) and has been listed as one of the most corrupt members of the U.S. Congress (Citizens for Responsibility and Ethics in Washington, Annual List of 20 Most Corrupt Members of Congress, *2008 Annual Report*).

40. "Mr. Waxman. Some have criticized you for lack of willingness to disclose your data and computer code. Could you briefly tell us how you have handled the availability of your research?

"Dr. Mann. . . . The statement was made earlier here that I didn't make my data available until 2004, and that is simply incorrect. . . . Our entire data set was available on the worldwide web several years before that. Now a code . . . is a different sort of thing. It is a matter of intellectual property . . . but as long as the algorithm is available then other[s] . . . can independently reproduce your work . . . that is what Dr. Wahl and Dr. Ammann have shown. They have independently implemented our algorithm in a different programming language that is available to anybody . . . over the past few years we have been making all of our codes available for all of the calculations that we do, and that is actually a standard that many others in our community . . . haven't really followed, so we are sort of leading the way there."

41. See chapter 2. A more detailed history behind the errors and ultimate correction of the Christy and Spencer temperature estimates is provided in chapter 12.

42. See hearing transcript at http://ftp.resource.org/gpo.gov/hearings/109h /31362.txt.

43. See, for example, Gavin Schmidt and Eric Steig, "Climate Code Archiving: An Open and Shut Case?" RealClimate.org. October 26, 2010, www .realclimate.org/index.php/archives/2010/10/climate-code-archiving-an -open-and-shut-case/.

44. See hearing transcript for details: http://ftp.resource.org/gpo.gov/ hearings/109h/31362.txt, beginning with "Mr. Waxman. Dr. Wegman isn't a climatologist, and I would like to give you the opportunity to respond to some of his statements from last week's hearing."

45. See hearing transcript for details: http://ftp.resource.org/gpo.gov /hearings/109h/31362.txt, beginning with "Mr. Waxman. He also said that global warming 'must be understood in the context which is that we have relatively speaking a Little Ice Age, which everybody seems to acknowledge, and so it is not so surprising that it is warming if we are coming out of a Little Ice Age.'"

46. See hearing transcript for details: http://ftp.resource.org/gpo.gov/ hearings/109h/31362.txt, beginning with "Mr. Waxman. Dr. Wegman testified he thought global warming 'is probably less urgent than some would have it be.'"

47. In written testimony submitted after the hearing in response to questions from the committee, I would point out that the National Science Foundation had launched an initiative with precisely this purpose more than a decade earlier, the NCAR Geophysical Statistics Project. I pointed out that I had participated in their inaugural workshop in Boulder as a graduate student back in 1994. The current director of that project, Doug Nychka, was one of the members of the NAS panel that examined our work. He was also an acknowledged consultant on the Wahl and Ammann study that had refuted the McIntyre and McKitrick claims that were now being repeated by Wegman (http://ftp. resource.org/gpo.gov/hearings/109h/31362.txt).

48. The exchange involved a humorous request for clarification by Cicerone (emphasis added):

"Mr. Inslee: . . . Dr. Cicerone, if Mr. and Mrs. Mann had never met and we never had the services of Dr. Mann, would that have varied the conclusion of the National Academy of Sciences on these fundamental questions?

"Dr. Cicerone. *You must be referring to his parents and not his wife.*

"Mr. Inslee. I am indeed."

49. Ritson was a coauthor of E. R. Wahl, D. M. Ritson, and C. M. Ammann, "Comment on 'Reconstructing Past Climate from Noisy Data,'" *Science*, 312 (2006): 529b, discussed in chapter 7.

50. Ritson's July 26, 2006, e-mail was discussed in, and provided as supplementary material in the form of an attachment to, the author's formal response to follow-up questions from the U.S. House Committee on Energy: http://ftp.resource.org/gpo.gov/hearings/109h/31362.txt. It is available in its entirety at www.meteo.psu.edu/~mann/house06. Ritson makes note in the letter of the error, first reported by Wahl and Ammann (discussed in chapter 9), involving the failure to remove deterministic trends from the proxy data before estimating the random component in the data. That error inflated their estimate of the natural low-frequency random fluctuations well beyond what would be observed for standard red noise and, as discussed in chapter 9, was at the center of their false claim that our procedures "manufacture hockey sticks." Ritson noted that the Wegman Report "makes no mention of this quite improper M&M procedure" and does not "provide any specification data for your own results that you contend confirm the M&M results."

51. E-mail to Wegman, from David Ritson to me and Gavin Schmidt, July 30, 2006.

52. The various claims by Wegman discussed in this paragraph were made in an e-mail he sent to Waxman's staff on September 1, 2006, which was subsequently forwarded to me.

53. Wegman claimed that his e-mail address was protected by a spam filter that cut out all e-mails referring to the "Wegman Report," "Mann," or other such key words, and that his coauthors did not check their e-mail very often.

54. The exact statement was: "It is not clear to me that before the journal peer review process is complete that we have an academic obligation to disclose the details of our methods."

55. Waxman followed up with a formal letter to Wegman on September 15, 2006: "Dr. Ritson assures us that his request for information is far from burdensome. Dr. Ritson appears to be simply requesting basic information on your methodological approach that is not transparent from the written report. In fact, some of the questions appear to be simple 'yes' or 'no' questions." "Your report noted that when 'methodology is not fully disclosed, peers do not have the ability to replicate the work and thus independent verification is impossible.' I hope that you will be able to provide Dr. Ritson with the requested information at your earliest convenience."

56. Ritson mailed spam-filter-proof versions of each e-mail, with any potentially offending key words removed. Ritson checked with Waxman's office

about whether Wegman had provided them with any response. The answer was no. Ritson expressed his frustrations in e-mail, dated September 2, 2007, subject: "Re Wegman," to NAS President Ralph Cicerone and NAS 2006 report chair Gerald North (forwarded to me by Ritson): "I can understand, but not sympathize with persons who, well aware of the weaknesses of their positions, simply fail to reply to queries. However in this instance Wegman has made a central point of the need for openness in science, and never, over my academic career, has anyone avoided actions by promising to implement a series of steps which they apparently had no intention of ever doing. This is written to you, not to involve you and the NAS in scientific arguments, but because such an incident is destructive of public and congressional trust in scientific objectivity."

57. E-mail with subject line: Wegman, from David Ritson to me and Gavin Schmidt, July 30, 2006.

12. Heads of the Hydra

1. In "Climate Chaos," *Panorama* radio special, BBC channel one, September 4, 2006, Luntz states: "It's now 2006 . . . I think most people will conclude that there is global warming taking place, and that the behavior of humans is affecting the climate."

2. The meeting was an IPCC conference held in Honolulu, Hawaii, in early 2009. It was a sort of retrospective on the state of the science with an eye forward toward the next assessment (AR5) to begin in 2010. It was also, at least informally, a reunion of sorts, a chance for the IPCC to take stock of what it had accomplished and for the various scientists involved to bask a bit in the glory of their shared prize.

3. S. Solomon, D. Qin, M. Manning, Z. Chen, M. Marquis, K. B. Averyt, M. Tignor, and H. L. Miller, eds., *IPCC, 2007: Climate Change 2007: The Physical Science Basis. Contribution of Working Group I to the Fourth Assessment Report of the Intergovernmental Panel on Climate Change* (Cambridge, UK, and New York, Cambridge University Press, 2007).

4. M. E. Mann and P. D. Jones, "Global Surface Temperatures Over the Past Two Millennia, *Geophysical Research Letters*, 30 (2003): 1820, doi:10.1029/2003GL017814, the "Mann and Jones" reconstruction, used a simple composite of high-quality proxy records from eight distinct regions across the Northern Hemisphere that spanned the past two thousand years.

5. Solomon et al., *IPCC, 2007,* 9.

6. Among the numerous of our detractors making such completely misleading claims was prominent Australian climate change denier Ian Plimer, *Heaven and Earth* (Lanham, MD, Taylor Trade Publishing, 2009).

7. George E. P. Box and Norman R. Draper, *Empirical Model-Building and Response Surfaces* (Oxford, Wiley, 1987), 424.

8. Some of the previous attacks against the instrumental temperature record were described in previous chapters. McKitrick and Michaels, in their purported bombshell paper of 2004 (see chapter 6), had claimed to demonstrate that instrumental thermometer measurements were contaminated by non-climatic factors. It didn't take long for other researchers to expose fatal flaws in their analysis, however. Then there was contrarian Chico, California, Fox News meteorologist Anthony Watts and his attempts to undermine confidence in the surface temperature record through his surfacestations.org project; that too, as we learned, didn't pan out (see chapter 5).

9. T. C. Peterson, "Assessment of Urban Versus Rural in Situ Surface Temperatures in the Contiguous United States: No Difference Found," *Journal of Climate*, 16 (2003): 2941–2959.

10. D. E. Parker, "Climate: Large-Scale Warming Is Not Urban," *Nature* 432 (2004): 290; D. E. Parker, "A Demonstration That Large-Scale Warming Is Not Urban," *Journal of Climate*, 19 (2006): 2882–2895.

11. Parker compared globally averaged land air temperatures when stratified with respect to whether contributing stations had experienced a windy or a calm night during any reporting period. Since the urban heat island effect is largely wiped out on windy nights by vertical atmospheric mixing, any urban heat island bias in the global record should show up in the form of warmer conditions for the calm night record.

12. K. E. Trenberth, P. D. Jones, P. Ambenje, R. Bojariu, D. Easterling, A. Klein Tank, D. Parker, F. Rahimzadeh, J. A. Renwick, M. Rusticucci, B. Soden, and P. Zhai, "Observations: Surface and Atmospheric Climate Change," in S. Solomon, D. Qin, M. Manning, Z. Chen, M. Marquis, K. B. Averyt, M. Tignor, and H. L. Miller, eds., *Climate Change 2007: The Physical Science Basis.* Contribution of Working Group I to the Fourth Assessment Report of the Intergovernmental Panel on Climate Change (Cambridge, UK, and New York, Cambridge University Press, 2007), section 3.2.

13. E.g., R. W. Spencer and J. R. Christy, "Precision and Radiosonde Validation of Satellite Gridpoint Temperature Anomalies, Part I: MSU Channel 2," *Journal of Climate*, 5 (1992): 847–857; R. W. Spencer and J. R. Christy, "Precision and Radiosonde Validation of Satellite Gridpoint Temperature Anomalies, Part II: A Tropospheric Retrieval and Trends 1979–90," *Journal of Climate*, 5 (1992): 858–866.

14. The interview was part of the segment "Three Views on Global Warming," May 12, 2004, with host Richard Harris, www.npr.org/templates/story /story.php?storyId=1893089.

15. J. W. Hurrell and K. E. Trenberth, "Spurious Trends in Satellite MSU Temperatures from Merging Different Satellite Records," *Nature*, 386 (1997): 164–167.

16. F. J. Wentz and M. Schabel, "Effects of Orbital Decay on Satellite-Derived Lower-Tropospheric Temperature Trends," *Nature*, 394 (1998): 661–664.

17. J. R. Christy, R. W. Spencer, and E. Lobl, "Analysis of the Merging Procedure for the MSU Daily Temperature Time Series," *Journal of Climate*, 5 (1998): 2016–2041.

18. Q. Fu et al., "Contribution of Stratospheric Cooling to Satellite-Inferred Tropospheric Temperature Trends," *Nature*, 429 (2004): 55–58.

19. C. A. Mears and F. J. Wentz, "The Effect of Diurnal Correction on Satellite-Derived Lower Tropospheric Temperature," *Science* (2005): 1548–1551.

20. J. R. Christy and R. W. Spencer, "Correcting Temperature Datasets," *Science*, 310 (2005): 972. The response by Mears and Wentz that followed the comment was unusually brief and to the point: "Once we realized that the diurnal correction being used by Christy and Spencer for the lower troposphere had the opposite sign from their correction for the middle troposphere sign, we knew that something was amiss. Clearly, the lower troposphere does not warm at night and cool in the middle of the day. We question why Christy and Spencer adopted an obviously wrong diurnal correction in the first place. They first implemented it in 1998 in response to Wentz and Schabel, which found a previous error in their methodology: neglecting the effects of orbit decay."

21. N. S. Keenlyside, M. Latif, J. Jungclaus, L. Kornblueh, and E. Roeckner, "Advancing Decadal-Scale Climate Prediction in the North Atlantic Sector, *Nature*, 453 (2008): 84–88.

22. E.g., David Whitehouse, "Has Global Warming Stopped?" *New Statesman*, December 19, 2007.

23. The authors expressed disapproval over the way their words were twisted in some press accounts. In their response to misleading media accounts of their work by Fox News and the UK *Daily Mail*, lead author Mojib Latif was quoted as saying, "I don't know what to do. They just make these things up." See Climate Progress, January 14, 2010, http://climateprogress.org/2010/01/14 /science-dr-mojib-latif-global-warming-cooling/.

24. The three assessments are (1) CRU, a joint product of CRU and the Hadley Centre of the UK Meteorological Office, e.g., P. Brohan, J. J. Kennedy, I. Harris, S. F. B. Tett, and P. D. Jones, "Uncertainty Estimates in Regional and Global Observed Temperature Changes: A New Dataset from 1850," *Journal*

of *Geophysical Research*, 111 (2006): D12106, doi:10.1029/2005JD006548; (2) NOAA, T. M. Smith et al., "Improvements to NOAA's Historical Merged Land-Ocean Surface Temperature Analysis (1880–2006)," *Journal of Climate*, 21 (2008): 2283–2293; (3) NASA/GISS, GISTEMP, J. Hansen, R. Ruedy, M. Sato, and K. Lo, "Global Surface Temperature Change," *Reviews of Geophysics*, 48 (2010): RG4004, doi:10.1029/2010RG000345.

25. These results are demonstrated in D. R. Easterling and M. F. Wehner, "Is the Climate Warming or Cooling?" *Geophysical Research Letters*, 36 (2009): L08706, doi:10.1029/2009GL037810.

26. Seth Borenstein, "Statisticians Reject Global Cooling" Associated Press, October 27, 2009.

27. David H. Douglass, John R. Christy, Benjamin D. Pearson, and S. Fred Singer, "A Comparison of Tropical Temperature Trends with Model Predictions," *International Journal of Climatology*, 28 (2008): 1693–1701; published online December 5, 2007, in Wiley InterScience, doi:10.1002/joc.1651.

28. B. D. Santer et al., "Amplification of Surface Temperature Trends and Variability in the Tropical Atmosphere," *Science*, 309 (2005): 1551–1555.

29. See "Tropical Tropospheric Trends," RealClimate, December 12, 2007.

30. The issue Douglass et al. purported to address was whether the observations fall within the range of the roughly twenty simulations. But they instead calculated the uncertainty in the average among all simulations, the so-called standard deviation of the mean. This number gets smaller and smaller as one averages more and more simulations, and the chances that any single case would fall within them becomes vanishingly small. But the observations represent just a single case—one possible realization of climate history—and the relevant question is whether that one realization falls within the scatter of the various models. The Douglass et al. comparison did not address that question.

31. B. D. Santer, P. W. Thorne, L. Haimberger, K. E. Taylor, T. M. L. Wigley, J. R. Lanzante, S. Solomon, M. Free, P. J. Gleckler, P. D. Jones, T. R. Karl, S. A. Klein, C. Mears, D. Nychka, G. A. Schmidt, S. C. Sherwood, and F. J. Wentz, "Consistency of Modelled and Observed Temperature Trends in the Tropical Troposphere," *International Journal of Climatology*, 28 (2008): 1703–1722, doi:10.1002/joc.1756.

32. The press conference, organized by the Science and Environmental Policy Project (SEPP), was entitled "Nature Rules the Climate: Human-Produced Greenhouse Gases Are Not Responsible for Global Warming: Therefore: Schemes to Control CO_2 Emissions Are Ineffective and Pointless, Though Very Costly," National Press Club, Washington, DC, December 14, 2007. The quotations in the text are attributed to comments at the press conference by lead author Douglass as reported in Fred Pearce, "Victory for Openness as

IPCC Climate Scientist Opens Up Lab Doors," *Guardian*, February 9, 2010, www.guardian.co.uk/environment/2010/feb/09/ipcc-report-author-data -openness.

33. As discussed in chapter 5; Dan Harris, Felicia Biberica, Elizabeth Stuart, and Nils Kongshaug, "Global Warming Denier: Fraud or 'Realist'?" ABC News, March 23, 2008.

34. J. D. McLean, C. R. de Freitas, and R. M. Carter, "Influence of the Southern Oscillation on Tropospheric Temperature," *Journal of Geophysical Research*, 114 (2009): D14104, doi:10.1029/2008JD011637.

35. David W. J. Thompson, John J. Kennedy, John M. Wallace, and Phil D. Jones, "A Large Discontinuity in the Mid-Twentieth Century in Observed Global-Mean Surface Temperature," *Nature*, 453 (2008): 646–649.

36. They had performed a linear regression relating the differences between consecutive temperature measurements to a standard index of El Niño. So-called first differences are equivalent to taking the derivative of the curve. This has the effect of removing any long-term trend in the data. It turns out, through a separate operation, the authors had also removed the short-term month-to-month fluctuations. That leaves only the intermediate year-to-year fluctuations.

37. G. Foster, J. D. Annan, P. D. Jones, M. E. Mann, B. Mullan, J. Renwick, J. Salinger, G. A. Schmidt, and K. E. Trenberth, "Comment on 'Influence of the Southern Oscillation on Tropospheric Temperature' by J. D. McLean, C. R. de Freitas, and R. M. Carter," *Journal of Geophysical Research*, 115 (2010): D09110, doi:10.1029/2009JD012960. Normally the original authors would be able to publish a rebuttal to the comment, but that too must pass peer review. In this case, McLean et al. were unable to offer a credible response, so the refutation was published without any rebuttal.

38. The publicity was coordinated by a New Zealand climate change denial group known as the New Zealand Climate Science Coalition, July 23, 2009.

39. C. Loehle, "A 2000-Year Global Temperature Reconstruction Based on Non-Tree Ring Proxies," *Energy and Environment*, 18 (2007): 1049–1058.

40. Gavin Schmidt detailed the problems in "Past Reconstructions: Problems, Pitfalls and Progress," RealClimate, December 7, 2007.

41. In fairness to the original authors of those records, they had never intended them to be used for such a purpose.

42. In one case, a chemical composition record was mistaken for the temperature estimates that the original authors had produced from that record. In other cases, oxygen isotope records that reflect a combination of climate influences were assumed to be simple "paleo-thermometers."

43. E.g., his error of confusing the mid-twentieth century for the end of the twentieth century in several records.

44. Of his eighteen total records, only five at most were potentially useful for comparing late-twentieth-century and medieval temperatures. Every careful study looking at the issue has indicated that one cannot estimate hemispheric, let alone global, temperatures—as claimed by Loehle—with such a paucity of records.

45. There were 1,209 proxy records around the globe that had the requisite annual-to-decadal temporal resolution. Of these, 59 extended back to A.D. 1000 and 36 back to A.D. 500. Of the 59 records dating back to A.D. 1000, 37 were from sources other than tree rings; and of the 36 available back to A.D. 500, 28 were from non-tree ring sources.

46. These tests were described in M. E. Mann, S. Rutherford, E. Wahl, and C. Ammann, "Testing the Fidelity of Methods Used in Proxy-Based Reconstructions of Past Climate," *Journal of Climate*, 18 (2005): 4097–4107; M. E. Mann, S. Rutherford, E. Wahl, and C. Ammann, "Robustness of Proxy-Based Climate Field Reconstruction Methods," *Journal of Geophysical Research*, 112 (2007): D12109, doi:10.1029/2006JD008272.

47. We had applied and tested this method of climate field reconstruction in numerous studies over the preceding six years: M. E. Mann and S. Rutherford, "Climate Reconstruction Using 'Pseudoproxies,'" *Geophysical Research Letters*, 29 (2002): 1501, doi:10.1029/2001GL014554; S. Rutherford, M. E. Mann, T. L. Delworth, and R. Stouffer, "Climate Field Reconstruction Under Stationary and Nonstationary Forcing," *Journal of Climate*, 16 (2003): 462–479; Z. Zhang, M. E. Mann, and E. R. Cook, "Alternative Methods of Proxy-Based Climate Field Reconstruction: Application to the Reconstruction of Summer Drought Over the Conterminous United States Back to 1700 from Drought-Sensitive Tree Ring Data," *Holocene*, 14 (2004): 502–516; S. Rutherford, M. E. Mann, T. J. Osborn, R. S. Bradley, K. R. Briffa, M. K. Hughes, and P. D. Jones, "Proxy-Based Northern Hemisphere Surface Temperature Reconstructions: Sensitivity to Methodology, Predictor Network, Target Season and Target Domain," *Journal of Climate*, 18 (2005): 2308–2329; M. E. Mann, S. Rutherford, E. Wahl, and C. Ammann, "Testing the Fidelity of Methods Used in Proxy-Based Reconstructions of Past Climate," *Journal of Climate*, 18 (2005): 4097–4107; M. E. Mann, S. Rutherford, E. Wahl, and C. Ammann, "Robustness of Proxy-Based Climate Field Reconstruction Methods," *Journal of Geophysical Research*, 112 (2007): D12109, doi:10.1029/2006JD008272.

48. M. E. Mann, Z. Zhang, M. K. Hughes, R. S. Bradley, S. K. Miller, S. Rutherford, and F. Ni, "Proxy-Based Reconstructions of Hemispheric and Global

Surface Temperature Variations Over the Past Two Millennia," *Proceedings of the National Academy of Science*, 105 (2008): 13252–13257.

49. M. E. Mann, Z. Zhang, S. Rutherford, R. S. Bradley, M. K. Hughes, D. Shindell, C. Ammann, G. Falugevi, and F. Ni, "Global Signatures and Dynamical Origins of the 'Little Ice Age' and 'Medieval Climate Anomaly,'" *Science*, 326 (2009): 1256–1260.

50. Our conclusions for the Southern Hemisphere and global average temperature were less definitive, owing to larger uncertainties arising from sparse available proxy data in the southern ocean.

51. The original hockey stick reconstruction and the dozen or so other reconstructions published in the scientific literature since largely fell within the error bars of our original reconstruction; that is to say, they were roughly compatible.

52. The official citation was S. McIntyre and R. McKitrick, "Proxy Inconsistency and Other Problems in Millennial Paleoclimate Reconstructions," *Proceedings of the National Academy of Science*, 106 (2009): E10. Letters to the editor were required to be under 250 words. They were published online only, along with the authors' response (subject to the same length constraints).

53. M. E. Mann, R. S. Bradley, and M. K. Hughes, "Reply to McIntyre and McKitrick, 'Proxy-Based Temperature Reconstructions Are Robust,'" *Proceedings of the National Academy of Science*, 106 (2009): E11.

54. They also claimed that the standard method of screening proxy records for a temperature signal (which was used in a restricted set of our analyses) "manufactures hockey sticks." This was again nonsense. There wasn't a single peer reviewed study to support the claim. They had instead cited a nonscientific document on the Internet to support the assertion. Screening simply selects those proxy records that are correlated with temperature records from the location of the proxy. If actual temperatures show a rapid increase, then a reliable proxy record must show the same thing. If temperatures show no increase, then a reliable proxy record must again also show the same thing. The importance of cross-validation procedures we and other practitioners use to test proxy reconstructions for reliability is that any record selected by screening has to continue to correlate with temperatures during an independent testing period.

55. M. W. Salzer, M. K. Hughes, A. G. Bunn, and K. F. Kipfmueller, "Recent Unprecedented Tree-Ring Growth in Bristlecone Pine at the Highest Elevations and Possible Causes," *PNAS*, 106 (2009): 20348–20353.

56. The authors examined three widely distributed sites in western North America near the upper elevation limit for tree growth, demonstrating

that ring growth in the late twentieth century was greater than in any other comparable period as far back as the records went (nearly four thousand years) across all three sites. Moreover, that pattern was independent of elevation and depended only on proximity to tree line—something that was inconsistent with CO_2 fertilization but consistent with climatic stress (which, after all, is what largely determines the position of the tree line in the first place). Using independent proxy records of temperature as well as high-elevation meteorological temperature observations, the authors demonstrated that increasing temperature was the primary factor behind the anomalous recent growth.

57. In a section entitled "Potential Data Quality Problems," we commented: "we also examined whether or not potential problems noted for several records . . . might compromise the reconstructions. These records include the four Lake Korttajarvi series used for which the original authors note that human effects over the past few centuries unrelated to climate might impact records." These, incidentally, are the records that McIntyre was apparently claiming were used "upside down." Yet there was no such thing as "upside down" in our methodology: In one of our methods (composite approach), proxy data were screened to determine if they possessed a local temperature signal, based on their correlation with modern instrumental data. In the other method (RegEM), the proxy data were used in a sophisticated multivariate regression, and again the relationship with climate was determined empirically, with no a priori assumptions made. Either the record was employed using these objective procedures, or it was thrown out.

58. Andrew Revkin, "Climate Auditor Challenged to Do Climate Science"; *New York Times*, "Dot Earth," October 5, 2009.

59. This is the phrase I used back in January 2005 on the RealClimate blog to emphasize that there wasn't just one hockey stick; all of the reconstructions revealed the warming of the past century to be anomalous.

60. M. N. Juckes et al., "Millennial Temperature Reconstruction Intercomparison and Evaluation," *Climate of the Past*, 3 (2007): 591–609.

61. D. S. Kaufman et al., "Recent Warming Reverses Long-Term Arctic Cooling," *Science*, 325 (2009): 1236.

62. F. C. Ljungqvist, "A New Reconstruction of Temperature Variability in the Extra-tropical Northern Hemisphere During the Last Two Millennia," *Geografiska Annaler*, 92 A (2010): 339–351.

13. The Battle of the Bulge

1. E-mail to a group of colleagues on February 29, 2009, discussing the coverage of climate change in recent high-profile articles and op-eds in the mainstream media.

2. I further advised my colleagues that "there needs to be a coordinated well thought out defense against this, but so far this has proven elusive, at least in my discussions with colleagues."

3. A little more than a month after the day (September 28) I sent this e-mail, CRU's e-mail server was criminally hacked into, and the illegally obtained messages formed the basis of the manufactured controversy that became known in media circles as "climategate," the subject of the next chapter. Ironically, this very e-mail was in the archive of the stolen CRU messages.

4. I couldn't help but see the irony, given that this was one day after I gave a talk entitled "Fighting a Strong Headwind: Challenges in Communicating the Science of Climate Change" at the annual meeting of the American Geophysical Union in San Francisco that focused, among other things, on flaws in the mass media's coverage of the topic of climate change.

5. That was hardly the end of Meyer's controversial on-air statements on matters involving climate change. On April 2, 2010, for example, he misleadingly conflated short-term predictability of weather and long-term change in climate as climate change deniers often seek to do: "there's no way we can get 15 years from now right, if we can't get 15 days from now right," and he made the absurd claim, echoing climate change deniers such as Pat Michaels, that climate scientists are driven to the results they obtain by money and somehow were more likely to do so if working for the government: "follow the money a little bit. Meteorologists aren't paid by the government, the ones on TV, the climatologists are. If there's nothing to talk about, will their jobs really be all that secure? So, follow the money a little bit." I couldn't help but notice that these CNN telecasts were regularly sponsored by fossil fuel interests such as the American Petroleum Institute, ExxonMobil, and a consortium of coal producers.

6. Lou Dobbs's radio show, February 24, 2010.

7. EPA, "EPA Finds Greenhouse Gases Pose Threat to Public Health, Welfare: Proposed Finding Comes in Response to 2007 Supreme Court Ruling," April 17, 2009, http://yosemite.epa.gov/opa/admpress.nsf/0/0ef7df675805295 d8525759b00566924.

8. The 2007 decision involved the case of *Massachusetts vs. EPA*, wherein the attorneys general of Massachusetts and eleven other states (California, Connecticut, Illinois, Maine, New Jersey, New Mexico, New York, Oregon,

Rhode Island, Vermont, and Washington) petitioned the court to require the EPA to regulate carbon dioxide and other greenhouse gases as pollutants under the provisions of the Clean Air Act. The court found in favor of the plaintiffs 5 to 4.

9. EPA, "EPA: Greenhouse Gases Threaten Public Health and the Environment: Science Overwhelmingly Shows Greenhouse Gas Concentrations at Unprecedented Levels Due to Human Activity," December 7, 2009, http://yo semite.epa.gov/opa/admpress.nsf/bd4379a92ceceeac8525735900400c27/08d1 1a451131bca585257685005bf252!OpenDocument.

10. Available at http://epa.gov/climatechange/endangerment/petitions.html.

11. Denial of Petitions for Reconsideration of the Endangerment and Cause or Contribute Findings for Greenhouse Gases Under Section 202(a) of the Clean Air Act, http://epa.gov/climatechange/endangerment/petitions.html.

12. John M. Broder, "Behind the Furor Over a Climate Change Skeptic," *New York Times*, September 25, 2009, A19.

13. Carlin appeared on the Glenn Beck show in a segment titled "Environmental Protection Agency Analyst Defends Controversial Climate Change Report" on Fox News, July 1, 2009, claiming, among other things, that temperatures had decreased between 2002 and 2008; www.foxnews.com/story /0,2933,529725,00.html.

14. Robin Bravender, "Free-Market Group Attacks Data Behind EPA 'Endangerment' Proposal," *E&E News*, October 7, 2009.

15. Rick Piltz was a senior associate in the U.S. Global Change Research Program (later the Climate Change Science Program) before he resigned in March 2005 over efforts by the Bush administration, via Phil Cooney and the CEQ, to water down various government reports on climate change. Piltz went on to found Climate Science Watch, a government watchdog organization with the aim of "promoting integrity in the use of climate science in government," sponsored by the Government Accountability Project, a whistleblower protection group.

16. S. Fred Singer, Hal Lewis, Will Happer, Larry Gould, Roger Cohen, and Robert H. Austin, "Petitioning for a Revised Statement on Climate Change," *Nature*, 460 (July 23, 2009): 457.

17. "The Climate Change Climate Change: The Number of Skeptics Is Swelling Everywhere," *Wall Street Journal*, June 26, 2009.

18. Marc Sheppard, "UN Climate Reports: They Lie," October 5, 2009, www.americanthinker.com/2009/10/un_climate_reports_they_lie.html.

19. D. S. Kaufman et al., "Recent Warming Reverses Long-Term Arctic Cooling," *Science*, 325 (2009): 1236.

20. On his climateaudit site, McIntyre posted five separate pieces attacking the work within the space of two weeks: "Kaufman and Upside-Down Mann," September 3, 2009; "Kaufman et al: Obstructed by Thompson and Jacoby," September 12, 2009; "The Kaufman Backstory," September 14, 2009; "Is Kaufman Robust," September 18, 2009; "Invalid Calibration in Kaufman 2009," September 18, 2009.

21. The data in question were the Lake Korttajarvi sediment records which, as discussed in chapter 12, were the subject of McIntyre's attack against our most recent work, the results of which were, once again, insensitive to whether or not these records were used at all.

22. McIntyre made the accusations in a series of posts on his climateaudit blog: "Fresh Data on Briffa's Yamal #1," September 26, 2009 (the URL article title was more inflammatory: climateaudit.org/2009/09/26/briffas-yamal-crack-cocaine-for-paleoclimatologists/), and "Yamal: A 'Divergence' Problem," September 27, 2009.

23. "Let the Backpeddling Begin," Deep Climate, October 7, 2009, http://deepclimate.org/2009/10/07/let-the-backpedalling-begin/.

24. See, e.g., Australian computer scientist Tim Lambert's Deltoid blog, "McIntyre Had the Data All Along," October 8, 2009, http://scienceblogs.com/deltoid/2009/10/mcintyre_had_the_data_all_alon.php.

25. We commented on the matter in the RealClimate article "Hey Ya! (mal)," September 30, 2009: "People have written theses about how to construct tree ring chronologies in order to avoid end-member effects and preserve as much of the climate signal as possible. Curiously no-one has ever suggested simply grabbing one set of data, deleting the trees you have a political objection to and replacing them with another set that you found lying around on the web."

26. "Hey Ya! (mal)," RealClimate, September 30, 2009.

27. This includes the hockey stick itself.

28. This had, in fact, been explicitly demonstrated for the recent Kaufman et al. reconstruction discussed earlier and for our *Proceedings of the National Academy of Sciences* (2008) temperature reconstruction, but it likely applied to just about all published reconstructions.

29. We commented further on the matter at RealClimate in "Hey Ya! (mal)."

30. Ross McKitrick, "Defects in Key Climate Data Are Uncovered," *National Post*, October 1, 2009.

31. Marc Morano, "Disgraced?! Michael Mann's Co-author 'Cherry Picked 10 Tree Ring Data Sets'—'Famous Temperature Hockey Stick Not Only

Disappears but Goes Negative,'" ClimateDepot.com, September 28, 2009, www.climatedepot.com/a/3103/Disgraced-Michael-Manns-coauthor-cherry -picked-10-tree-ring-data-sets—Famous-temperature-Hockey-Stick-not- only-disappears-but-goes-negative.

32. Anthony Watts, "Broken Hockey Stick Fallout: Leading UK Climate Scientists Must Explain or Resign," Watts Up with That, September 29, 2009, http://wattsupwiththat.com/2009/09/29/leading-uk-climate-scientists-must -explain-or-resign/.

33. Thomas Fuller, "A Tale of Tree Rings, Global Warming and Fangorn's Ents," *Examiner,* October 2, 2009, www.examiner.com/environmental-policy -in-national/a-tale-of-tree-rings-global-warming-and-fangorn-s-ents.

34. James Delingpole, "How the Global Warming Industry Is Based on One Massive Lie," *Telegraph,* September 29, 2009, http://blogs.telegraph.co.uk /news/jamesdelingpole/100011716/how-the-global-warming-industry -is-based-on-one-massive-lie/.

35. Chris Horner, "Mann-Made Warming Confirmed," *National Review,* September 28, 2009.

36. Andrew Bolt, "An Inconvenient Truce as Green," *Herald Sun,* October 7, 2009.

37. Fred Pearce, "Climate Change Special: State of Denial," *New Scientist,* November 2006.

38. The precise wording of the demand was as follows: "This is a request under the Freedom of Information Act. Santer et al, Consistency of modelled and observed temperature trends in the tropical troposphere (Int J Climatology, 2008), of which NOAA employees J. R. Lanzante, S. Solomon, M. Free and T. R. Karl were co-authors, reported on a statistical analysis of the output of 47 runs of climate models that had been collated into monthly time series by Benjamin Santer and associates. I request that a copy of the following NOAA records be provided to me: (1) any monthly time series of output from any of the 47 climate models sent by Santer and/or other coauthors of Santer et al 2008 to NOAA employees between 2006 and October 2008; (2) any cor- respondence concerning these monthly time series between Santer and/or other coauthors of Santer et al 2008 and NOAA employees between 2006 and October 2008 . . ."

39. Stephen McIntyre filed or helped coordinate the filing of numerous FOIA demands against CRU scientists beginning in 2007. The campaign esca- lated dramatically during 2009. According to Richard Girling, "The Leak Was Bad, Then Came the Death Threats," Times (UK) Online, February 7, 2010, CRU received sixty UK FOI requests from across the world in July 2009 alone. Jones is quoted in the article: "We were clearly being targeted." According to the

article, only 22 percent of the filings were from the United Kingdom, 39 percent were from abroad, and 39 percent were untraceable. Many of the submissions made use of a template provided at McIntyre's Web site, http:// climateaudit.org/2009/07/24/cru-refuses-data-once-again/#comment-188529 (archived May 3, 2011): "FOI_09–97 / I hereby make a EIR/FOI request in respect to any confidentiality agreements) restricting transmission of CRUTEM data to non-academics involing [sic] the following countries: [insert 5 or so countries that are different from ones already requested]. / 1. the date of any applicable confidentiality agreements; / 2. the parties to such confidentiality agreement, including the full name of any organization; / 3. a copy of the section of the confidentiality agreement that 'prevents further transmission to non-academics.' / 4. a copy of the entire confidentiality agreement." A typo in the template (*involing* in place of *involving*) could be found in many of the submissions, and in one case, the submitter left in the boilerplate "insert 5 or so countries" rather than actually filling in the names of the countries. For further details, see the listing of FOI demands received by CRU in an official response from the University of East Anglia's Information Services Directorate dated January 21, 2010 (available as of May 3, 2011 at www.whatdotheyknow .com/request/25032/response/66822/attach/2/Response%20letter%20199%20 100121.pdf).

40. E. J. Steig, D. P. Schneider, S. D. Rutherford, M. E. Mann, J. C. Comiso, and D. T. Shindell, "Warming of the Antarctic Ice Sheet Surface Since the 1957 International Geophysical Year," *Nature*, 1457 (2009): 459–463.

41. This contrarian talking point is itself based on cherry-picking the evidence. Parts of Antarctica have cooled in recent decades, but only areas of East Antarctica's interior, during certain seasons, and as a result of change in the pattern of atmospheric circulation known as the Antarctic oscillation or AAO. See Eric Steig and Gavin Schmidt, "Antarctic Cooling, Global Warming," RealClimate, December 4, 2004, www.realclimate.org/index.php/archives /2004/12/antarctic-cooling-global-warming/.

42. See, e.g., H. Goosse et al., "Consistent Past Half-Century Trends in the Atmosphere, the Sea Ice and the Ocean at High Southern Latitudes," *Climate Dynamics*, 33 (2009): 999–1016; B. E. Barrett et al., "Rapid Recent Warming on Rutford Ice Stream, West Antarctica, from Borehole Thermometry," *Geophysical Research Letters*, 36 (2009): L02708, doi:10.1029/2008GL036369; A. J. Orsi and J. Severinghaus, "Evidence of Recent Warming in Polar Latitudes from Borehole Temperature," Abstract C24A-04, presented at 2010 Fall Meeting, AGU, San Francisco, California, December 13–17, 2010.

43. Christopher Booker, "Despite the Hot Air, the Antarctic Is Not Warming Up," *Telegraph*, January 27, 2009.

44. "Eric Steig Wears No Clothes," Air Vent, February 2, 2009, http://noconsensus.wordpress.com/2009/02/01/eric-steig-wears-no-clothes/ (archived May 3, 2011).

45. Stephen McIntyre, "Steig Professes Ignorance," climateaudit, August 14, 2009.

46. Eric Steig, "On Overfitting," RealClimate, June 1, 2009, www.realclimate.org/index.php/archives/2009/06/on-overfitting/.

47. Darrell Kaufman, "Climate Science Not Falsified," *Arizona Republic*, December 12, 2009.

48. The piece first appeared on townhall.com, October 15, 2009, http://townhall.com/columnists/pauldriessen/2009/10/15/none_dare_call_it_fraud (archived May 3, 2011). It later appeared on the Post Chronicle site, October 27, 2009, www.postchronicle.com/cgi-bin/artman/exec/view.cgi?archive=171&num=262493 (archived July 18, 2011).

49. See Sourcewatch page on Paul Driessen, www.sourcewatch.org/index.php?title=Paul_Driessen (archived May 3, 2011).

50. Center for a Constructive Tomorrow (CFACT), Center for the Defense of Free Enterprise, Frontiers of Freedom, and Atlas Economic Research Foundation have each been funded by some combination of ExxonMobil, Scaife Foundation, and Koch Industries. See Sourcewatch.org page on Committee for a Constructive Tomorrow, www.sourcewatch.org/index.php?title=Committee_for_a_Constructive_Tomorrow (archived May 3, 2011); "Koch Industries: Still Fueling Climate Denial," Greenpeace USA, April 2011; Sourcewatch.org page on Center for the Defense of Free Enterprise, www.sourcewatch.org/index.php?title=Center_for_the_Defense_of_Free_Enterprise (archived May 3, 2011).

51. According to Media Transparency, the Landmark Legal Foundation received a total of more than $6.3 million from the Sarah Scaife and Scaife Family Foundations between 1986 and 2009, greatly exceeding the sum from all other listed contributors; http://mediamattersaction.org/transparency/organization/Landmark_Legal_Foundation/funders.

52. According to Media Transparency, the Southeastern Legal Foundation received a total of more than $1.5 million from Scaife Family Foundations between 1985 and 2009, greatly exceeding the contributions from any other listed contributors and nearly equaling the sum of all other listed contributions; http://mediamattersaction.org/transparency/organization/Southeastern_Legal_Foundation/funders.

53. Http://climaterealists.com/index.php?id=6818 (archived May 3, 2011) provides a copy of a letter, posted on December 9, 2010, submitted to the UK House of Commons Science and Technology Committee:

From: D.J. Keenan

To: Science and Technology Committee (UK Commons)

Sent: 1 December 2010 16:34

Subject: Reviews into the Climatic Research Unit's E-mails at the University of East Anglia

Pursuant to the reviews of the Climatic Research Unit at the University of East Anglia, oral evidence was heard from Lord Oxburgh, on 8 September 2010, and from Sir Muir Russell, Vice Chancellor Edward Acton, and Pro Vice Chancellor Trevor Davies, on 27 October 2010. Each hearing considered the fraud allegation against CRU Professor Phil Jones. The allegation was made by me. The following describes some issues that pertain to the allegation, and proposes a means of resolution . . .

54. E-mail sent to me by doug.keenan@informath.org on December 3, 2009, copied to Phil Jones, Wei-Chyung Wang, and Steve McIntyre. In the e-mail, Keenan stated: "I reported you to the FBI in 2006, for your fraud with the hockey stick. The FBI decided that the case was not appropriate for them to pursue. . . . Then I found out that the proper policing authority for such crimes is the Office of the Attorney General of the state."

55. E-mail sent to doug.keenan@informath.org by stephen.mcintyre@uto ronto.ca on December 3, 2009, copied to me, Phil Jones, and Wei-Chyung Wang. The e-mail read "Please do not copy me on these emails. I was not a party to these complaints and do not wish to be involved in this. Regards, Steve McIntyre."

56. E-mail on April 20, 2007, from D. J. Keenan to Wei-Chyung Wang, copied to Phil Jones (later sent by Jones to me).

57. The two papers in question are W-C. Wang, Z. Zhaomei, and T. R. Karl, "Urban Heat Islands in China," *Geophysical Research Letters*, 17 (1990): 2377–2380; P. D. Jones, P. Ya. Groisman, M. Coughlan, N. Plummer, W-C. Wang, and T. R. Karl, "Assessment of Urbanization Effects in Time Series of Surface Air Temperature Over Land," *Nature*, 347 (1990): 169–172.

58. Douglas J. Keenan, "The Fraud Allegation Against Some Climatic Research of Wei-Chyung Wang," *Energy and Environment*, 18 (2007): 985–995.

59. H. McCulloch, "Irreproducible Results in Thompson et al., 'Abrupt Tropical Climate Change: Past and Present' (PNAS 2006)," *Energy and Environment*, 20 (2009): 367–373.

60. For those who are interested in the technical details, the matter involved the statistical concept of standardized Z scores. These are the series of

numbers that result from taking an original dataset, subtracting off the average, and dividing by the standard deviation. This yields a convenient new version of the dataset whose average is zero and standard deviation is equal to one. That latter property only holds for the full dataset. If one takes some subset of the data, the average will not in general be zero and the standard deviation will not in general be one. McCulloch's error essentially amounts to having assumed that the original Z scores defined by Thompson and colleagues, over the full data interval A.D. 0–2000, would still have average zero and standard deviation of one over a much shorter interval of A.D. 1610–1970. Because of this error, McCulloch ended up using a different weighted average of ice core data than the simple uniform weighted average used in the original Thompson et al. paper, which was the only reason he was unable to reproduce the Thompson et al. result. The error was all McCulloch's. How McCulloch and all reviewers of the paper could have missed something as basic as this is rather bewildering, even more so since McCulloch could have simply walked over to the other side of campus to ask Thompson; they're both faculty members at Ohio State University.

61. McCulloch wrote to *Nature* alleging that Steig had plagiarized his work. The claim related to the 2009 Steig et al. *Nature* article on Antarctic temperature trends, which, as we have seen, was subject to a dizzying assault by climate change deniers in the months following its publication. While Steig was on a field campaign in Antarctica and not receiving e-mail—or reading blogs, for that matter—McCulloch "published" a piece on the climateaudit blog criticizing the Steig et al. analysis—correctly, as it turned out—saying that the estimates of statistical significance of trends cited in the article were in error because they did not account for the presence of what is known as autocorrelation. Autocorrelation is a property, present in many climate data, that reduces the significance of a trend in the data below what it otherwise would be. McCulloch's claim was in fact correct; through an oversight, the effects of autocorrelation, contrary to what was stated in the paper, had not been taken into account. Once Steig was able to confirm that such an error had been made, he recalculated the trend significances correctly. The impact was found to be minimal, changing none of the conclusions of the study. Steig and colleagues nonetheless submitted a "corrigendum" to *Nature* correcting the trend significances. When it was published in August 2009 (*Nature*, 457: 459–462), McCulloch contacted *Nature*. McCulloch complained that Steig had appropriated his own finding. Yet it is self-evident that Steig et al. were aware of the need for the autocorrelation correction, since the paper explicitly stated (albeit, it turns out, in error) that it had been made. Had McCulloch

notified Steig of the error when he first discovered it, or had he submitted a formal comment to *Nature* identifying the error, he would have received credit and acknowledgement. He chose, however, to do neither of these things. To suggest that Steig's correction of an error in his own work, using standard methods, could constitute plagiarism was simply absurd. *Nature* rejected McCulloch's allegation. Meanwhile, however, Roger Pielke Jr., with some help from disinformation specialist Marc Morano, had already blasted the plagiarism allegation against Steig throughout the climate change denial echo chamber.

14. Climategate: The Real Story

1. The hackers had access to the materials in early October 2009, but held off releasing them until mid-November 2009, apparently to inflict maximum damage to the Copenhagen climate summit in early December 2009. See, e.g., Ben Webster, "Climate E-mail Hackers 'Aimed to Maximise Harm to Copenhagen Summit,'" *Sunday Times*, December 3, 2009.

2. There were four downloads via that link before we disabled RealClimate.

3. See Charles Arthur, "Hacking into the Mind of the CRU Climate Change Hacker," *Guardian* (environment blog), February 5, 2010, www.guardian.co.uk /environment/blog/2010/feb/05/cru-climate-change-hacker.

4. While noting that "McIntyre insists he had no role in the hack," the *Guardian* pointed to a "strikingly similar" incident involving McIntyre several months earlier. See Fred Pearce, "Search for Hacker May Lead Police Back to East Anglia's Climate Research Unit," *Guardian*, February 9, 2010, which states: "On 24 July, McIntyre says he received a big FOI refusal from CRU. . . . The next day McIntyre announced that he had got a mass of data. In November, there was a big FOI refusal, and again within days the 'FOIA2009.zip' files was [sic] all over the web. McIntyre was behind the first leak, though he initially was coy about it, talking about a 'mole.' . . . This was a tease. There was no human 'mole' in the sense of someone deliberately leaking material. Just a security breach."

5. Lisa Lerer, "Hacking Into the Mind of the CRU Climate Change Hacker," *Guardian*, February 5, 2010.

6. Lisa Lerer, "Saudi Arabia Calls for 'Climategate' Investigation," *Politico*, December 7, 2009.

7. The connections with Saudi Arabia are intriguing, if speculative. It was, after all, the Saudi Arabian delegation that first objected when the IPCC draft

report in 1995 claimed to have established an appreciable human influence on climate. Saudi royal family members, moreover, were substantial shareholders, second only to Rupert Murdoch, in News Corp. According to Kenneth Li, "Alwaleed Backs James Murdoch," *Financial Times*, January 22, 2010, Prince Alwaleed bin Talal al-Saud of the Saudi royal family, through his Kingdom Holding Company, owned 7 percent of News Corp's shares, making the holding company the second largest shareholder, after Rupert Murdoch and family, who own a controlling interest of more than 30 percent of the shares. News Corp is the parent company of several of the British tabloids, Fox News, and the *Wall Street Journal* that were most active in promoting the climategate charges. Moreover, a News Corp subsidiary had been implicated in similar acts of corporate espionage earlier that year. A federal case was brought against News America, a division of News Corp, when, according to the complaint filed, it had "illegally accessed plaintiff's computer system and obtained proprietary information" and "disseminated false, misleading and malicious information about the plaintiff" (David Carr, "Troubles That Money Can't Dispel," *New York Times*, July 17, 2011). News Corp settled the case for $30 million and then bought the company days later. A possible News Corp role in climategate was the topic of discussion on Current TV's Keith Olbermann show ("Is a Murdoch Henchman Responsible for Climate-Gate?" July 20, 2011, http://current.com/shows/countdown/videos/is-a-murdoch-henchman-responsible-for-climate-gate).

8. Richard Black, "Climate E-mail Hack 'Will Impact on Copenhagen Summit,'" BBC News, December 3, 2009.

9. According to a comment posted by Gavin Schmidt at RealClimate (www.realclimate.org/index.php/archives/2009/11/the-cru-hack-context/comment-page-4/#comment-143886): "This archive appears to be identical to the one posted on the Russian server except for the name change. Curiously, and unnoticed by anyone else so far, the first comment posted on this subject was not at the Air Vent [a site that had previously posted the e-mails], but actually at ClimateAudit (comment 49 on a thread related to stripbark trees, dated Nov 17 5.24am (Central Time I think)). The username of the commenter was linked to the FOIA.zip file at realclimate.org. Four downloads occurred from that link while the file was still there (it no longer is)."

10. That there was coordination (and perhaps collusion) between the architects of climategate and various fringe media and public relations allies is suggested by a comment posted by Charles Rotter, the moderator of the most prominent climate change denial Web site, "Watts Up with That": "A lot is happening behind the scenes. . . . Much is being coordinated among major

players and the media. . . . You will notice the beginnings of activity on other sites now. Here soon to follow."

11. For further discussion of these and other claims, see Morgan Goodwin, "Climategate: An Autopsy," DeSmogBlog, March 30, 2010, www.desmog blog.com/climatgate-autopsy.

12. See "Newtongate: The Final Nail in the Coffin of Renaissance and Enlightenment 'Thinking,'" Carbon Fixated Blog, November 21, 2009, http://carbonfixated.com/newtongate-the-final-nail-in-the-coffin-of-renaissance-and -enlightenment-thinking/.

13. David Wright made precisely that erroneous claim in an ABC News report that ran on December 9, 2009. A similar falsehood had earlier been promoted in various conservative media venues, e.g., "he's talking about a trick that another scientist previously used in a peer-reviewed journal to apparently hide the decline in temperatures—incredible" (Glen Beck, Fox News, November 23, 2009) and "Jones speaks of the 'trick' of filling in gaps of data in order to hide evidence of temperature decline" (Editorial, *Investor's Business Daily*, November 23, 2009).

14. This particular falsehood had been promoted recently by venues such as Fox News, e.g., Bill Hemmer on Fox's *America's Newsroom*, December 3, 2009: "Recently leaked emails reveal that scientists use, quote, 'tricks' to hide evidence of a decline in global temperatures over the past, say, few decades."

15. These were regional effects related to an ongoing El Niño event, with help from the negative state of the North Atlantic oscillation (NAO) atmospheric pattern.

16. Based on preliminary data for 2009 (issued on December 8, 2009, after the official November numbers had come in), the WMO concluded that "The 2000–2009 decade will be the warmest on record, with its average global surface temperature about 0.96 degree F above the 20th century average. This will easily surpass the 1990s value of 0.65 degree F" (www.wmo.int/pages/me diacentre/press_releases/pr_869_en.html).

17. The event took place in February of that winter. See, e.g., Emily Heil and Elizabeth Brotherton, "Heard on the Hill: Global Warming Snow Job," *Roll Call*, February 9, 2010.

18. Michael E. Mann, "E-mail Furor Doesn't Alter Evidence for Climate Change," *Washington Post*, December 18, 2009.

19. "Climatologists Under Pressure," *Nature*, 462 (December 3, 2009): 545.

20. K. R. Briffa, F. H. Schweingruber, P. D. Jones, T. J. Osborn, S. G. Shiyatov, and E. A. Vaganov, "Reduced Sensitivity of Recent Tree-Growth to Temperature at High Northern Latitudes," *Nature*, 391 (1998): 678–682.

21. See, e.g., R. D'Arrigo et al., "On the 'Divergence Problem' in Northern Forests: A Review of the Tree-Ring Evidence and Possible Causes," *Global and Planetary Change*, 60 (2008): 289–305.

22. In "Singer on Climategate Parliamentary Inquiry," an April 3, 2010, editorial on Anthony Watts's climate change denial blog (archived April 19, 2011), http://wattsupwiththat.com/2010/04/03/singer-on-climategate-parlia mentary-inquiry/, Singer stated of the phrase "hide the decline": "I believe that it refers to Michael Mann's 'trick' in hiding the fact that his multi-proxy data did not show the expected warming after 1979. So he abruptly cut off his analysis in 1979 and simply inserted the thermometer data supplied by Jones." Note that Singer also wrongly identifies the ending year of our reconstruction as 1979 when it was in fact 1980.

23. This was explained clearly in the methods section of MBH98: "the training interval is terminated at 1980 because many of the proxy series terminate at or shortly after 1980."

24. The differences are due in part, for example, to how missing data are treated, e.g., whether they are essentially just ignored (as in HadCRUT) or interpolated from neighboring data (as in NASA).

25. Michaels had since left the University of Virginia and was now employed by the Cato Institute.

26. According to Media Transparency, http://mediamattersaction.org /transparency/organization/The_Cato_Institute/funders, the Cato Institute received more than $2 million from the Sarah Scaife Foundation and more than $4 million from the David H. Koch Foundation between 1985 and 2009, the combined amount accounting for roughly one-third of the total listed contributions.

27. Patrick J. Michaels, "How to Manufacture a Climate Consensus," *Wall Street Journal*, December 17, 2009.

28. I made these points in a response to Michaels's piece that ran as a letter to the editor in the *Wall Street Journal* a couple weeks later; Michael E. Mann, "Science Journals Must Be Unpolluted by Politics," *Wall Street Journal*, December 31, 2009.

29. K. E. Trenberth, "An Imperative for Climate Change Planning: Tracking Earth's Global Energy," *Current Opinion in Environmental Sustainability*, 1 (2009): 19–27, doi:10.1016/j.cosust.2009.06.001.

30. The specific request was to delete e-mail exchanges with Keith Briffa, who was a lead author of the AR4 IPCC report. Briffa was apparently being harassed with FOIA demands for any correspondence he had had with various other scientists. Among the individuals mentioned in Jones's e-mail were me, Caspar Ammann, and Eugene Wahl. Jones had apparently misplaced Wahl's

e-mail address and asked me to make the request of Wahl. I did not comply, though I forwarded the e-mail to Wahl as I thought he should be aware of a discussion concerning him that was taking place. Indeed, to satisfy my employer, Penn State University, that I myself had not deleted any e-mail in response to Jones's request, I printed the messages out directly from my e-mail archives. There were only a handful of correspondences that I'd had with Briffa during the time period in question. They simply pointed to recent papers that were relevant to the sections of the IPCC report that Briffa was involved with.

31. James Hoggan, "Mainstream Media Misdirected in Stolen Email Story," DeSmogBlog, December 14, 2009.

32. Sarah Palin, "The Hacker Case Verdict," Facebook, April 30, 2010.

33. Richard Brenne, posted as a comment on Joe Romm, "Contest: Rename the Scandal Formerly Known as Climategate," Climate Progress, April 4, 2010.

34. These terms were offered in an online contest by Joe Romm of the Center for American Progress in "Contest: Rename the Scandal Formerly Known as Climategate."

35. Goodwin, "Climategate: An Autopsy."

36. Ibid.

37. Calls for an investigation began with Chris Horner of CEI demanding a lawsuit "against the obviously dishonest tactics and claims by the global warming industry under the Racketeer Influence and Corrupt Organization Act (RICO)" on November 21 and UK climate change denier Lord Nigel Lawson calling on November 23 for an "independent inquiry into claims that leading climate change scientists manipulated data to strengthen the case for man-made global warming." Soon enough, Inhofe was calling for an investigation, as reported by Tony Romm's *The Hill* blog and the Drudge Report (November 23) and the *Washington Times* and Fox News (November 24). Almost immediately, allegations were made of a whitewash (CNBC's *Kudlow Report*, November 24). The whitewash theme was echoed in the days ahead by the *Telegraph* (James Delingpole's blog and Christopher Booker in the print edition on November 27).

38. Goodwin, "Climategate: An Autopsy."

39. Fox News, *Fox Nation*, November 23.

40. I corresponded frequently with CNN producer Nadia Kounang and found both her and the CNN correspondent on the story, Brooke Baldwin, highly professional and a pleasure to work with.

41. E.g., "American Morning: Cloud Over Climate Summit," December 7, 2009; "Campbell Brown: Global Warming: Trick or Truth?" December 7, 2009; "Anderson Cooper 360 Degrees: Climate Conspiracy," December 7, 2009.

42. Roberts and I must have exchanged a dozen text messages and e-mails over one particular weekend, and I was convinced that he was engaged in a good faith effort to get to the bottom of the various allegations swirling about.

43. I became somewhat exasperated with Roberts in mid-April 2010. The UK House of Commons had issued a report clearing climategate scientists of any wrongdoing. Roberts had asked me to go on CNN to discuss the report. It was a particularly busy time, as classes were just finishing up at Penn State, but I agreed—at least initially. I rescinded my offer when it emerged, quite late in the game, that I was to appear side-by-side with self-appointed "climate auditor" Stephen McIntyre. The story at hand was one of exoneration of scientists. There weren't two sides to that story, and I saw no reason to share my time on camera with someone expressing an "other view" that was completely bankrupt.

44. Wright's piece earned him the dubious title of "worst person in the world" on MSNBC's *Countdown with Keith Olbermann* that evening (December 9, 2009).

45. The segment, which ran on the February 4, 2010, *CBS Evening News* with Katie Couric was entitled "Mistakes in Climate Report Fuel Skepticism" (www.cbsnews.com/stories/2010/02/04/eveningnews/main6175058 .shtml?tag=mncol;lst;1). The correspondent, Mark Phillips, reported that "The subject of the spoof is Michael Mann of Penn State University, who was accused of tampering with climate data to produce his famous hockey stick graph which shows that the rise in man-made greenhouse gasses corresponds to a rise in world temperatures. An academic board today cleared Mann, saying his science holds up—but the damage may have already been done."

46. The piece appeared online on February 22 and in print in the March 1 issue of *Newsweek*.

47. The full correction appeared online on March 5, 2010, and read: "Editor's note: This story was corrected to acknowledge that Al Gore's slide show used data from a hockey-stick study other than Michael Mann's, that Mann made all the data germane to the study available online, and that Mann's study data stopped at 1980." Not all of the corrections could be made in time for them to appear in the print version of the article, which appeared on March 29.

48. Michael Mann, "Iceberg Ahead," "Letters: March 15, 2010," *Newsweek*, March 29, 2010; appeared online on March 18, 2010.

49. Guterl appeared to me to be an honest journalist who was hoodwinked by others while researching his story, didn't check his facts carefully,

and was perhaps biased by a desire to produce a winning narrative that he could sell to *Newsweek*'s editors. That interpretation seemed even more likely when *Newsweek* published a denialist editorial that easily could have been written by a fossil fuel industry advocate: "Uncertain Science: Bickering and Defensive, Climate Researchers Have Lost the Public's Trust," *Newsweek*, May 28, 2010. The editorial attempted to position its stance as neutral, e.g., "Very few scientists dispute a link between man-made CO_2 and global warming," but its true policy agenda was revealed in statements like "[T]here are excellent reasons to limit emissions and switch to cleaner fuels. . . . At the moment, however, certainty about how fast—and how much—global warming changes the earth's climate does not appear to be one of those reasons."

50. Fred Guterl, "It's Gettin' Hot in Here: The Big Battle Over Climate Science," *Discover*, March 2010.

51. This was despite the pressure and circuslike atmosphere that groups such as the Commonwealth Foundation—a Scaife-funded free market think tank based in Pennsylvania—attempted to generate in the region. Initially, student reporters for Penn State's *Daily Collegian* had played up any controversy they could find, airing accusations by groups like the Commonwealth Foundation no matter how misleading or defamatory. I don't blame the student reporters. They learn from watching the professionals (the student editor in chief of the *Collegian* had served as a summer intern for the Scaife-owned *Pittsburgh Tribune-Review*). Over time, I believe the paper's staff and reporters became better at recognizing and ignoring obvious attempts by outsiders to manufacture fake controversy.

52. David Wright and Max Culhane, "Many Television Weather Forecasters Doubt Global Warming," *Nightline*, April 22, 2010.

53. Ibid.

54. Wright here was responding to my admission that climate scientists weren't doing as good a job as they could in communicating the science.

55. The piece (Dan Harris and Christone Brower, "Climate Scientists Claim 'McCarthy-Like Threats,'" *ABC Evening News*, May 23, 2010) included interviews with me and other climate scientists that further served to shift the narrative from the content of the stolen e-mails and toward the topic of the increased vilification, harassment, and intimidation of climate scientists.

56. Even ExxonMobil, which had claimed to have discontinued its funding of climate change denial, had a crucial role to play in promoting the faux scandal. According to a July 19, 2010, story that ran on the front page of the *Times* of London: "Several [of the Exxon-funded groups] made outspoken attacks on climate scientists at the University of East Anglia and argued their

leaked e-mails showed that the dangers of global warming had been grossly exaggerated. . . . The scientists were exonerated this month by an independent inquiry but groups funded by Exxon have continued to lambast them. The Media Research Centre, which received $50,000 last year from Exxon, called the inquiry a 'whitewash' and condemned 'climate alarmists.'"

57. "Koch Industries: Secretly Funding the Climate Denial Machine," Greenpeace USA, March 2010.

58. See, e.g., Janet Raloff, "IPCC Admits Himalayan Glacier Error," *Science News* Web edition, January 20, 2010, www.sciencenews.org/view/generic/id /55455/title/Science_%2B_the_Public__IPCC_admits_Himalayan_glacier _error.

59. See, e.g., Robin McKie, "Climate Scientists Admit Fresh Error Over Data on Rising Sea Levels," *The Observer*, February 14, 2010, www.guardian. co.uk/environment/2010/feb/14/benny-peiser-houghton-ipcc-apology.

60. Leake's repeated fabrications of alleged IPCC errors were eventually themselves collectively referred to as "Leakegate"; see, e.g., Tim Lambert's Deltoid blog, "Leakegate: The Case for Fraud," February 14, 2010.

61. Christopher Booker, "Pachauri: The Real Story Behind the Glaciergate Scandal," *Telegraph*, January 23, 2010.

62. S. L. Rao, "Behind the Attacks on Pachauri," *Business Standard*, February 4, 2010.

63. Stefan Rahmstorf, "IPCC Errors: Facts and Spin," RealClimate.org, February 14, 2010.

64. See, e.g., Gavin Schmidt, "The Guardian Disappoints," February 23, 2010. Schmidt notes numerous such examples, but one of the more salient instances involves Pearce's discussion of the hockey stick. Quoting Schmidt, "Some of the more egregious confusions and errors were in the third part of the series. In this part, a number of issues that were being discussed among the paleo-community in 1999 were horribly mixed up. For instance, there was a claim that arguments on the zeroth-order draft of the 2001 IPCC report were based on Briffa's reconstruction showed the 11th century as being almost as warm as the 20th century, while Mann's graph found little sign of the earlier warming. But this is simply untrue since at the time Briffa's curve only went back to 1400 A.D. (not the 11th Century) and the discussions had nothing to do with the medieval warm period, but rather the amount of multi-decadal variability in the three different reconstructions then available. . . . That discussion was conflated with a completely separate April 1999 issue based on a disagreement about a perspectives piece in *Science* (which appeared as Briffa and Osborn, 1999) and which was in any case amicably resolved." See also the guest article by Ben Santer, "Close Encounters of the Absurd Kind," February

24, 2010. We later published a response from the *Guardian's* environmental editor James Randerson, "The Guardian Responds," March 24, 2010. Many of our readers took the opportunity to express their own criticisms of the *Guardian's* coverage in the comments.

65. The interview was "Q & A: Professor Phil Jones," BBC, February 13, 2010, http://news.bbc.co.uk/2/hi/8511670.stm. The BBC interviewer, Roger Harrabin, indicated that several of his questions had been "gathered from sceptics."

66. E-mail from Lindzen leaked by Watts on his blog, wattsupwiththat .com/2008/03/11/a-note-from-richard-lindzen-on-statistically-significant -warming: "Look at the attached. There has been no warming since 1997 and no statistically significant warming since 1995. Why bother with the arguments about an El Niño anomaly in 1998? . . . Best wishes, Dick."

67. Jones's response was "Yes, but only just. I also calculated the trend for the period 1995 to 2009. This trend (0.12C per decade) is positive, but not significant at the 95% significance level. The positive trend is quite close to the significance level. Achieving statistical significance in scientific terms is much more likely for longer periods, and much less likely for shorter periods."

68. Jonathan Leake, "World May Not Be Warming, Say Scientists," *Sunday Times,* February 14, 2010, www.timesonline.co.uk/tol/news/environment /article7026317.ece.

69. Jonathan Petre, "Climategate U-Turn as Scientist at Centre of Row Admits: There Has Been No Global Warming Since 1995," *Daily Mail,* February 14, 2010, www.dailymail.co.uk/news/article-1250872/Climategate-U-turn-Astonishment-scientist-centre-global-warming-email-row-admits-data-organised.html?ITO=1490.

70. On the February 16, 2010, edition of *Fox News and Friends,* the following exchange between Fox News commentators Steve Doocy and Glenn Beck took place:

Doocy: "Let's talk real quickly. There's a report out of the British tabloids—newspapers yesterday that said that apparently Phil Jones is a professor over in England who has been overseeing a lot of this data and in fact is famous for the so-called hockey stick chart that shows that the earth has got a fever. Apparently he doesn't actually have the paperwork that supports it, and there's been no global warming for apparently 15 years."

Beck: "15 years. And it's now cooling. He says it's cooling in the last few years. I mean, I don't know why anyone believes this, but you'll notice that all of the supporters will all say, well it doesn't matter anyway. It doesn't matter anyway. If this was truly about science, especially at this critical time in our economic history, we'd be saying whoa, whoa whoa. We'd be doing what India's doing. Back off. Wait a minute, wait a minute. This whole thing is falling apart. We're not going to do this."

71. The characterization was provided by an observer in the comment thread of a blog post by Deep Climate: "Round and Round We Go with Lindzen, Motl and Jones," March 2, 2010, http://deepclimate.org/2010/03/02/round -and-round-we-go-with-lindzen-motl-and-jones/.

72. From a comment in blog post, Tim Lambert, "James M. Taylor Hides the Decline," Deltoid Blog, May 22, 2010, http://scienceblogs.com/deltoid/2010 /05/james_m_taylor_hides_the_decli.php.

73. Johann Hari, "Climategate Claptrap, II," The Nation Online, April 15, 2010.

74. The article in question was M. Siddall, T. F. Stocker, and P. U. Clark, "Constraints on Future Sea-Level Rise from Past Sea-Level Change," *Nature Geoscience*, 2 (2009): 571–575; originally published online on July 26, 2009.

75. M. Siddall, T. F. Stocker, and P. U. Clark, "Retraction: Constraints on Future Sea-Level Rise from Past Sea-Level Change," *Nature Geoscience*, 3 (2010): 217; published online on February 21, 2010.

76. Bret Baier, "Special Report with Bret Baier: More Questions About Validity of Global Warming Theory," February 22, 2010, www.foxnews.com /story/0,2933,587179,00.html.

77. The impact of the errors in question was discussed on RealClimate in Stefan Rahmstorf, "Sealevelgate," March 11, 2010.

78. P. W. Gething et al., "Climate Change, and the Global Malaria Recession," *Nature*, 465 (2010): 342–345.

79. Climate change is only one of the factors that could affect changes in the spread of malaria, and nothing in the *Nature* article or in the IPCC assessment claimed otherwise. The influence of climate on malaria development and spread is complex and often subtle, requiring sophisticated approaches combining information from climate models with models of disease dynamics. As it happened, I had been collaborating with a Penn State scientist, Matt Thomas, who was widely quoted making essentially this point. Morano couldn't resist the opportunity to attack Thomas, me, and Penn State University simultaneously in his post: "MalariaGate: 'Discredited' Penn State Prof. Matt Thomas 'has been doing overtime trying to salvage remnants of discredited man-made global warming theory,'" ClimateDepot.com, May 23, 2010, www.climatedepot .com/a/6675/MalariaGate-Discredited-Penn-State-Prof-Matt-Thomas -has-been-doing-overtime-trying-to-salvage-remnants-of-discredited -manmade-global-warming-theory (archived May 5, 2011).

80. For a timeline of the attacks, see Lee Fang, "A Case of Classic Swift-Boating: How the Right-Wing Noise Machine Manufactured 'Climategate,'" The Wonkroom, Center for American Progress, December 9, 2009.

81. The letter, sent to various journalists, was obtained by the site Think-Progress, http://wonkroom.thinkprogress.org/2009/11/23/vitter-climategate-fraud/.

82. Tony Romm, "Inhofe to Call for Hearing into CRU, U.N. Climate Change Research," The Hill Online, November 23, 2010, http://thehill.com/blogs/blog-briefing-room/news/69141-inhofe-to-call-for-hearing-into-cru-un-climate-change-research.

83. A letter addressed to me dated November 24, 2010, announces: "The Minority Staff . . . is conducting an investigation. . . . Your name has surfaced in either the emails or the documents. . . . This letter is to notify you that through either the Freedom of Information Act ("FOIA") or other information disclosure laws, we may be asking for copies of any documents or records you have. . . . Please note that there are severe civil and criminal penalties, federal and statutes, for the destruction of certain materials."

84. See Keith Johnson and Gautam Naik, "Lawmakers Probe Climate Emails," *Wall Street Journal*, November 24, 2009.

85. Seth Borenstein, "Obama Science Advisers Grilled Over Hacked E-mails," Associated Press, December 3, 2009. Holdren had refuted the claim that scientists had attempted to "hide the decline" in temperatures and corrected Sensenbrenner's repetition of the erroneous denier talking point that the NAS had discredited the hockey stick.

86. United States Senate Committee on Environment and Public Works, Minority Staff, "'Consensus' Exposed: The CRU Controversy," February 2010.

87. Dan Lashof, National Resources Defense Council blog, March 1, 2010, http://switchboard.nrdc.org/blogs/dlashof/are_you_now_or_have_you_ever_b.html.

88. Douglas Fisher and the Daily Climate, "Cyber Bullying Intensifies as Climate Data Questioned," *Scientific American*, March 1, 2010.

89. Glenn Beck radio program, February 10, 2010, www.glennbeck.com/content/articles/article/198/36153/. The precise statement was "There's not enough knives. If . . . the IPCC had been done by Japanese scientists, there's not enough knives on planet Earth for hara-kiri that should have occurred. I mean, these guys have so dishonored themselves, so dishonored scientists."

90. Media Matters, November 29, 2009, http://mediamatters.org/blog/200911290004.

91. Fisher and the Daily Climate, "Cyber Bullying Intensifies."

92. Who Is Behind the Climate Change and Carbon Trading Scams?" Stormfront.org, March 30, 2010, www.stormfront.org/forum/showthread .php?p=7961482.

93. Clive Hamilton, "Who Is Orchestrating the Cyber-Bullying?" Australian Broadcasting Corporation (ABC), The Drum, February 23, 2010.

94. Fisher and the Daily Climate, "Cyber Bullying Intensifies."

95. This was covered by *ABC Evening News with Diane Sawyer* in Harris and Brouwer, "Climate Scientists Claim 'McCarthy-Like Threats'": "Recently, a white supremacist website posted Mann's picture alongside several of his colleagues with the word 'Jew' next to each image."

96. According to Media Transparency, the National Center for Public Policy Research received more than $260,000 from the Sarah Scaife Foundation between 1986 and 2008, amounting to about 10 percent of its total listed received funds (http://mediamattersaction.org/transparency/organization/National_Center_for_Public_Policy_Research/funders).

97. "Economic Stimulus Funds Went to Climategate Scientist: Funds Should Be Returned to U.S. Treasury, Says National Center for Public Policy Research," press release, January 14, 2010.

98. The vast majority of scientists at major research universities are supported by the university during the nine-month academic year, but are expected to cover the three summer months with research funding at their regular academic year monthly salaries. NSF had benefited from a major infusion of economic stimulus money during 2009, and roughly one in three NSF grants that year was funded off stimulus funds. I was the beneficiary of a little more than one month of summer salary per year from the two NSF grants in question. Most of the grant money went toward funding students, postdoctoral researchers, travel, and publication costs.

99. The incident was gleefully reported thusly by James Delingpole of the *Telegraph* on January 3, 2010 (http://blogs.telegraph.co.uk/news/jamesdelingpole/100021135/climategate-michael-manns-very-unhappy-new-year/): "I am so glad to report that Michael Mann—creator of the incredible Hockey Stick curve and one of the scientists most heavily implicated in the Climategate scandal—is about to get a very nasty shock. When he turns up to work on Monday, he'll find that all 27 of his colleagues at the Earth System Science Center at Penn State University have received a rather tempting email inviting them to blow the whistle on anyone they know who may have been fraudulently misusing federal grant funds for climate research. . . . Under US law, regardless of whether or not a prosecution results, the whistleblower stands to make very large sums of money: it is based on a percentage of the total government funds which have been misused, in this case perhaps as much as

$50 million." The incident was also reported in Colleen Boyle, "Agent Looking for 'Climategate' Insiders," *Daily Collegian*, January 11, 2010.

100. "Economic Stimulus Funds Went to Climategate Scientist."

101. Mitchell Anderson, "The Latest Smear Campaign Against Michael Mann," DeSmogBlog, January 15, 2010, www.desmogblog.com/latest-smear-campaign-against-michael-mann.

102. Www.archive.org/details/tobacco_kym52d00.

103. The video was produced for Philip Morris employees. It purports to provide the "state of science on environmental tobacco smoke" and attempts to refute the scientific findings of the Environmental Protection Agency demonstrating just such a health threat. This movie is part of the collection UCSF Tobacco Industry Videos.

104. Www.cato.org/people/deepak-lal.

105. Deepak Lal, "Man-Made Global Warming: The Climate-Change 'Science' Continues to Unravel," *Business Standard*, January 30, 2010.

106. H. M. Mamudu, R. Hammond, and S. Glantz, "Tobacco Industry Attempts to Counter the World Bank Report Curbing the Epidemic and Obstruct the WHO Framework Convention on Tobacco Control," *Social Science and Medicine*, 67 (2008): 1690–1699.

107. Http://live.psu.edu/story/44327.

108. Ed Barnes, "Penn State Probe into Mann's Wrongdoing a 'Total Whitewash,'" Fox News, February 5, 2010.

109. "Congressman Slams Grant to Embattled Climate Researcher," *Washington Times*, February 3, 2010.

110. Ben Geman, "State Dept. Defends Climate Science as GOP Lawmakers Seek Review of Researcher," *The Hill*, February 17, 2010.

111. See the various FOIA and similar legal demands filed by these groups at http://climscifoi.blogspot.com/.

112. Richard Mellon Scaife is widely held to be the individual behind the "vast right-wing conspiracy" referred to by former first lady Hilary Clinton in characterizing a malicious campaign to discredit former President Bill Clinton. See, e.g., Brooks Jackson, "Who Is Richard Mellon Scaife?" CNN, April 27, 1998, www.cnn.com/ALLPOLITICS/1998/04/27/scaife.profile/. A similar case has been made for the Koch brothers of Koch Industries; see, e.g., Jane Mayer, "Covert Operations: The Billionaire Brothers Who Are Waging a War Against Obama," *New Yorker*, August 30, 2010.

113. According to Media Transparency, the Commonwealth Foundation for Public Policy Alternatives received more than $2 million from the Sarah Scaife and Scaife Family Foundations between 1988 and 2009, larger than any other contribution, and amounting to more than 50 percent of the total listed

funding (http://mediamattersaction.org/transparency/organization/Common wealth_Foundation_for_Public_Policy_Alternatives/funders).

114. Letter "RE: Allegations of Intellectual Fraud by PSU Professor" from Matthew J. Brouillette, president and CEO of the Commonwealth Foundation, to Governor Ed Rendell and seven Pennsylvania state senators, copied to Penn State President Graham Spanier, November 30, 2009: "as you continue to debate funding for higher education institutions in Pennsylvania, including PSU, I trust that the Rendell Administration, leaders of the General Assembly, and chairmen of the Senate and House Education Committees would thoroughly investigate these apparently inappropriate actions by Dr. Mann; and, if necessary, withhold state funds in response to any failures to adequately address any acts of intellectual fraud committed by Dr. Mann. I thank you in advance for your attention to this matter, and I look forward to your public addressing of this troubling situation. Sincerely, Matthew J. Brouillette President & CEO" (letter available at www.scribd.com/doc/23383109/CF-Letter-RE -Michael-Mann).

115. Letter to Penn State President Graham Spanier from state senator Jeffrey E. Piccola, December 3, 2009, available at www.scribd.com/doc/23612181 /Sen-Piccola-Letter-on-PSU-Prof-Michael-Mann.

116. These investigations include the Independent Climate Change Email Review, announced on December 3, 2009, called for and funded by the University of East Anglia. Its work and findings are wholly independent, conducted by a team of scientific experts led by Sir Muir Russell of the United Kingdom. There was also the House of Commons Science and Technology Committee Inquiry into the disclosure of climate data from the Climatic Research Unit at the University of East Anglia, announced on January 22, 2010. Finally, there was the Royal Society panel, set up in consultation with CRU, which consisted of Lord Ron Oxburgh of the Royal Society and seven other leading international scientists and academics, whose charge was to "assess the integrity of the research published by the Climatic Research Unit in the light of various external assertions."

117. "News Release: Investigate 'Mann-Made' Global Warming; Commonwealth Foundation Calls for Independent Investigation of Penn State Professor in Eye of Climategate Storm," January 12, 2010.

118. "Paid Advertisement: Will Penn State's Investigation of Climategate Be a [picture of a whitewashing of a fence]: Academic Integrity Rally 12:00 Noon," *Daily Collegian*, January 27, 2010. It ran a similar round of attacks ads during the first week of classes in August-September 2010.

119. On February 3, 2010, after the completion of its initial inquiry phase, it issued a finding dismissing any and all charges against me of suppressing or

falsifying data; deleting, concealing, or destroying data or information; and misusing privileged or confidential information. On a final, more diffuse charge of whether I had engaged in "actions that seriously deviated from accepted practices within the academic community for proposing, conducting, or reporting research or other scholarly activities," while finding that there was no evidence to support the charge, the committee nonetheless judged it best to remand the matter to a separate committee of academic peers who would formally investigate the charge and report back later.

120. One individual even gained unauthorized access to the Penn State staff e-mail listserv to circulate malicious allegations about me to all of the faculty and staff in my wife's college.

121. E.g., "Michael Mann 'Accused of Fraud and Perpetrating a Hoax,'" ClimateDepot.com, January 15, 2010. Who was Morano quoting? None other than the Commonwealth Foundation.

122. Interview with Morano on *Fox News America Live*, hosted by Megyn Kelly, April 27, 2010.

123. The ads were sponsored by Milloy's Green Hell Blog.

124. The No Cap and Trade Coalition sponsored a publicity campaign attacking me over the video, as described below in the text.

125. Jeff Tollefson, "An Erosion of Trust," *Nature*, 466 (June 30, 2010): 24–26, began with a description of the video: "Last November, a catchy music video popped up on YouTube and attracted thousands of fans. Called 'Hide the Decline,' the video featured a caricature of climate researcher Michael Mann admitting that he had committed fraud while creating his famous 'hockey-stick' graph of temperatures over the past millennium. Accompanied by a kitten playing the guitar, the cartoon image of Mann joyfully sings, 'Making up data the old hard way, fudging the numbers day by day.'"

126. According to investigators with the New Jersey law firm Cozen O'Conner, Balgaar (as of February 14, 2010) had registered the Web page through his company, MacMagicians Inc. (http://macmagicians.com). Balgaar described himself as a Web designer, expert graphic designer, video designer, musician, and "true Christian believer." He gave an interview related to a previous anti-global warming YouTube video he had made, which was pulled after Van Morrison filed a complaint of infringement with YouTube (http://blip.tv /file/1742749, as archived February 14, 2010). The minnesotansforglobalwarming Web site, as of February 14, 2010, was hosted by two network servers owned by ThePlanet.com Internet Services, located in Houston, Texas. A Google search on the Macmagicians Web page performed by Cozen O'Conner revealed a cache of material linked with the Republican Party, including a Republican National Committee Web page attacking Al Gore during the 2000 election.

127. Antonio Regalado and Dionne Searcey, "Where Did That Video Spoofing Gore's Film Come From?" *Wall Street Journal*, August 3, 2006.

128. Jake Tapper and Max Culhane, "Al Gore YouTube Spoof Not So Amateurish, Republican PR Firm Said to Be Behind 'Inconvenient Truth' Spoof," ABC News, August 4, 2006.

129. The image's copyright holder was a photographer named Tom Cogill, who had taken the photo for the University of Virginia alumni magazine. The photo was available on my Web site with the explicit instructions "Permission to reprint must be obtained from Tom Cogill," which linked to Cogill's Web site and contact information.

130. According to the group's Web page (www.nocapandtrade.com/about/) as archived on April 18, 2011, the No Cap and Trade Coalition is an alliance of a large number of groups that, as it turns out, are heavily funded by Koch Industries and the Scaife Foundations as detailed previously in the book: e.g., Americans for Prosperity, the Commonwealth Foundation, Competitive Enterprise Institute, FreedomWorks, Heartland Institute, and the National Center for Policy Analysis.

131. "Media Advisory: Leading Climate Scientist Threatens Legal Action for Satirical YouTube Video, No Cap and Trade Coalition and Minnesotans for Global Warming Respond," April 19, 2010, advertising a press conference involving "Minnesotans for Global Warming, Dr. Patrick Michaels, Myron Ebell, and members of the No Cap and Trade Coalition," to be held 10 A.M. on April 20, 2010, at the National Press Club, Washington, D.C.

132. *Neil Cavuto Show*, Fox News, April 28, 2010.

133. Some examples of messages left on my voice mail: "I pray you really do sue those who made the parody about your work. That way there will be full discovery, and your intentional fraud, along with the collusion of your so called peers, will become part of the court record and will allow you to be charged with perhaps the biggest fraud of the century. I think about twenty years in the federal pen would look very good on you." and "Please, please sue the parody video folks. Maybe we can finally get a look at all your source data. Plus all that other great stuff the discovery process exposes."

134. Indeed, the obnoxious refrain of one later *Pittsburgh Tribune-Review* editorial attack against me was "Mann Up" (May 15, 2010).

135. Amanda Little, "The King and I: An Interview with Sir David King, Britain's Top Scientist and Climate Crusader," *Grist*, February 17, 2006: "You've been heckled at your lectures by U.S. climate skeptics who argue that climate science is 'in its infancy'—an opinion that has also been voiced by members of the Bush administration." King responded: "I've been chased around by several

people funded by the Competitive Enterprise Institute who make these kind of statements that simply fly in the face of all the evidence."

136. An individual named Jeffrey Steigerwalt (who is active in local tea party politics) led the event and had registered the protesters. He identified himself to PennFuture director Jan Jarrett and vice president Heather Sage as a funder and organizer for the Commonwealth Foundation. The only coverage of the protest in the U.S. media was on the Commonwealth Foundation Web site in a posting by Elizabeth Stelle, a "research associate" with the Commonwealth Foundation, "Climate Change 'Deniers' Protest in Pittsburgh," May 3, 2010, www.commonwealthfoundation.org/policyblog/detail/climate-change-deniers-protest-in-pittsburgh. Ms. Stelle was identified by PennFuture staff as having been present throughout the protests and among the most active members of the small group, wearing one of the "Mann-Made Global Warming" T-shirts that the group was selling and posing for several of photographs of the event linked from the Commonwealth Foundation Web site.

137. See, e.g., the January 11, 2011, coverage at the DeSmogBlog, www.desmogblog.com/cia-vet-stalking-hockey-stick-author-mike-mann, and at Climate Progress, http://climateprogress.org/2011/01/11/cia-vet-kent-clizbe-stalking-hockey-stick-author-mike-mann/.

138. Their latest videos were produced and hosted by the No Cap and Trade Coalition and featured on the Heartland Institute Web site.

139. The video was available as of May 11, 2011, and archived February 3, 2011. The Heartland Institute, which hosted the video on its Web page (www.heartland.org/environmentandclimate-news.org/YouTubevideo.html) during February 2011, provided this introduction: "It's a parody of 'I'm a Believer' written by Neil Diamond and performed by the Monkees. This version was written by Elmer Beauregard and Brian D. Smith and performed by Elmer and the M4GW players. This song is in honor of all the new Republican Freshman entering Congress and the Senate most of whom are Deniers and proud of it."

140. Eric Alterman, "The Chutzpah Hall of Fame," *The Nation*, February 3, 2011, www.thenation.com/blog/158269/chutzpah-hall-fame.

141. CFACT ran a booth at the well-known annual conservative event CPAC, at which passers-by were encouraged to hurl eggs at photos of both Al Gore and me. Bradford Plumer, "In the Belly of the Beast: CPAC's Carnival of Groups Offers a Glimpse Into the Conservative Universe," New Republic Online, February 12, 2011, www.tnr.com/article/politics/83367/cpac-nra-safari-club-carnival, reported: "I saw one girl chuck an egg so vehemently that she [had] to leap back to avoid the splatter." The "girl," it turned out, was a CFACT employee named Christina Wilson. She placed a photo showing her in front of

the egg-throwing display on her Twitter page (archived February 11, 2011) at http://twitpic.com/3yporx. According to the CFACT Web page, she is a "development officer" for CFACT: www.cfactcampus.org/about/staff/ (archived May 11, 2011).

142. Such was the conclusion of a March 2010 Gallup poll: Frank Newport, "Americans' Global Warming Concerns Continue to Drop: Multiple Indicators Show Less Concern, More Feelings That Global Warming Is Exaggerated," March 11, 2010, www.gallup.com/poll/126560/americans-global -warming-concerns-continue-drop.aspx.

15. Fighting Back

1. Www.parliament.uk/parliamentary_committees/science_technology /s_t_cru_inquiry.cfm.

2. David Adam, "Climate Scientist Admits Sending 'Awful Emails' but Denies Perverting Peer Review," *Guardian*, March 1, 2010.

3. Www.uea.ac.uk/mac/comm/media/press/CRUstatements/SAP; additional related materials can be found at www.uea.ac.uk/mac/comm/media /press/CRUstatements/independentreviews.

4. "Investigation of Climate Scientist at Penn State Complete," July 1, 2010. Full report at http://live.psu.edu/fullimg/userpics/10026/Final_Investigation_ Report.pdf. We covered the development at RealClimate, www.realclimate.org /index.php/archives/2010/07/penn-state-reports/. My response to the finding is available at www.essc.psu.edu/essc_web/news/MannInquiryStatementFi nal.html.

5. I was mildly scolded in the report for being careless on a couple of occasions in sharing unpublished manuscripts provided by colleagues of mine who had not given explicit consent to further distribute, though the committee noted that this was done with the tacit approval of those colleagues.

6. The August 19, 2011, close-out memorandum is available at the NSF Inspector General Web site, www.nsf.gov/oig/search/ (the case code is 'A09120086'). It has also been uploaded at: http://thinkprogress.org/wp-con tent/uploads/2011/08/NSF-Mann-Closeout.pdf.

7. Www.cce-review.org/pdf/FINAL%20REPORT.pdf. We reported on the findings at RealClimate, www.realclimate.org/index.php/archives/2010/07/ the-muir-russell-report/.

8. Www.cce-review.org.

9. "Cleared on Climategate," *Reliable Sources with Howard Kurtz*, CNN, July 11, 2010. Kurtz complained, for example, "Glenn Beck didn't report on

this at all. Last fall, when the e-mails were leaked, he called global warming a big hoax and he said, 'Why has no network covered this global warming fix?' Why has Glenn Beck and others not revisited it?"

10. CNN aired a segment "Climategate: Trick or Truth" on *The Situation Room* on July 8 in which I was interviewed by Mary Snow. Quoting from the transcript: "an independent review has concluded there was no evidence to question the rigor and honesty of the climate scientists in those hacked e-mails including meteorology professor, Michael Mann of Penn State. But Professor Mann and others have already paid a personal price. He tells us he's received threatening e-mails and phone messages."

11. Wyatt Andrews, "An End to Climategate? Penn State Clears Michael Mann," CBS News, July 1, 2010.

12. The *New York Times* ran both an editorial and a news article reporting on the vindication of the scientists. Positive coverage was given in numerous other major newspapers and magazines, including *USA Today, Washington Post, Los Angeles Times, Philadelphia Inquirer, The Hill,* the Associated Press, the *Guardian,* and *Newsweek,* as well as numerous other news outlets both within and outside the United States.

13. The *Wall Street Journal* ran an op-ed by Patrick Michaels, "The Climategate Whitewash Continues," July 12, 2010, parroting many of the usual climategate myths. The paper published my response—"It's the critics of warming who lack credibility"—in the July 16 edition.

14. The press release, issued by PennFuture on July 1, 2010, read: "PennFuture hails vindication of Dr. Michael Mann at Penn State; Time for Matt Brouillette of the Commonwealth Foundation to 'man up and apologize' for pillorying vindicated climate scientist, says PF's Jan Jarrett."

15. The affair was covered in David Adam, "Climate Emails Inquiry: Energy Consultant Linked to Physics Body's Submission," *Guardian,* March 4, 2010. According to the *Guardian,* "Evidence from a respected scientific body to a parliamentary inquiry examining the behavior of climate-change scientists, was drawn from an energy industry consultant who argues that global warming is a religion, the *Guardian* can reveal." The individual in question was Peter Gill, an IOP official who was head of a company called Crestport Services. According to Deep Climate, another climate change denier on the subcommittee was Terri Jackson, who had been, for example, an invited speaker at the Heartland Foundation (www.heartland.org/events/2010Chicago/speakers.html). Jackson appears to have been working closely with Gill to advance a contrarian agenda within the IOP Energy Group (see http://deepclimate.org/2010/03/18/iop-energy-group-founder-featured-speaker-at-upcoming-heartland-confer ence/).

16. Several critical IOP members, including prominent members of other committees, are quoted in the article cited in the previous note.

17. On July 22, the following message was sent out to IOP members in an e-mail: "Following the meeting of the Science Board on 17 June 2010, it is with regret that I announce that the Energy Sub-group is to be disbanded, immediately. This, as you can imagine, is a direct consequence of the Climategate affair."

18. Louise Gray, "'Hockey Stick' Was Exaggerated," *Telegraph*, April 14, 2010.

19. On April 19, 2010, the following addendum was added to the posted report at www.uea.ac.uk/mac/comm/media/press/CRUstatements/SAP: "For the avoidance of misunderstanding in the light of various press stories, it is important to be clear that neither the panel report nor the press briefing intended to imply that any research group in the field of climate change had been deliberately misleading in any of their analyses or intentionally exaggerated their findings. Rather, the aim was to draw attention to the complexity of statistics in this field, and the need to use the best possible methods."

20. It is also worth noting that a sixth, independent, and rather exhaustive review of all of the climategate allegations by the EPA provided further exoneration of the various climate scientists involved in the affair. This review was developed in conjunction with the EPA's denial on July 29, 2010, of the petitions filed by various climate denier organizations and public officials for a reconsideration of the EPA's endangerment finding that would support the regulation of greenhouse gas emissions. The response and supporting reports can be found at http://epa.gov/climatechange/endangerment/petitions.html.

21. "Hackers Attempt to Access Canadian Government Centre for Climate Modeling and Analysis," Marketwire press release, December 5, 2009, reported that, according to University of Victoria climate scientist Andrew Weaver, the thieves stole a computer of his related to his job as an editor of *the Journal of Climate*. The *Journal of Climate*, among other things, had published a number of papers discrediting the various claims of McIntyre and McKitrick.

22. An up-to-date archive of the more than twenty accumulated FOIA and legal demands by these and other organizations is available at http://climscifoi.blogspot.com/.

23. First, *USA Today* covered the harassment of climate scientists (me in particular) in a front page article: Brian Winter, "Questions About Research Slow Efforts to Tackle Climate Change," March 10, 2010. That was followed by increased focus on the funding of denialists after Greenpeace released a report—"Koch Industries: Secretly Funding the Climate Denial Machine"—at the end

of March 2010. The report was given prominent coverage by MSNBC commentator Rachel Maddow. Renewed focus on the harassment of climate scientists came with an *ABC Evening News* piece, Dan Harris and Christine Brouwer, "Climate Scientists Claim 'McCarthy-Like Threats,' Say They Face Intimidation, Ominous E-Mails," *ABC Evening News*, May 23, 2010.

24. See, e.g., Julian Walker, "Cuccinelli Opts for More Modest Virginia State Seal," *Virginian-Pilot*, May 1, 2010.

25. Forced by the court to explain his reason for suspecting fraud in the use of state funds, all Cuccinelli could muster in his court filing was that in one internal University of Virginia grant dealing with the topic of land-atmosphere interaction in the African savanna, of which I was a minor coinvestigator, the CV I provided in support of the grant proposal included references to the original MBH hockey stick papers—which in the view of professional climate change deniers was, of course, the result of fraud, despite the fact that every investigation of the hockey stick found no evidence of impropriety of any kind.

26. In an odd conflict of interest, the office of legal counsel at the university directly answers to the attorney general himself.

27. Rosalind Helderman, "McDonnell Can't Recall Issuing Civil Demand Similar to Cuccinelli's Request to U-Va.," Virginia Politics Blog, *Washington Post*, May 5, 2010.

28. Rosalind Helderman, "Moran Weighs in on Cuccinelli Subpoena," Virginia Politics Blog, *Washington Post*, May 13, 2010.

29. Dan Vergano, "Science Group: Climate Science 'Witch Hunt' Underway in Virginia," USA Today Online, "Science Fair" section, May 3, 2010.

30. Dahlia Lithwick, "Suing Science," *Slate*, May 4, 2010.

31. "U-Va. Should Fight Cuccinelli's Faulty Investigation of Michael Mann," May 7, 2010; "University of Virginia Should Fight the Va. Attorney General's Inquiry," May 13, 2010; "U-Va. Admirably Resists Mr. Cuccinelli's Fishing Expedition," May 29, 2010; "A Judge Puts a Damper on Mr. Cuccinelli's Witch Hunt," August 31, 2010; "Ken Cuccinelli Seems Determined to Embarrass Virginia," October 6, 2010.

32. Among them were "First They Came for the Climatologists," *Richmond Times Dispatch*, May 11, 2010; "Cuccinelli, U.Va and Academic Freedom," *Lynchburg News and Advance*, May 30, 2010. The editorial boards of some right-leaning papers predictably provided cover for Cuccinelli: The *Washington Times* (May 12, 2010) and Scaife's *Pittsburgh Tribune Review* (May 15, 2010) both wrote editorials defending Cuccinelli and his witch hunt.

33. "Cuccinelli, U.Va and Academic Freedom." The editorial ends this way: "We've been supportive of the attorney general's effort to determine the

constitutionality of health care reform. . . . But in his environmental jihad against UVa and its former faculty member, the attorney general has ventured into an area where no politician—liberal or conservative—should tread. Thrusting partisan politics into academics is wrong."

34. Position Statement on Attorney General's Investigation of Dr. Michael Mann, University of Virginia Faculty Senate Executive Council, May 5, 2010.

35. Mashey, among other things, invented the Mashey Shell, a key innovation in the early development of the UNIX operating system, while at Bell Labs in the mid-1970s. He is listed in the online Museum of Computer History, www.computerhistory.org/trustee/John,Mashey/.

36. Http://deepclimate.org/2009/12/22/wegman-and-rapp-on-tree-rings-a -divergence-problem-part-1/; http://deepclimate.org/2010/01/06/wegman- and-rapp-on-proxies-a-divergence-problem-part-2/; http://deepclimate.org/2010 /02/08/steve-mcintyre-and-ross-mckitrick-part-2-barton-wegman/; http://deep climate.org/2010/08/03/what-have-wegman-and-said-done-lately/; http://deep climate.org/2010/10/24/david-ritson-speaks-out/; http://deepclimate.org/2010/09 /15/wegman-report-update-part-2-gmu-dissertation-review/; http://deepclimate .org/2010/04/22/wegman-and-saids-social-network-sources-more-dubious -scholarship/; http://deepclimate.org/2010/07/29/wegman-report-update-part-1- more-dubious-scholarship-in-full-colour/; http://deepclimate.org/2010/09/26 /strange-scholarship-wegman-report/; http://deepclimate.org/2010/10/25/the -wegman-report-sees-red-noise/; http://deepclimate.org/2010/11/16/replica tion-and-due-diligence-wegman-style/; http://deepclimate.org/2010/12/02 /wegman-et-al-miscellany/. See also these related posts at DeSmogBlog: www .desmogblog.com/plagiarism-conspiracies-felonies-breaking-out-wegman -file; www.desmogblog.com/wegmans-report-highly-politicized-and-fatally -flawed; www.desmogblog.com/deep-climate-exposes-more-cheating-team -wegman.

37. John Mashey, *Strange Scholarship in the Wegman Report (SSWR): A Facade for the Climate Anti-science PR Campaign*, http://deepclimate .org/2010/09/26/strange-scholarship-wegman-report/.

38. Some of this evidence came to light by virtue of a PowerPoint presentation by Wegman's student Yasmin Said in 2007, which was accidentally left on a public server at George Mason University: "Experiences with Congressional Testimony: Statistics and the Hockey Stick," George Mason University, Data and Statistical Sciences Colloquium Series, September 7, 2007. It's original location was www.galaxy.gmu.edu/stats/colloquia/AbstractsFall2007/Talk Sept7.pdf. It was removed from this location soon after its discovery was publicized by Deep Climate and Mashey. Said's presentation included the following revelations:

1. Wegman was originally approached with an offer to serve as author of the report by a third party, Dr. Jerry Coffey, on September 1, 2005. Coffey is a fellow member with Wegman of the Washington branch of the American Statistical Society, as well as a member of the Tea Party Patriots (www.rpvnetwork.org/profile/DrJerryL Coffey; accessed June 3, 2010) who has expressed strongly contrarian views about global warming. Deep Climate identified postings by Coffey at the site PersonalLiberty.com. In one, he refers to the "Gore global warming boondoggle." In another, he recommends contrarian reading materials on global warming: "My favorite short read on global warming is Lawrence Solomon's 'The Deniers.' I particularly enjoyed the chapter on Ed Wegman since I had a ringside seat when Ed's analysis got started. Others [sic] books you might enjoy are the last couple by Patrick Michaels; Fred Singer and Dennis Avery on the 1500 year cycle; and [Roy] Spencer's latest."

2. After the initial contact with Coffey, Wegman received materials and a visit from Joe Barton's chief of staff, Peter Spencer. Spencer asked Wegman if he would take on the task of pursuing "criticism of the paleoclimate temperature reconstruction published by Dr. Michael Mann" (presumably by McIntyre) that was "not being taken seriously within the climate change community."

3. Wegman accepted the offer and picked a current graduate student, Said, and a previous graduate student, David W. Scott (now at Rice University) as coauthors. One other unspecified participant apparently "later dropped out."

4. Wegman and his crew were "warned that [they] should be prepared for criticism" and "should have thick skins."

5. Over the following nine months, Barton's chief of staff sent Said and the others "a daunting amount of material for us to review."

39. See Dan Vergano, "Experts Claim 2006 Climate Report Plagiarized," *USA Today*, November 22, 2010; Dan Vergano, "Climate Study Gets Pulled After Charges of Plagiarism," *USA Today*, May 16, 2011; additional details were provided in Dan Vergano, "Retracted Climate Critics' Study Panned by Expert," USA Today Online, May 16, 2011, http://content.usatoday.com/com munities/sciencefair/post/2011/05/retracted-climate-critics-study -panned-by-expert-/1.

40. Raymond S. Bradley, *Paleoclimatology: Reconstructing Climates of the Quaternary*, 2nd ed. (Burlington, MA, Academic Press, 1999).

41. One representative example is as follows:

Bradley: "Trees growing near to the latitudinal or altitudinal treeline are mainly under growth limitations imposed by temperature and hence ring-width variations in such trees contain a strong temperature signal."

Wegman: (emphasis denotes key change): "Trees growing near to their ecological limits either in terms of latitude or altitude show growth limitations imposed by temperature and thus ring width variations in such trees contain a *relatively* strong temperature signal."

In some cases, the changes in wording were such that the point Bradley originally made was rendered wrong or even nonsensical. For example, Deep Climate notes that Bradley's statement, "However, optimum climatic reconstructions may be achieved by using both ring widths and densitometric data to maximize the climatic signal in each sample (Briffa et al., 1995)," which correctly recognizes that it is possible to combine different types of tree ring reconstructions to get a better climate reconstruction, is contorted by Wegman to imply incorrectly that only temperature signals are being sought, and that it is actually the norm to directly combine both types of tree-ring information (it is not): "Both tree ring width and density data *are used* in combination to extract the maximal climatic temperature signal" (emphasis added).

42. The passage is: "The resulting 'regional curve' provided a target for deriving a mean growth function, which could be applied to all of the individual core segments regardless of length (Fig. 10.13). Averaging together the core segments, standardized in this way by the regional curve . . . has far more low frequency information than the record produced from individually standardized cores . . . and retains many of the characteristics seen in the original data."

43. As originally noted by Deep Climate: "Wegman (and Rapp) on Tree Rings: A Divergence Problem (Part 1)," DeepClimate.org, December 22, 2010, http://deepclimate.org/2009/12/22/wegman-and-rapp-on-tree-rings-a-divergence-problem-part-1/.

44. Their alternative statement read: "Because the early history of tree rings confounds climatic signal with low frequency specimen specific signal, tree rings are not usually effective for accurately determining low frequency, longer-term effects."

45. Http://deepclimate.org/2009/12/22/wegman-and-rapp-on-tree-rings-a-divergence-problem-part-1/.

46. "Wegman and Said on Social Networks: More Dubious Scholarship," DeepClimate.org, April 22, 2010, http://deepclimate.org/2010/04/22/wegman-and-saids-social-network-sources-more-dubious-scholarship/.

47. Stanley Wasserman and Katherine Faust, *Social Network Analysis: Methods and Applications* (New York, Cambridge University Press, 1994), section 1.3.

48. Wouter de Nooy, Andrej Mrvar, and Vladimir Batagelj, *Exploratory Social Network Analysis with Pajek* (New York, Cambridge University Press, 2005).

49. Deep Climate provides this remarkable example:

Opening of section 2.3 in Wegman et al.: "A social network is a mathematical structure made of nodes, which are generally taken to represent individuals or organizations. . . . . Social network analysis (also called network theory) has emerged as a key technique and a topic of study in modern sociology, anthropology, social psychology and organizational theory."

Wikipedia: "A social network is a social structure between actors, mostly individuals or organizations. . . . . Social network analysis (also sometimes called network theory) has emerged as a key technique in modern sociology, anthropology, Social Psychology and organizational studies, as well as a popular topic of speculation and study."

As Deep Climate notes, the Wegman et al. variations belie any actual understanding of the concepts being discussed. The change in the first sentence suggests that Wegman et al. do not understand the distinction between a structure and its representation. In the second sentence, Wegman et al. have blurred the key distinction drawn in the Wikipedia article between social network analysis as a tool and as a topic for study in its own right.

50. Mashey, *Strange Scholarship in the Wegman Report.*

51. Several of the papers included were completely irrelevant to the narrow remit of Wegman et al. to assess the hockey stick. Nearly three pages of summary were devoted to two papers by Carl Wunsch dealing with topics such as the Dansgaard-Oeschger oscillations of the last glacial period and ocean observations. These papers appear to have been included simply because they might superficially appear to cast doubt on climate modeling. Wunsch's work has been misrepresented by climate change deniers in the past (see discussion in chapter 5). My Ph.D. thesis and signal detection work related to it were included, even though they have no connection whatsoever with my later hockey stick work. A number of papers cited in the references were pulled from the dubious gray literature and had no discernable connection with the field of paleoclimatology, but did serve to launder fringe denialist claims.

52. According to Mashey's estimates, thirty-one of the ninety pages of the WR contained entirely or largely plagiarized material taken from three books and two Wikipedia pages.

53. See version 1.2 of the Mashey report, September 26, 2010, http://deep climate.files.wordpress.com/2010/09/strange-scholarship-v1–02.pdf, especially, pages 4 (executive summary), 23, 68, and 70. Mashey shows that Wegman, in a

talk he gave at the National Center for Atmospheric Research (NCAR) in 2007, lifted, without attribution, three slides directly from my September 17, 2007, public lecture, "The Science of Climate Change," at Penn State University, which was available on my public lecture archive at www.meteo.psu.edu/~mann /Mann/lectures/ScienceOfClimateChange07.ppt. Wegman used these slides in an October 27, 2007, talk at an invited workshop on statistics and climate science at NCAR, www.image.ucar.edu/public/Workshops/ASAclimate/wegma nASA.htm.

54. Any alleged misconduct is made all the more serious in this case because both Wegman and Said acknowledged government grant support for the paper. The paper formally acknowledged support from the National Institutes of Health and the U.S. Army Research Office: "The work of Dr. Yasmin Said was supported in part by the National Institutes on Alcohol Abuse and Alcoholism under grant 1 F32 AA015876–01A1. The work of Dr. Edward Wegman was supported in part by the Army Research Office under contract W911NF-04–1-0447. The work of Dr. Said and Dr. Wegman was also supported in part by the Army Research Laboratory under contract W911NF-07–1-0059. The content is solely the responsibility of the authors and does not necessarily represent the official views of the National Institute on Alcohol Abuse and Alcoholism or the National Institutes of Health."

55. The paper, "Social Networks of Author–Coauthor Relationships," was published in July 2007 in the journal *Computational Statistics and Data Analysis*. The authors were all associated with Wegman and George Mason University. The lead author was Wegman's student Yasmin Said, while Wegman was the second author. The third author, John Rigsby—a Wegman graduate student at the time—was attributed as performing the original social network analysis of me for the Wegman report. The fourth author, Walid Sharabati—a former Wegman student—supplied an analysis of Wegman's own coauthor network in follow-up questions from the House Energy and Commerce Committee hearings of summer 2006. The paper was accepted without revision in just six days, contrasting sharply with all other articles in the same issue of the journal. The article had few references. For further details, see "Wegman and Said on Social Networks."

56. The abstract begins this way: "Wegman et al. (2006) [i.e. the Wegman report] undertook a social network analysis of a segment of the paleoclimate research community. This analysis met with considerable criticism in some circles, but it did clearly point out a style of co-authorship that led to intriguing speculation about implications of peer review. Based on this analysis and the concomitant criticism, we undertook to examine a number

of author–coauthor networks in order to see if there are other styles of authorship."

57. As described originally in the WR.

58. Based, ironically, on his former student Sharabati's analysis of the co-author network of his own mentor—Wegman.

59. Comment by "Lotharsson," April 23, 2010, http://deepclimate.org/2010 /04/22/wegman-and-saids-social-network-sources-more-dubious -scholarship/#comment-3370.

60. "Wegman and Said on Social Networks."

61. Vergano, "Climate Study Gets Pulled After Charges of Plagiarism"; additional details were provided in Vergano, "Retracted Climate Critics' Study Panned by Expert." In the latter, Vergano reports: "So, how did the paper get published? The journal shows the manuscript was submitted July 8, 2007 and accepted July 13, 2007, for publication. This is a very fast review of a paper. Most take months and require review by two-three outside experts." And "In response to a Freedom of Information Act request last year, Wegman sent USA TODAY two emails detailing the paper's review, to and from his friend, the journal editor, Stanley Azen of the University of Southern California." Vergano quotes the e-mail exchange, which suggests that Azen accepted the paper on the spot, without any peer review at all.

62. Vergano, "Retracted Climate Critics' Study Panned by Expert."

63. Vergano, "Climate Study Gets Pulled After Charges of Plagiarism"; ibid.

64. See again Mashey's report. While the extent of any potential coordination among Spencer, McIntyre, and the Wegman panel members is still unknown, the report provides evidence that there were likely face-to-face meetings, and we now know that Wegman panel members worked closely with McIntyre in conducting their statistical analyses.

65. Some degree of collaboration was hinted at earlier in the curious pattern of repeated errors that David Ritson had discovered shortly after the report was published (see chapter 11).

66. See http://ftp.resource.org/gpo.gov/hearings/109h/31362.txt, specifically the following exchange between Rep. Stupak and Wegman:

Mr. Stupak: "Did you or your co-authors contact Mr. McIntyre and get his help in replicating his work?"

Dr. Wegman. "Actually, no . . ."

67. "Replication and Due Diligence Wegman Style," DeepClimate, November 16, 2010, http://deepclimate.org/2010/11/16/replication-and-due -diligence-wegman-style.

68. Figure 4.4 of the WR.

69. See the related discussion in chapter 11. As Deep Climate notes in the November 16, 2010, post: "both the source code and the hard-wired 'hockey stick' figures clearly confirm what physicist David Ritson pointed out more than four years ago, namely that McIntyre and McKitrick's 'compelling' result was in fact based on a highly questionable procedure that generated null proxies with very high auto-correlation and persistence. All these facts are clear from even a cursory examination of McIntyre's source code, demonstrating once and for all the lack of competence and lack of due diligence exhibited by the Wegman report authors."

70. As noted by Deep Climate in the same November 16, 2010, post: "It turns out that the sample leading principal components (PC1s) shown in two key Wegman et al. figures were in fact rendered directly from McIntyre and McKitrick's original archive of simulated 'hockey stick' PC1s."

71. Deep Climate in the November 16, 2010, post, discusses "the astonishing fact that this special collection of 'hockey sticks' is not even a random sample of the 10,000 pseudo-proxy PC1s originally produced in the GRL study. Rather it expressly contains the very top 100—one percent—having the most pronounced upward blade. Thus, McIntyre and McKitrick's original Fig 1–1, mechanically reproduced by Wegman et al., shows a carefully selected 'sample' from the top 1% of simulated 'hockey sticks.' And Wegman's Fig 4–4, which falsely claimed to show 'hockey sticks' mined from low-order, low-auto-correlation 'red noise,' contains another 12 from that same 1%!"

72. The context of this quotation is a more thorough description by Mashey of the cherry-picking that had been done in the WR. Mashey provided this explanation in the comment thread of Deep Climate's "Replication and Due Diligence" post:

To me, the real zinger is showing that WR Figure 4.4

a) Was produced by McIntyre code whose output hadn't been published.

b) Was effectively presented as a sample of 12 from 10,000 that showed that MBH "mined" random data for positive hockey sticks.

c) Whereas it was a sample of 12 from the 100 (1%) sorted as having the highest positive hockey sticks . . . MM screwed up various places in the sequence of steps to generate that, but this takes cherry-picking to a new height. Does anyone have any idea how serious statisticians might view this?

d) In the US, this would be like declaring the average male height to be 6'6", without bothering to mention the sample was taken on an NBA basketball court.

73. The longer context of Mashey's statement, from the comment thread of Deep Climate's "Replication and Due Diligence" post, was as follows: "I had resolved to avoid the real statistics analysis in the WR, but eventually realized there was none. Using McIntyre's code and help, they reproduced [McIntyre]-like charts. . . . . The WR contains no actual new statistical analysis of MBH itself, just reworks of MM material . . . the legal term is culpable ignorance."

74. In "Steve McIntyre and Ross McKitrick, Part 2: The Story Behind the Barton-Whitfield Investigation and the Wegman Panel," DeepClimate, February 8, 2010, http://deepclimate.org/2010/02/08/steve-mcintyre-and-ross-mckitrick-part-2-barton-wegman/, Deep Climate summarizes the affair thusly: "So there you have it. This supposedly 'independent' panel began with a sounding out by a rabid Republican partisan [Jerry Coffey] and convinced climate 'skeptic.' And Wegman agreed to a process that not only excluded climate scientists, but also involved [Barton staffer] Peter Spencer as a key conduit and gatekeeper providing climate science documentation and commentary. And all this was done by a House committee that had refused to even acknowledge the offer of a proper scientific review from the National Academy of Sciences."

75. The investigation was announced in a letter from Roger R. Stough, vice president for research and economic development at George Mason University to Ray Bradley, copied to George Mason University (GMU) president Alan Merten, April 8, 2010. A letter from James S. Coleman, vice provost for research at Rice University, to Ray Bradley absolved report coauthor David Scott of any role, stating that "During the Inquiry, persuasive evidence was obtained that one of the other authors, Dr. Edward J. Wegman, has taken full responsibility for preparing the allegedly plagiarized text. . . . " GMU's conduct of the investigation, which still has not come to completion as of July 2011, has itself been strongly criticized; see John Mashey, *Strange Inquiries at George Mason University*, www.desmogblog.com/gmu-paralyzed-plagiarism-investigation, as well as these DeepClimate posts: http://deepclimate.org/2010/10/08/wegman-under-investigation-by-george-mason-university/; http://deepclimate.org /2010/12/23/george-mason-universitys-endless-inquiry/.

76. Vergano, "Experts Claim 2006 Climate Report Plagiarized," 6A.

77. Among other mainstream media reports were UPI, "Plagiarism Charged in Congress Report," November 22, 2010; Alex Pareene, "Joe Barton's Climate Report Was Plagiarized," Salon, November 22, 2010; Kate Sheppard, "Smoky Joe Strikes Again," *Mother Jones*, November 23, 2010; "Influential Climate Change Report 'Was Copied from Wikipedia,'" *Daily Mail*, November 23, 2010. The story was covered by numerous prominent online sources such as

Huffington Post, DailyKos, Talking Points Memo, and Think Progress, as well as the *Washington Post*'s online Virginia news section.

78. See Richard Littlemore, "Wegman's Report Highly Politicized—and Fatally Flawed," DeSmogBlog, February 8, 2010, www.desmogblog.com/weg mans-report-highly-politicized-and-fatally-flawed. Littlemore noted that it is a criminal offense to mislead Congress while testifying under oath, and that a congressional investigation into the misconduct by Wegman and his coauthors might therefore be warranted.

79. Stories continued to appear well after the initial *USA Today* article, e.g., Sara Afzal, "4 Recent Cases of Plagiarism Charges in the Headlines," *Christian Science Monitor*, December 8, 2010. *USA Today* followed up with additional articles through spring 2011, as noted earlier.

80. For example, Molly Davis, "Allegations of Plagiarism, Bias, Misinformation Surround Barton Climate Report," appeared in the *Texas Independent* on December 9, 2010, while Joe Barton was still under consideration to take the chairmanship of the House Energy and Commerce Committee when the House of Representatives readjourned under Republican leadership in January 2011. It is possible that the additional bad PR for Barton emerging from the Wegman scandal in November–December 2010 was the final nail in the coffin of his unsuccessful bid to again head this influential committee.

81. This is from the comments of Scott A. Mandia, a professor of meteorology at SUNY-Suffolk who in fall 2010 helped found the Climate Science Rapid Response Team, a network of one-hundred-plus scientists ready to engage with the media in response to any attacks or smears (www.climaterapi dresponse.org/). The comment was made in the thread of our November 21, 2010, RealClimate postmortem on climategate: "One Year Later." Mandia's full comment was: "Climategate was a tactical error by those that wish to deny the science or to delay action. Their illegal action woke up a sleeping bear. Scientists are now engaging with the media . . . and they will not stand by watching one of their own being attacked. They have the truth, they have the passion, and now they have the numbers" (www.realclimate.org/index.php/archives/2010/11 /one-year-later/comment-page-2/#comment-191452).

82. "Climate Change and the Integrity of Science," letter to *Science*, 328 (May 5, 2010): 689–690.

83. For example, Will Happer, who testified for the Republicans at the May 18 hearing of the House Select Committee on Energy Independence and Global Warming on Climate Science in the Political Arena, indicated that he considered Cuccinelli's investigation inappropriate when asked specifically by Representative Jay Inslee (D-WA).

84. "VA Attorney General's Misguided Investigation," press release, Union of Concerned Scientists, May 7, 2010.

85. "Science Subpoenaed: The University of Virginia Should Fight a Witch-Hunt by the State's Attorney General," *Nature*, 465 (May 12, 2010): 135–136.

86. The editorial noted that "Mann's research has been upheld by the US National Academy of Sciences, and an investigation by Pennsylvania State University into the e-mails also cleared Mann of any misconduct."

87. Joint letter to President John Casteen III from the American Meteorological Society and University Corporation for Atmospheric Research, May 14, 2010.

88. Statement of the AAAS Board of Directors Concerning the Virginia Attorney General's Investigation of Prof. Michael Mann's Work While on the Faculty of University of Virginia, May 18, 2010.

89. Letter to Attorney General Cuccinelli, copied to Virginia Governor Robert F. McDonnell, signed by over eight hundred "scientists and academic leaders living and working in Virginia," May 18, 2010; later updated to over nine hundred, May 26, 2010.

90. Preliminary statement contained in filing by the rector and visitors of the University of Virginia's petition to set aside civil investigative demands issued to the University of Virginia, delivered to Honorable Debra Shipp, Clerk, Albemarle Circuit Court, Charlottesville, Virginia, May 27, 2010.

91. Asked by a journalist about my thoughts on the development, I too invoked Thomas Jefferson, noting with some degree of pride that he was in fact one of our country's very first climate scientists; Paul Guinnessy, "UVa Decides to Fight AG's Demands," Physics Today Online, May 28, 2010, http://blogs.physicstoday.org/mt/mt-tb.cgi/4887. The text of my comment was: "I was pleased to see U.Va stand up for the principles that Thomas Jefferson's great university was founded upon, in particular, the principle of academic freedom. Unbeknownst to many . . . Jefferson . . . was in fact . . . one of the very first climate scientists. He collected some of the earliest climate observations in America at Monticello, and he had James Madison collect similar observations up at Montpelier Virginia to the north. When I was at U.Va, my students and I analyzed those observations and published an article detailing what they told us about past summer droughts in the region (D. Druckenbrod, M. E. Mann, D. W. Stahle, M. K. Cleaveland, M. D. Therrell, and H. H. Shugart, 'Late 18th Century Precipitation Reconstructions from James Madison's Montpelier Plantation,' *Bulletin of the American Meteorological Society*, 84 [2003]: 57–71)."

92. See, e.g., Darren Samuelsohn, "Global Warming Critic Plots Revenge," Politico, September 9, 2010, which states "Wisconsin Rep. Jim Sensenbrenner wants to keep the Select Committee on Energy Independence and Global Warming alive so it can investigate climate science" and "Issa told Politico last week that he would use his panel to conduct inquiries into the stolen 'Climategate' e-mails that global warming skeptics say show collusion among prominent scientists, although independent reviews have found nothing of the sort."

93. Michael E. Mann, "Get the Anti-science Bent out of Politics," *Washington Post*, October 8, 2010, A17.

94. Sherwood Boehlert, "Can the Party of Reagan Accept the Science of Climate Change?" *Washington Post*, November 19, 2010.

95. A press release announcing the effort was sent out on November 22, 2010.

96. More information is available at www.climaterapidresponse.org.

97. AGU had recently announced its own climate Q&A service. The AGU effort was the official launch, begun on December 7, 2010, of an earlier pilot version of the program, shortly following the CRU e-mail hack affair. As described on the AGU Web site, the AGU service involved "over 700 Ph.D.-trained scientists" willing to take part in this "novel form of scientist/journalists," the mission of which is "to enable high-quality climate science reporting by connecting the media with an email service staffed by expert climate scientists with quick turnaround and peer collaboration."

98. In Greg Sargent, "GOP Leadership Cool to Hearings Into 'Scientific Fraud' Underlying Global Warming," Washington Post Online, November 8, 2010, http://voices.washingtonpost.com/plum-line/2010/11/eis_gop_really _holding_hearing.html, Barton's staff purportedly indicated that "that global warming science won't be the focus of upcoming hearings. Rather, Barton wants to hold hearings to try to get the Environmental Protection Agency to study the impact action on global warming will have on jobs." (Barton did not get the Energy and Commerce Committee chairmanship anyway.) Stephen Stromberg, "Will the GOP House Attack Science, After All?" Washington Post online, January 3, 2011, http://voices.washingtonpost.com/postpartisan /2011/01/will_the_gop_house_attack_scie.html, reported that "It appears that Rep. Darrell Issa (R-Calif.), the incoming chairman of the House government reform committee and the Republicans' new chief investigator, is backing off of climate issues."

99. The Energy and Commerce Committee held a hearing on Climate Science and EPA's Greenhouse Gas Regulations on March 8, 2011 (http:// energycommerce.house.gov/hearings/hearingdetail.aspx?NewsID=8304).

The hearing had as many mainstream climate scientists as contrarians testifying. There were no fireworks and hardly any media coverage. The House Science Committee later held a hearing on climate science, Climate Change: The Processes Used to Create Science and Policy, on March 31, 2011 (see, e.g., the coverage at www.agu.org/sci_pol/hearing_summaries/112th_con gress/commerce_justice.shtml). While some of the panel members and witnesses attempted to malign mainstream climate science and climate scientists, the main news generated by the hearing was that one of the Republican's own witnesses (retired astrophysicist Richard Muller) had actually recently confirmed the instrumental record of global warming. Those expecting the sort of fireworks promised by Issa, Sensenbrenner, and Barton during the fall 2010 congressional campaigns were likely sorely disappointed with the first session of the new Congress.

100. See, e.g., "On First Day of New Congress, Koch Operatives Met with GOP Chairman Planning to Gut the Clean Air Act," Alternet.org, February 11, 2011, www.alternet.org/newsandviews/article/464218/on_first_day_of_new _congress_koch_operatives_met_with_gop_chairman_planning_to_gut_ the_clean_air_act/; Evann Gastaldo, "Koch Brothers Wield Influence in New Congress," Newser.com, February 6, 2011, www.newser.com/story /111376/koch-brothers-wield-influence-in-new-congress.html.

101. In an interview, John Shimkus (R-IL), who ran the new Environment and Economy subcommittee of the Energy and Commerce committee, generated headlines in November when he quoted the Bible to argue that God would not destroy Earth through global warming; Darren Samuelsohn, "John Shimkus Cites Genesis on Climate Change," *Politico*, November 10, 2010.

102. See, e.g., Jim Nolan, "Judge Rules Against Cuccinelli in U.Va Case," *Richmond Times-Dispatch*, August 30, 2010.

103. From "Ken Cuccinelli Seems Determined to Embarrass Virginia," *Washington Post*, October 6, 2010.

104. In February 2011, a group known as the American Tradition Institute (ATI) headed by Chris Horner of CEI filed a Virginia FOIA demand that the University of Virginia turn over my private documents and e-mails. The wording of the demand was nearly identical to that by Cuccinelli. At the time of the writing of this book, the university was in the process of challenging a court order that specifies how exempt materials will be handled. ATI's actions were condemned in an editorial by the *Washington Post* ("Harassing Climate-Change Researchers," May 29, 2011) that stated: "Freedom of information laws are critical tools that allow Americans to see what their leaders do on their behalf. But some global warming skeptics in Virginia are showing that even the best tools can be misused." ATI's actions were also criticized in a statement by

the American Association for the Advancement of Science (Statement of the Board of Directors of the American Association for the Advancement of Science Regarding Personal Attacks on Climate Scientists, approved by the AAAS Board of Directors, June, 28, 2011). In its press release, AAAS specifically referred to the ATI FOIA against the University of Virginia to obtain my personal e-mails, and a similar FOIA ATI issued to NASA to obtain e-mails of NASA climate scientist James Hansen.

105. For example, a paper by two economists launching yet more flawed attacks against the hockey stick (and the field of paleoclimatology more generally) and laundering several of the McIntyre/Wegman talking points, was published online in a statistics journal in August 2010. The journal also published a number of scathing critiques of this paper as a follow-up, however. Further details and coverage are provided at "Responses to McShane and Wyner," Real-Climate, December 13, 2010, www.realclimate.org/index.php/archives/2010/12/responses-to-mcshane-and-wyner/.

106. In January 2011, I was forwarded a press release that is evocative of the infamous Seitz letter, discussed elsewhere in this book; the press release gave the false apparent imprimatur of the National Academy of Sciences (NAS) to a bogus petition manufactured by climate change deniers. The forwarded press release (www.nas.org/polPressReleases.cfm?Doc_Id=1729) contained the lead "40% of scientists doubt manmade global warming." It was formatted to look as if it bore the insignia of the Academy. Instead, it was issued by an organization that calls itself National Association of Scholars, which derives its funding from the Scaife Foundations and other such entities (see, e.g., www.sourcewatch.org/index.php?title=National_Association_of_Scholars). The basis of the claim advertised by the press release: more dubious claims by S. Fred Singer. See DeSmogBlog, www.desmogblog.com/liars-damn-liars-and-dr-s-fred-singer.

107. Indeed, the attacks have spread as the climate change policy debate has moved forward in other countries. In summer 2011 as the Australian government was preparing to vote on a carbon pricing scheme, Australian climate scientists were at the receiving end of widespread threats to their safety and lives. In June, Rosslyn Beeby, "Climate of Fear: Scientists Face Death Threats," *Canberra Times*, June 4, 2011, reported that "Australia's leading climate change scientists are being targeted by a vicious, unrelenting email campaign that has resulted in police investigations of death threats." In July, a climate scientist giving a lecture at a climate conference in Melbourne was "confronted with a death threat" and presented with a hangman's rope by a man in the front row (Brendan Nicholson and Lauren Wilson, "Climate Anger Dangerous, Says German Physicist," *The Australian*, July 16, 2011).

108. One particularly amusing example was provided by the *Telegraph*'s James Delingpole on October 26, 2010 (http://blogs.telegraph.co.uk/news /jamesdelingpole/100055500/global-cooling-and-the-new-world-order/), who included my name with that of Al Gore, George Soros, Bill Gates, Carol Browner, John Holdren, Barack Obama, David Cameron, Ted Turner, Robert Redford, and numerous other luminaries in a list of those "guilty" of using human-caused climate change to establish a "new world order." A less amusing, more recent example dates from March 2011, when James Inhofe, with help from CEI's Christopher Horner and from Stephen McIntyre, sought to spread a malicious smear about me. Inhofe's office apparently selectively leaked snippets from a transcript of a NOAA inspector general's report to falsely suggest that I had told another scientist (Eugene Wahl) to delete e-mails. Wahl quickly put out a statement indicating that the claim was false. The matter was covered in detail by Center for American Progress's Joe Romm (http://cli mateprogress.org/2011/03/09/inhofe-watts-horner-mcintyre -michael-mann-email/).

Epilogue

1. NASA's James Hansen made this argument in "Twenty Years Later: Tipping Points Near on Global Warming," an opinion piece he wrote for the *Guardian*, June 23, 2008, www.guardian.co.uk/environment/2008/jun/23/cli matechange.carbonemissions: "CEOs of fossil energy companies know what they are doing and are aware of long-term consequences of continued business as usual. In my opinion, these CEOs should be tried for high crimes against humanity and nature." Don Brown, a Penn State colleague and lawyer who served as a program manager for sustainability at the EPA under the Clinton administration, and is currently the director of the Collaborative Program on Ethical Dimensions of Climate Change in the Rock Ethics Institute of Penn State University, also made this argument in "A New Type of Crime Against Humanity," October 24, 2010, http://rockblogs.psu.edu/climate/2010/10 /a-new-kind-of-vicious-crime-against-humanity-the-fossil-fuel-industrys- disinformation-campaign-on-cl.html.

2. This is a seasonally adjusted estimate, subtracting the effects of the annual cycle that leads to higher-than-average concentrations in the Northern Hemisphere winter and lower-than-average concentrations in the Northern Hemisphere summer due to the "breathing of the biosphere." Up-to-date numbers are available from NOAA at www.esrl.noaa.gov/gmd/ccgg /trends/.

3. Jared Diamond, *Collapse: How Societies Choose to Fail or Succeed* (New York, Viking Press, 2005).

4. Some of the deniers of the ozone hole problem back then, such as S. Fred Singer, are today denying the threat of human-caused climate change (see chapter 5).

5. Public polling by one leading expert, Jon Krosnick of Stanford University, suggests that the impact on public opinion was short-lived, with the percentage of the public that believe in the scientific reality of human-caused climate change rebounding to pre-climategate levels by June 2010 (http://woods.stanford.edu/research/americans-support-govt-solutions-global-warming.html).

6. This was true as late as 2003; thus my reticence to comment on policy-related matters during my exchanges with Senator Inhofe during the July 2003 Senate hearing: "I am not a specialist in public policy and I do not believe it would be useful for me to testify on that."

7. Emanuel made the statement in the course of his testimony at a March 31, 2011, hearing of the House Science Committee. See press release by the American Geophysical Union, "House Science Committee Holds Climate Science Hearing," April 13, 2011, www.agu.org/sci_pol/hearing_summaries/112th_congress/commerce_justice.shtml.

8. "Climate of Fear," *Nature*, 464 (March 11, 2010): 141.

9. The U.S. National Academy of Sciences recently weighed in on this issue in "America's Climate Choices," May 2011, http://dels.nas.edu/Report/Americas-Climate-Choices/12781: "Most people rely on secondary sources for information, especially the mass media; and some of these sources are affected by concerted campaigns against policies to limit CO_2 emissions, which promote beliefs about climate change that are not well-supported by scientific evidence. U.S. media coverage sometimes presents aspects of climate change that are uncontroversial among the research community as being matters of serious scientific debate. Such factors likely play a role in the increasing polarization of public beliefs about climate change, along lines of political ideology, that has been observed in the United States."

10. On May 13, 2011, the *Washington Post* ran an editorial, "Climate Change Denial Becomes Harder to Justify," discussing the implications of the May 2011 National Academy study described in the preceding note and echoing several of the themes of this epilogue. "Climate-change deniers," the editorial stated, ". . . are willfully ignorant, lost in wishful thinking, cynical or some combination of the three. And their recalcitrance is dangerous . . . because the longer the nation waits to respond to climate change, the more catastrophic the planetary damage is likely to be—and the more drastic the needed

response." It finishes with this prescription: "Every candidate for political office in the next cycle, including for president, should be asked whether they disagree with the scientific consensus of America's premier scientific advisory group, as reflected in this report; and if so, on what basis they disagree; and if not, what they propose to do about the rising seas, spreading deserts and intensifying storms that, absent a change in policy, loom on America's horizon." A day later (May 14, 2011), *USA Today* issued its own editorial, "Our View: America, Pick Your Climate Choices": "[Climate change] deniers [are] in the same position as the 'birthers,' who continue to challenge President Obama's American citizenship—a vocal minority that refuses to accept overwhelming evidence." The editorial ended: "The latest scientific report provides clarity that denial isn't just a river in Egypt. It paves a path to a future fraught with melting ice caps, rising sea levels, shifting agricultural patterns, droughts and wildfires."

Selected Bibliography

Bradley, R. S. *Global Warming and Political Intimidation: How Politicians Cracked Down on Scientists as the Earth Heated Up.* Amherst, MA, University of Massachusetts Press, 2011.

———. *Paleoclimatology: Reconstructing Climates of the Quaternary,* 2nd ed. Burlington, MA, Academic Press, 1999.

Ehrlich, P. R., and A. H. Ehrlich. *Betrayal of Science and Reason: How Antienvironmental Rhetoric Threatens Our Future.* Washington, DC, Island Press, 1996.

Gelbspan, R. *Boiling Point: How Politicians, Big Oil and Coal, Journalists, and Activists Have Fueled a Climate Crisis—And What We Can Do to Avert Disaster.* New York, Basic Books, 2004.

———. *The Heat Is On: The Climate Crisis, the Cover-up, the Prescription.* New York, Perseus Books, 1997.

Gould, S. J. *The Mismeasure of Man,* rev. ed. New York, W.W. Norton, 1996.

Hoggan, J., and R. Littlemore. *Climate Cover-up: The Crusade to Deny Global Warming.* Vancouver, BC, Greystone Books, 2009.

Mann, M. E., and L. R. Kump. *Dire Predictions: Understanding Global Warming.* New York, Pearson/DK, 2008.

Mathez, E. A. *Climate Change: The Science of Global Warming and Our Energy Future.* New York, Columbia University Press, 2009.

Mooney, C. *The Republican War on Science.* New York, Basic Books, 2005.

Mooney, C., and S. Kirshenbaum. *Unscientific America: How Scientific Illiteracy Threatens Our Future.* New York, Basic Books, 2010.

Oreskes, N., and E. M. Conway. *Merchants of Doubt: How a Handful of*

Scientists Obscured the Truth on Issues from Tobacco Smoke to Global Warming. New York, Bloomsbury Press, 2010.

Powell, J. L. *The Inquisition of Climate Science.* New York, Columbia University Press, 2011.

Sagan, C. *The Demon-Haunted World: Science as a Candle in the Dark.* New York, Random House, 1996.

Schmidt. G., and J. Wolfe. *Climate Change: Picturing the Science.* New York, W.W. Norton, 2009.

Schneider, S. H. *Science as a Contact Sport: Inside the Battle to Save Earth's Climate.* Washington DC, National Geographic, 2009.

Acknowledgments

I owe great thanks for the help and support many individuals have provided over the years. First and foremost are my family: my wife Lorraine, my daughter Megan, my parents Larry and Paula, my brothers Jay and Jonathan, and the rest of the Manns, Sonsteins, Finesods, and Santys.

Special thanks to the crew at Columbia University Press, including Science Publisher Patrick Fitzgerald, Assistant Editor Bridget Flannery-McCoy, Senior Manuscript Editor Roy Thomas, Senior Designer Milenda Lee, and Publicity Director Meredith Howard. Thanks also to my developmental editor Jonathan Cobb and to Debbie Masi at Westchester Book Group for production management. My deep appreciation to all who provided invaluable comments on various drafts of the book including Mark Cane, Tom Cronin, Aaron Huertas, Philip Kitcher, Naomi Oreskes, Dan Schrag, and Seth Stein. An extra special thanks to Gavin Schmidt not only for the invaluable feedback on the final book draft but for his friendship, support, and sage advice over the years.

I want to thank some of the teachers and mentors at various stages of my education who helped instill in me a curiosity about the natural world, a love of science, and an understanding of how to do it. Among them are Raymond Bradley, Charles Camp, Didier de Fontaine, Tony Haymet, Malcolm Hughes, Jonathan Lees, Jeffrey Park, Bill Ruddiman, Barry Saltzman, and Ron Smith. I also want to acknowledge those who provided special inspiration during the toughest of times: Paul Ehrlich, Donald Kennedy, Walter Munk, Bill Nye, Steve Schneider, and Steven Soter.

I am greatly indebted to policy makers on both sides of the aisle who stood up against powerful interests to defend my colleagues and me against politically motivated attacks: Sherwood Boehlert, Jerry Brown, Al Gore, Mark Herring, Jay Inslee, Jim Jeffords, Edward Markey, John McCain, Jim Moran, Harry Reid, and Henry Waxman. I am similarly indebted to the tireless efforts of staff members Greg Dotson, Sue Fiering, David Goldston, Edith Holleman, Johannes Loschnigg, Chris Miller, John Mimikakis, Dan Pearson, Cliff Rechtschaffen, Janill Richards, Alexandra Teitz, and Ana Unruh-Cohen.

I am much obliged to various other individuals who have provided support and legal advice over the years: Pete Altman, Kert Davies, Sarah Faulkner, Pete Fontaine, Peter Frumhoff, Francesca Grifo, Jan Jarrett, Kalee Kreider, Roger McConchie, Pete Meyers, Rick Piltz, David Silbert, Scott Stapf, Todd Stern, David Vladeck, and Nikki Vo.

I want to thank and commend journalists who have strived so hard to cover the science and the events and circumstances surrounding it with nuance and accuracy at a time when uncritical coverage is all too common: Neela Banerjee, Richard Black, Seth Borenstein, Juliet Eilperin, Faye Flam, Justin Gillis, Eli Kintisch, George Monbiot, Kate Sheppard, Paul Thacker, Dan Vergano, and Brian Winter.

I want to thank President Graham Spanier, Vice Presidents for Research Eva Pell and Hank Foley, Dean Bill Easterling, and Department Chair Bill Brune at Penn State, and President John Casteen III and Department Chair Patricia Wiberg at the University of Virginia for their firm and principled stance in defense of academic integrity. I would also like to thank the members of Penn State's RA10 Inquiry and Investigatory Committees, Sarah Assman, Welford Castleman, Mary Jane Irwin, Nina Jablonski, Alan Scaroni, Fred Vondracek, Candice Yekel, for their diligence, patience, and wisdom.

While there are too many other colleagues, collaborators, friends, and supporters I wish to acknowledge and thank than I can possibly list, some that come to mind are John Abraham, John Albertson, Richard Alley, Caspar Ammann, Palaniswamy Ananthakrishnan, James and Edna Anderson, Steven Andrew, Rick Anthes, David Archer, Vicki Arroyo, Ugo Bardi, Michael Berube, Tim Bralower, Gregory Bowman, Tom Bowman, Sue Brantley, Keith Briffa, Ronn Brourman, Frank Cartieri, Dave Clarke, Amy Clement, Kim Cobb, Ford Cochran, Ed Cook, John Cook, William Connolley, Jason Cronk, Jen Cronk, Tom Crowley, Heidi Cullen,

Fred Damon, Brendan Demelle, Andrew Dessler, Steve D'Hondt, Henry Diaz, Kerry Emanuel, Howie Epstein, Carla Esposito, Jenni Evans, Grant Foster, Jose Fuentes, Ross Gelbspan, Nellie Gorbea, Kevin Grandia, David Graves, Jay Gulledge, James Hansen, Susan Joy Hassol, Bill Hay, Gabi Hegerl, James Hoggan, Greg Holland, Jan Jarrett, Phil Jones, Jim Kasting, Bill Keene, Klaus Keller, Sheril Kirshenbaum, Adam Kissel, Chip Knappenberger, Lee Kump, Deb Lawrence, Sukyoung Lee, Alan Leshner, Rachel Levinson, Richard Littlemore, Scott Mandia, John Mashey, Scott McConnell, Bill McKibben, Chris Miller, Sonya Miller, Becky Mock, Chris Mooney, Ellen Mosely-Thompson, Ray Najjar, Cary Nelson, Gerald North, Randy Olson, Michael Oppenheimer, Naomi Oreskes, Tim Osborn, Jonathan Overpeck, Rajendra Pachauri, Chuck Pavloski, Ray Pierrehumbert, Sharon Pillar, Emma Pullman, Alan Robock, Stefan Rahmstorf, Joe Romm, Scott Rutherford, Heather Sage, Ben Santer, Keith Seitter, Don Shelby, Hank Shugart, Stein Siggurdson, Peter Sinclair, Dave Smith, Tony Socci, Susan Solomon, Richard Somerville, Eric Steig, Sean Sublette, Larry Tanner, Betsy Taylor, Anne Thompson, Lonnie Thompson, Kim Tingley, Kevin Trenberth, Fred Treyz, Nancy Tuana, Dave Verardo, Eugene Wahl, Bud Ward, Peter Webster, Ray Weymann, Joel Wiles, Jay Zieman, and Herm Zimmermann.

Index

"Mann number," 165

margin of error, 41, 47

Markey, Edwin, 246

Mashey, John, 239, 240, 243, 312n4

Massachusetts vs. EPA, 327–328n8

Massey West Virginia coal mine disaster, 236

mathematical models. *See* climate models

Maunder, Edward W., 40

"Maunder minimum," 40, 41

Max Planck Institute for Meteorology, 184

MBH98: expanding level of research, 50–53; methods, 170, 172; publication of, 48–50, 211–212; source code used in, 129–130, 158

MBH99: methods, 170, 172; publication of, 52–53; reactions in scientific community, 56–58; in Third Assessment Report, 53–56, 98; *See also* hockey stick

MCA (medieval climate anomaly), 36, 289–290n29. *See also* medieval warm period

Mears, Carl, 182–183, 315n20

media: collision of science and, 87–89; coverage of Climate Research Unit hacking, 217–220, 234–235; improvements in climate change coverage, 192; initial response to MBH98, 49; initial response to MBH99, 53; op/ eds in conservative newspapers, 68; participants in climate change denial, 69–73; propagation of disinformation in, 64, 184–186; *See also specific media sources*

medieval climate anomaly (MCA), 36, 283–284n29, 369

medieval warm period, 34–36, 36*f,* 261–262; in Climate Research Unit hacking scandal, 214; and contrarians, 56, 57; early research on, 34–35; in Wegman Report, 239–240

Mencken, H. L., 237

Merchants of Doubt (Oreskes and Conway), 60

Meyers, Chad, 193–194

Michaels, David, 59

Michaels, Patrick, 3, 82, 94–95, 106, 141, 213

microwave sounding units (MSUs), 181–183

"Mike's trick," 6, 8, 29, 209–211, 338n22

Milloy, Steven J., 70, 230, 283n53

Mirowski, Philip, 276n4

The Mismeasure of Man (Gould), 131, 135

molecular change, 7

Monastersky, Richard, 115, 116

Monbiot, George, 70

Monckton, Christopher, 71, 284n58

Monsanto Corporation, 74

Monte Carlo method, 7

Montreal Protocol, 251

Mooney, Chris, 60, 61

Morano, Marc, 73–74, 187, 199, 220, 224, 225, 230

MSUs (microwave sounding units), 181–183

"MTM-SVD," 270n3

Muller, Richard, 140

Munk, Walter, 125

Murdoch, Rupert, 336n7

Murray, Cherry, 197

[